# College Geometry
# Using The Geometer's Sketchpad®

**Barbara E. Reynolds, SDS**
*Cardinal Stritch University*

**William E. Fenton**
*Bellarmine University*

**Key College Publishing**
Innovators in Higher Education

www.keycollege.com

Barbara E. Reynolds, SDS
Department of Mathematics and Computer Science
Cardinal Stritch University
Milwaukee, WI 53217

William E. Fenton
Department of Mathematics
Bellarmine University
Louisville, KY 40205

Key College Publishing was founded in 1999 as a division of Key Curriculum Press® in cooperation with Springer, New York. We publish innovative texts and courseware for the undergraduate curriculum in mathematics and statistics as well as mathematics and statistics education. For more information, visit us at www.keycollege.com.

Key College Publishing
1150 65th Street
Emeryville, CA 94608
(510) 595-7000
info@keycollege.com
www.keycollege.com

This material is based upon work supported by the National Science Foundation under award numbers DUE #01-25130 and DUE #03-38301. Any opinions, findings, and conclusions or recommendations expressed in this publication are those of the author(s) and do not necessarily reflect the views of the National Science Foundation.

**Activity and Figure Sketches CD-ROM**

Development Editor: Allyndreth Cassidy
Production Director: McKinley Williams
Project Manager: Eric Houts
Production Coordinator: Ken Wischmeyer
Text Designer: Adriane Bosworth
Copyeditor: Tara Joffe
Proofreader: Andrea Fox
Indexer: Victoria Baker
Compositor: Interactive Composition Corporation
Art Editor/Photo Research: Jason Luz
Cover Concept: Charley Harper
Cover Designer and Coordinator: Jensen Barnes
Printer: Alonzo Printing

Editorial Director: Richard J. Bonacci
General Manager: Mike Simpson
Publisher: Steven Rasmussen

Photo credits:
Page 189 © Corbis
Page 190 © Royalty-Free/CORBIS
Page 195 © M.C. Escher's "Circle limit IV" © 2005 The M.C. Escher Company—Baarn—Holland. All rights reserved.
Page 243 © Foto Marburg/Art Resource, NY

Printed in the United States of America
10 9 8 7 6 5 4 3 2 1   09 08 07 06 05

The cover art, designed by Jensen Barnes, was inspired by a design created by Charley Harper titled, *Skipping School*. Mr. Harper uses mathematical tools to create his artwork such as T-square and French curve, common draftsman's tools. This piece was chosen because it looks like something that could have been designed with The Geometer's Sketchpad. Mr. Harper has written several titles including *Beguiled By The Wild* featuring his own collection of writing and artwork. His studio is located in Cincinnati, Ohio.

# BRIEF CONTENTS

# CONTENTS

# OUR MOTIVATION, PHILOSOPHY, AND PEDAGOGY

College geometry serves many purposes. Some students taking the course will be introduced to mathematical proofs for the first time. Many will be preparing for teaching geometry at the high-school or middle-school level. Still others will be experimenting with technology for the first time. Even the level of mathematics background of the students will differ from course to course. Due to the varied audience for this course, we chose to write a book that would make full use of technology through The Geometer's Sketchpad to introduce rigorous mathematical proofs and prepare future teachers. A course based on this text can be taught with high-school geometry and college algebra as the minimal prerequisites. If calculus and linear algebra are additional prerequisites (as is the case in some institutions), students will be able to cover more of the text and explore ideas in greater depth.

When we started this project, we surveyed many existing college geometry texts before deciding on the content of our text. Unlike the standard calculus curriculum, an undergraduate course in geometry does not have a standard table of contents. Our search led us to find much variation and a few common threads. We talked with colleagues who teach this course at a variety of institutions, and we found that everyone has a personal list of favorite topics. To be honest, each of us had our own list, which although they intersected, were not identical. We realized very early that we would have to decide which topics to include. Eventually we noticed that we were using four simple questions to determine whether to include a topic:

1. Does this topic lend itself to explorations and conjecture with The Geometer's Sketchpad?

2. Does this topic allow us to examine some interesting questions in geometry—and connect the study of geometry to the larger tapestry of mathematics?

3. Does this topic allow explorations that lead students to make and test their own conjectures?

**4.** Is this topic useful content for future middle-school and high-school mathematics teachers while leading to important ideas that reach considerably beyond the content of a high-school geometry course?

Those topics for which we answered "yes" to all four questions are included.

We wanted to develop a coherent text in which topics for each chapter would relate to each other, and larger ideas and themes would emerge that would run through the course. We offer several distinct approaches to the study of geometry: a *synthetic approach* (Chapters 1–3), an *analytic approach* (Chapter 4, Appendix A), and a *transformational approach* (Chapters 6–8). While we do not follow a strictly *axiomatic* treatment, we address the importance of the choice of an axiom system by introducing Euclid's postulates early in the course (Chapters 1–3), and later contrasting these with the hyperbolic axioms of non-Euclidean geometry (Chapter 9). In the chapter on Taxicab Geometry (Chapter 5), we develop axioms for a metric space and explore the impact that changing the metric has on the shape of some familiar curves. Chapter 10 is a culminating chapter in which we examine the real projective plane axiomatically, analytically, and through projective transformations, thus bringing together these different approaches to the study of geometry.

An important goal of this course is to teach students to write good mathematical proofs. At the core of our pedagogical approach is a belief that students tend to write in a way that reflects their understanding of the underlying concepts and ideas. That is, if students understand the mathematical concepts, they tend to write correct proofs; and when a student does not quite understand an important concept, the weaknesses or errors in the proof reflect this. Thus, we approach the development of proofs from two directions: First, in the activities that open each chapter, we ask students to spend a lot of time working with Sketchpad diagrams, observing, making conjectures, and talking to each other about their ideas. We give students lots of room to form their own ideas about particular geometric situations, all the while listening to their conversations. Then in the ensuing class discussion, we talk about their conjectures, affirming what they have seen where it is correct and helping them to reshape their ideas where their reasoning is weak. The discussion in the text illustrates specific proof strategies, giving examples and explaining accepted patterns of logical reasoning for developing proofs.

Throughout this text, we use the dynamic power of Sketchpad to engage students in explorations leading to conjecture. To construct a diagram in a Sketchpad worksheet, the students have to think about certain geometric ideas. Our experience has been that if students understand what is going on in a particular geometric situation—if they see the relationships made visible in their Sketchpad diagram and if they engage in discussions about what they see in the diagram—they will find words to express what they see. Through both small-group and whole-class discussion, we guide students toward accepted mathematical language for expressing these ideas. In the discussion following their explorations with Sketchpad, we engage students in conversation, listen to what they are saying, and shape their language toward generally accepted (mathematically correct) expressions of these ideas.

Keeping in mind future teachers and their needs, we have included questions at the end of each exercise set that are particularly designed to be answered by mathematics education majors. These questions are based on the recommendations presented in the NCTM *Principles and Standards for School Mathematics*. Through class discussion, we emphasize connections between the concepts and highlight issues that are important for future teachers.

As we wrote these materials, we envisioned a course taught in a cooperative learning environment. In our own classes, we form groups with three or four students and have these groups work together in a computer lab on the activities that open each chapter. Our pedagogical approach is to guide students toward foundational insight and conceptual understanding through their group work on the activities. In our experience, groups who have diverse backgrounds often have fruitful discussions as they work on the activities. Each person sees things in a different way and brings a different viewpoint to the small-group discussion. We see our role as facilitators, guiding students to formalize their mathematical ideas as they develop correct proofs of conjectures that arise from their work on the activities. While we have found that teaching geometry in this manner is quite effective, each instructor adapts these materials to her/his preferred teaching style.

## CHAPTER DEPENDENCIES

Chapters 1–3 draw on experiences from high school geometry. Some of our students (particularly returning adult students who took high-school geometry more than ten years earlier) find their recollection of high-school geometry is rather dim, but by the end of Chapter 3 they have recalled the elements of geometry that they need to draw on throughout the rest of the course. Yet there are several theorems in these chapters that go sufficiently beyond the high-school curriculum to engage everyone in substantive discussions. The elementary geometry presented in these first three chapters goes well beyond a mere review of high-school geometry and is worth the time taken to review this material.

Chapter 4 (Analytic Geometry) and Appendix A (Trigonometry) assume familiarity with college algebra and at least minimal exposure to basic right-triangle trigonometry.

Once Chapters 1–4 and Appendix A (if needed) have been covered, students are prepared for Chapters 5, 6, or 9. Preservice elementary and middle school teachers find Chapters 5 (Taxicab Geometry) and 6 (Transformational Geometry) particularly appealing, while Chapter 9 (The Hyperbolic Plane) is important for mathematics majors and future high-school teachers who need to develop deeper understanding of the importance of axiom systems.

Chapters 7 (Isometries and Matrices) and 8 (Symmetry in the Plane) both depend on Chapter 6. Chapter 7 requires linear algebra, and can be omitted if your students have not taken a prior course in linear algebra. Appendix B reviews matrix calculations for those students who might have forgotten these things.

Some of our students have suggested that an instructor not wait until the end of the course to cover Chapter 9. A few students suggested that we consider covering Chapter 9 following Chapter 6—and then go back to Chapter 7 or 8. They said that Chapter 9 really challenged everything they ever knew about the shape of the world, and they would have preferred to have had more time to digest the ideas of hyperbolic geometry before the course ended. On the other hand, those class testers who covered Chapter 5 in their course have indicated that Chapter 5 provides a good conceptual lead-in for Chapter 9.

Chapter 10 (Projective Geometry) draws together themes that run through nearly every chapter and serves as a capstone to the course.

## SUPPLEMENTS

There are two supplements for this course. One is a CD for students and the other is a CD for instructors. The CD for students is packaged in the back of each new book. If your students have purchased used books, encourage them to order the Used Book CD Package to accompany this text (ISBN 1-597570-31-1). The CD provides Sketchpad sketches for all activities and figures in the text.

For instructors, we are providing Instructor Resources on CD (ISBN 1-597570-00-1). This CD provides an overview of the text, suggestions for teaching in collaborative learning groups, tips for teaching with technology, sample solutions to all activities, and lesson plans for each chapter.

For more information, please contact your Key College Publishing sales representative at **888-877-7240** or visit **www.keycollege.com**.

## ACKNOWLEDGMENTS

Writing a book is not something that one does alone. Looking back over the five years that this book has been in development, we are surprised to realize how many people have shared this journey with us, and we are grateful for their ongoing support.

We gratefully acknowledge the contributions of students who have worked with us as we developed this text. Students in our geometry classes at both Cardinal Stritch University (in Milwaukee, Wisconsin) and Bellarmine University (in Louisville, Kentucky) have given us feedback on early drafts of these materials. Their feedback challenged us to refine and clarify what we have written. A number of student assistants have worked closely with us, testing the activities and critiquing their effectiveness in introducing the concepts that are discussed in each chapter. Vanessa Sowinski, Lindsey Blue, Lindsay Bronson, Katherine Kubicek, Howard Fahje, and Ken Bellinger worked with Sister Barbara Reynolds at Cardinal Stritch University, while Ryan Church and Jon Lamkin worked with Bill Fenton at Bellarmine University. Additionally, two of Sister Barbara's students were integral in the creation of the supplements. Katherine Kubicek created the Sketchpad documents for the activities and figures on the student CD, and Bryna Goeckner developed the sample solutions and lesson plans for the instructor CD.

A number of people at Key College Publishing were also helpful in the process of getting this book published. Early in this project, Richard Bonacci and Steve Rasmussen made helpful suggestions as we developed a proposal for an NSF grant, Nick Jackiw provided extensive comments after teaching from a draft of the manuscript, and Scott Steketee offered technical assistance with Sketchpad. It has been a delight to work closely with Allyndreth Cassidy, Senior Development Editor. In addition to arranging for math checking, reviews, and class testing, she provided enthusiastic encouragement along the way.

We received many helpful suggestions from reviewers throughout the development process, which helped us to improve the manuscript. Our reviewers include:

Thomas Banchoff—Brown University, Rhode Island

Anne Brown—Indiana University–South Bend, Indiana

Joseph Fiedler—California State University–Bakersfield, California

Catherine Gorini—Maharishi University of Management, Iowa

Sarah Greenwald—Appalachian State University, North Carolina

Ronald Milne—Goshen College, Indiana

Bruce O'Neill—Milwaukee School of Engineering, Wisconsin

Tamas Szabo—Weber State University, Utah

In addition, we are pleased to acknowledge the careful math-checking work of Lisa Lister at Bloomsburg University, Pennsylvania, and Adam Massey, a talented mathematics student at Brown University, Rhode Island.

It has been affirming that a number of instructors have been interested in using drafts of our manuscript in their own geometry classes. Feedback from these class testers and their students has been valuable in revising the text. Our class testers include:

Jorgen Berglund—California State University–Chico, California

Anne Brown—Indiana University–South Bend, Indiana

Jim Cottrill—Illinois State University, Illinois

Joseph Fiedler—California State University–Bakersfield, California

Kenneth Forsythe—Silver Lake College, Wisconsin

Thomas Fox—University of Houston–Downtown, Texas

Patricia Giurgescu—Pace University, New York

David Gove—California State University–Bakersfield, California

Mike Hall—Arkansas State University–Jonesboro, Arkansas

Robert Klein—Ohio University, Ohio

Thomas Mattmann—California State University–Chico, California

Nathalie Sinclair and Nick Jackiw—Michigan State University, Michigan

Andrius Tamulis—Cardinal Stritch University, Wisconsin

Bill Whitmire—Francis Marion University, South Carolina

This project was supported in part by grants from the National Science Foundation (DUE #01-25130 and DUE #03-38301). We would like to thank those who served on our NSF Advisory Board: Thomas Banchoff, Anne Brown, Robert Megginson

(University of Michigan), and Draga Vidakovic (Georgia State University). In addition, Susan Pustejovsky (Alverno College, Wisconsin) and Jack Bookman (Duke University, North Carolina), our NSF Evaluation Team, gathered feedback from our class testers and their students, which was very helpful in improving the manuscript.

Finally, Bill would like to thank his family, Ann Jirkovsky and Billy Fenton, and Sister Barbara would like to thank the Sisters of the Divine Savior for their ongoing personal support over the five years we have spent writing and refining this book.

We invite you on a journey—a journey with your mind! The course for which this textbook is designed will lead you to explore geometric worlds, visually at first, and then using both deductive and inductive reasoning processes. We invite you to explore geometry through computer-based investigations, to make observations and conjectures about what you see, and then to develop proofs or disproofs to support or refute your conjectures.

Playing can be a gateway to new ideas. In this course, we ask you to play with geometric figures, to explore their properties, and to observe relationships and interactions among those figures. As you play, you will be asked to make conjectures about what you see happening. Although you might have heard it said that "seeing is believing," mathematicians tend to be a bit skeptical in this regard. Once mathematicians think they see something, they often ask, "Is this really true, or does it just appear to be so?"

The Geometer's Sketchpad is a tool that supports this kind of investigative learning. Throughout this course, you will be asked to use Sketchpad to construct various geometric figures—and to play with them, observing what is happening. One of the nice features of Sketchpad is that you can *construct* figures that have certain properties. Constructed figures retain their properties as they are manipulated. If you merely draw figures, geometric properties will not hold.

Your instructor might ask you to work in a cooperative learning group throughout this course. If so, it will be important for you to develop a good working relationship with the members of the group. For a cooperative learning group to be effective for every member, each member of the group must be regularly engaged in the group work. Each must be committed to keeping up with her or his own individual study of the course materials. Your group will need to meet regularly to share ideas about the problems you are working on, and each of you must keep up with your own study.

When we asked our own students what suggestions they would give to others who use this textbook, they asked us to share the following strategies, which they feel contributed to their success as students.

The activities that open each chapter are an introduction to the concepts covered in the chapter. Take time to do these activities, and to reflect with each other on your observations. Try to decide what the crucial idea is in each activity.

You might not get the answers right at first, but keep trying to understand the ideas. When you work on the activities, ask yourselves, "Why are they asking us that question here?" The purpose of the activities is to prepare you to understand the concepts when you read the discussion in the text, and this does work. If your group has done a good job on the activities, the discussion in the text will be much easier to read.

Play and make observations. Construct robust figures. Move the objects around, and observe what is happening. Take notes while you are working. Type notes directly into your Sketchpad worksheets. That way, you can articulate your ideas while you are working together. At the end of the working session, you can save and print a copy of your Sketchpad worksheets with notes for each member of your group.

Talk with your colleagues. Trying to tell someone else what you observe helps to solidify the ideas in your own mind. Formulate conjectures about what you see happening. Test your conjectures. Try to find the extremes at which they hold. Talk to your colleagues about why you think your conjecture holds.

Prepare ahead of time for each working session with your group. Read over the activities before coming to the computer lab. Think about what you will have to do to solve a particular problem, and perhaps even read ahead into the chapter to see what topics are discussed there. Your work with your small group will be much more rewarding if everyone comes prepared.

Finally, make use of the CD packaged at the back of your new book. This CD has Sketchpad files for each chapter. You will find it helpful to open the activities sketches when you first start working on the activities. Most of the activities are presented on at least two pages in the Sketchpad documents for a set of activities— the first page simply repeats the statement of the activity as it is presented in your textbook, and the second gives a summary of the activity statement with plenty of room for you to work through the problem. If you use the activities document provided for each chapter, you will not need to have your book open while you are working at the computer. Also, each figure in the text is included on the CD so that you can interact with the diagrams. These sketches bring to life the static images in the book.

Keep in mind throughout the course that for your learning to be effective, your play needs to be reflective. Pay attention to what you see, ask lots of questions, and think about the meaning of the answers you find.

## CHAPTER ONE

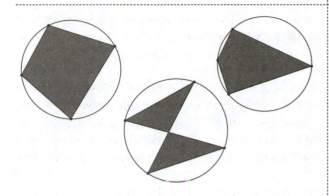

An important objective of this course is to expand your ability to visualize and reason about geometric ideas. The Geometer's Sketchpad® is a dynamic and powerful geometry tool. Sketchpad™ makes it easier to see, literally, what happens to various geometric objects—points, lines, segments, circles, and so on—when you move them around the plane. As you work on these activities, think about what it means to *construct* an object rather than to *draw* it. The geometric ideas and constructions introduced in these activities will be used throughout the course.

# Using The Geometer's Sketchpad: Exploration and Conjecture

## PART I: GETTING STARTED WITH SKETCHPAD

In this first chapter, you are invited to experiment with some of the capabilities of The Geometer's Sketchpad. Let's get started. Sit in front of a computer, and launch Sketchpad. Read through this discussion of icons as you experiment with them in Sketchpad.

The Geometer's Sketchpad is a dynamic tool for geometry that you can use to create very elaborate geometric figures. Fortunately, even a beginner can do many interesting things right away. As the course progresses, you will become more and more sophisticated with Sketchpad, perhaps rivaling your instructor!

Let's start by taking a look at the basics. At the top of the screen are the commands for the menus that Sketchpad has available. Look through the various commands on these menus. In the first few chapters, we use only a few of these menus, saving the more specialized ones for later. On the left side of the screen is the Toolbox, containing the basic tools that you will use over and over. It is worth discussing these tools individually.

Use the **Arrow** tool to select and move objects in the sketch. Click on any object—a point, a line, a segment, a circle, an arc—to select it; the object will be highlighted. Click on the object again to deselect it. You can drag selected objects across the sketch, or you can delete them. Also, the color of a selected object can be changed by using commands on the Display menu. (If you hold down the **Arrow** tool, you will see three options. For now, we will use only the basic selection arrow.)

Use the **Point** tool to construct points in the sketch. The new point will be shown at the end of an arrow. When you move this point over another object, notice that the object is highlighted, showing that you can construct the new point on this object—this is useful for putting points on lines or circles.

Use the **Compass** tool to construct circles, just as you would with a mechanical compass. Select two points by clicking at the center, then on the circle. Or click at the center and drag out to construct the circle.

The **Straightedge** tools are used for constructing straight objects—lines, segments, or rays. If you hold down this button, you will see the three options available.

Use the **Text** tool to label objects in the sketch. Simply click on the object, and a label will appear. If you double-click (on a Mac) or right-click (on a PC) and select **Properties**, you can change the label. Labels move to stay with their objects.

The **Text** tool is also used to create *captions,* which are useful for putting names and numbers on your diagrams, for describing your figures, and for answering the questions in the activities. To open a text box in which you can type your caption, double-click in a blank area. Notice that the **Text** palette also appears in the sketch.

This allows you to change the font, size, color, and style of the text in a caption. There is also a button on the **Text** palette that allows you to use mathematical notation, such as subscripts and special symbols, in your caption.

 Finally, we have the **Custom** tools icon. Sometimes you may need special tools, either tools supplied with Sketchpad or tools you create yourself. (We will work with some of these custom tools in later chapters.) This tool lets you select any custom tools that have been created or opened for the sketch.

Before printing your work, you should always choose **Print Preview** from the File menu. Sometimes a large Sketchpad diagram will print on two or more pages. If you use **Print Preview**, you will be able to choose **Scale To Fit Page**. Sometimes the most interesting part of the Sketchpad diagram will fall over the page break; using **Scale To Fit Page** will save you the frustration of reprinting pages that don't appear as you expect.

Now you are ready to do some geometry. We encourage you to attempt these activities with one or two colleagues so that you can talk about what you see as you experiment with Sketchpad diagrams.

## 1.2 ACTIVITIES

Do the following activities, writing your explanations clearly in complete sentences. Include diagrams whenever appropriate. You will be able to answer many of these questions by typing a sentence or two directly into your Sketchpad diagram.

You should get into the habit of saving your work for each activity, as later work sometimes builds on earlier work. You will find it helpful to read ahead into the chapter as you work on these activities.

1. Construct an arbitrary quadrilateral in Sketchpad and find the midpoints of its sides. When these midpoints are connected (in clockwise order), what kind of figure is formed? Drag the vertices of your quadrilateral to help you decide whether your conjecture is always correct. Can you explain why your conjecture is correct?

2. Start with an arbitrary quadrilateral, *WXYZ*. Draw its diagonals *WY* and *XZ*. (The diagram will be easier to analyze if you use different colors for some of the lines.)
   a. Try to manipulate the quadrilateral so that the diagonals don't intersect each other (that is, they don't intersect at a point interior to *WXYZ*). What do you observe about the shape of the quadrilateral?
   b. What if the diagonals of *WXYZ* bisect each other? Can you construct a quadrilateral with this property?

3. Draw line $\ell$, and construct line *m* parallel to $\ell$. Draw a third line, *t*, that intersects $\ell$ and *m*. These three lines will form eight angles. Use Sketchpad to measure the size of each angle. What do you observe? Express your observation as a conjecture, clearly indicating your hypothesis and your conclusion.

**4.** Is it possible to draw a line that intersects exactly one side of a given triangle (without passing through a vertex)? Why or why not? Is it possible to draw a line that intersects exactly two sides of the triangle? Exactly three sides? Explain your answers.

**5.** If one side of a triangle is extended at one vertex, the angle it creates with the other side at that vertex is called an *exterior angle* of the triangle. Construct an example of this in Sketchpad. Measure the exterior angle. Compare this measurement with each of the interior angles at the other two vertices. What do you observe? Express your observation as a conjecture. Is your conjecture still true if you vary the triangle?

**6.** Construct an equilateral triangle using Sketchpad. (Don't just estimate; use circles and intersection points to guarantee that your triangle is equilateral.) Once you have constructed this triangle, choose **Hide** from the Display menu to clear any auxiliary construction objects from the screen.

   a. Create point *P* in the interior of the triangle. Construct line segments from *P* that are perpendicular to each side. Then hide the extra objects you used to do this. Measure the lengths of the three segments, and have Sketchpad calculate the sum.

   b. Drag *P* around your picture. Your segments should remain perpendicular to the three sides. What do you observe? Drag *P* onto one of the vertices of your triangle. What do you observe? Make a conjecture about the sum of these three perpendicular segments.

**7.** Construct a circle with center point *C*. Then construct three points *P*, *Q*, and *R* on this circle.

   a. Draw segments *PQ* and *QR*. The angle *PQR* is called an *inscribed angle* for the circle. Measure ∠*PQR*. Also measure ∠*PCR*, which is called the *central angle*. (Using different colors may make your diagram easier to analyze.) What do you observe? Vary the points *P*, *Q*, and *R* so that you examine many different angles. Does your observation still hold? Express your observation as a conjecture. Can you explain why your conjecture is correct?

   b. Construct chord *PR*. Drag point *Q* around the circle. What do you observe about the measure of ∠*PQR*? What happens if you drag *Q* past point *R*? What happens if you change the radius of the circle? The family of angles that you are looking at as *Q* moves around the circumference of the circle are referred to as *angles subtended by the chord PR*. Express your observations in the form of a conjecture. Explain why your conjecture is true.

**8.** Construct a circle with diameter *PR*. Make sure you construct your circle in such a way that *PR* is certain to be a diameter, not merely a chord, of the circle.

   a. Construct point *Q* on the circle, and measure ∠*PQR*. What do you observe? How does the situation change if *Q* moves past point *R*?

   b. Make a conjecture about this situation. Justify your conjecture.

9. A quadrilateral is called *cyclic* if its four vertices lie on a common circle. Construct an example of this, and measure the four angles of your quadrilateral. What do you observe about the opposite angles? Express your observation as a conjecture. Does your conjecture still hold if you move the vertices? Prove your conjecture.

10. Create two lines. Construct points $A$, $C$, and $E$ on one of these lines and points $B$, $D$, and $F$ on the other. Label these six points. Connect the points by six lines—$\overleftrightarrow{AB}$, $\overleftrightarrow{BC}$, $\overleftrightarrow{CD}$, $\overleftrightarrow{DE}$, $\overleftrightarrow{EF}$, and $\overleftrightarrow{FA}$. (Using different colors for some of the lines will make the diagram easier to analyze.)

   In a general sense, these six points and six lines form a hexagon. (The hexagon might be easier to see if you have Sketchpad construct its interior. To do this, select the hexagon's six vertices in order, then choose an option from the Construct menu.) Drag the vertices to be sure that you see the six sides of the hexagon.

   The points where opposite sides intersect are called the *diagonal points* of the hexagon. In your diagram, these occur at $\overleftrightarrow{AB} \cap \overleftrightarrow{DE}$, $\overleftrightarrow{BC} \cap \overleftrightarrow{EF}$, and $\overleftrightarrow{CD} \cap \overleftrightarrow{FA}$. Construct and label these diagonal points.

   What do you observe about the diagonal points? Express your observation as a conjecture. Is your conjecture still true if you move some of the vertices or move the two lines?

## 1.3  DISCUSSION

## PART II:  OBSERVATION → CONJECTURE → PROOF

The ancient Greek philosophers—Plato, Euclid, and others—felt that all of mathematics rested on geometric foundations. They even used geometry to do some basic arithmetic, such as addition, multiplication, and square roots. Many arithmetic computations can be done geometrically, but not all. In the history of geometry, this proved to be a major issue.

According to the ancient Greek philosophers, the only tools allowed for geometry were the straightedge (which is not the same as a ruler) and the compass. In other words, you can draw only lines and circles. You are not allowed to measure things using the calibrations on a ruler, because such measuring is considered imprecise. However, you are allowed to compare lengths, and even to copy lengths, using a compass.

In this course, you will use a tool that was not available to the ancient Greeks—The Geometer's Sketchpad. This tool is more powerful than the straightedge and compass, though it uses the same fundamental objects: points, lines, and circles. The power of Sketchpad lies in its dynamic capability. You construct a diagram in Sketchpad in the same way you would with a straightedge and compass, but then you can experiment by altering and varying the diagram. By dragging objects around your diagram, you can explore many different possibilities very quickly.

Throughout this course, you will be asked to construct geometric figures using Sketchpad and to observe what is happening as you move some of the elements of the figure around. You will be asked to formulate conjectures about your observations, stating what you think is happening in a given situation. You then will be asked to test your conjectures to see if they hold up in different circumstances. In stating a conjecture, there are certain conditions that are given; these form the *hypothesis* of your conjecture. Your *conjecture* is an assertion (or a claim) that if the conditions required in the hypothesis are met, then the *conclusion* will hold. Only after you have proven that your conjecture always holds (is valid) are you allowed to call it a *theorem*.

## SOME SKETCHPAD TIPS

Here is one way you might have approached Activity 1. A quadrilateral consists of four line segments connected to form a closed figure. Use one of the **Straightedge** tools to construct one segment, then another segment connected to it, then a third segment connected to the second, and finally a fourth segment connected to the third and first segments. (This is easier to do than to describe.) Once the quadrilateral is constructed, use the **Arrow** tool to select one of the sides. Choose an option from the Construct menu to find the midpoint of that side. Repeat this for the other sides. (It is also possible to find the midpoints of all the desired segments at once. Simply select the segments you want, and go to the Construct menu. If the **Midpoint** command is not available in that menu, be sure that you have selected only segments.) After the four midpoints have been constructed, use one of the **Straightedge** tools again to connect them in clockwise order.

Using different colors can help focus your attention on certain portions of a figure as you experiment with a construction. For instance, select the four line segments that join the midpoints of the original quadrilateral, and go to the Display menu. The **Color** command gives you many choices. Pick a new color. Look again at the quadrilateral that you constructed by connecting the midpoints. Because this new quadrilateral is now in a contrasting color, it will be easier to see why it is interesting. Use the **Arrow** tool to drag one of the original vertices around the sketch and watch what happens.

There are many other ways you might have done Activity 1. For instance, you might visualize a quadrilateral as four vertices connected by line segments. Your construction would begin with four points. Then you would select these points two at a time and choose **Segment** from the Construct menu. (It is also possible to construct all four of the desired segments at once. Simply select the four points in the order you want to connect them, and go to the Construct menu. If the **Segment** command is not available in that menu, be sure that you have selected only points.) As you become familiar with the tools and the menus available, you will think of many approaches to any particular construction.

# QUESTIONS, QUESTIONS, QUESTIONS!

As you worked on Activity 1, what kind of figure did you observe when you connected the midpoints of your quadrilateral? You might have conjectured that the figure is a parallelogram. (This is Varignon's Theorem, established in 1731.) What is a parallelogram? Try to state precisely what it means for a figure to be a parallelogram. Why does this construction seem to lead to a parallelogram? There must be parallel segments involved, but how can you be sure that the segments you see really are parallel? Perhaps you can add something to the picture that will show why the segments are related. Would it help to create triangles in the sketch? What other facts in the problem could be useful?

Here is another question about Activity 1: As you dragged a vertex around your sketch, you most likely saw a situation where two edges of the quadrilateral crossed each other. This is not what most people think of as a quadrilateral, but it is one nevertheless. Is your conjecture still true in this situation? Does your explanation still work?

How did you construct the equilateral triangle you worked with in Activity 6? It is not enough to simply draw three line segments that look congruent; you must use extra objects to force the segments to be congruent. If your construction is *robust*, you will be able to select any of the vertices or edges of the triangle and move them around without losing the property that the figure is an equilateral triangle. Is your construction of an equilateral triangle robust? *Hint:* The radii of any particular circle are all congruent. So, if the base of your triangle is a radius, each of the sides can be a radius too, though maybe not for the same circle. Once you get a robust construction that produces an equilateral triangle, you should ask yourself why that construction works. In other words, how can you be sure that the three sides are truly congruent? (If your construction is not a robust construction, what can you do to make it so?)

As you worked on Activity 6, something interesting should have happened as you moved $P$ around inside the triangle. What conjecture did you make about the sum of the lengths of the three segments you constructed? Viviani's Theorem states that this sum is always the same and, furthermore, that this sum is equal to an important value for the triangle. Does your conjecture say something similar?

Activity 6 raises some issues about geometric language. What does it mean for a point to be *on* a triangle or *on* a circle? Try to write a sentence or two explaining this. What does it mean for a point to be *interior to* a particular triangle or circle? This is a bit harder to explain. (Just to complicate things, imagine that a large triangle is drawn on the surface of a sphere. Where is the interior now?) What does it mean for a point to be *exterior to* a triangle or circle?

Although at first you may have put the point $P$ interior to the equilateral triangle, Sketchpad might allow you to move $P$ beyond the boundary of the triangle (depending on how you constructed the figure). Does your conjecture still hold when $P$ is on the triangle? Does it hold if $P$ is outside the triangle? If not, it may be possible to modify your conjecture so that it still holds even when $P$ is *on* or *outside of* the triangle.

You can take this activity in many directions. Does something similar happen for isosceles triangles? What if the figure is a square instead of an equilateral triangle? What about a pentagon? Perhaps you could investigate segments from P to the vertices instead of from P to the sides. Let your imagination guide you to new discoveries.

The previous few paragraphs raise many questions! You may have answered some of these questions already as you worked on the activities. We will ask many more questions in this book. As you read, pause and try to answer each question, or at least make sure you understand what the question is asking and why we are asking it. As you work the activities and read the discussions in this book, try to ask your own questions: "Is that always true? Why might it be true? Will it still be true if I change some part of the problem? Can I explain why it works?" The question *why* is a major theme of mathematics, and mathematicians are constantly searching for explanations.

Throughout this course, we intend to engage you in quite a few exploratory activities, and we hope that you will ask a lot of questions about what you are observing. Your questions are important, so take time to ponder them. Experiment with your Sketchpad diagrams to discover what insights they can provide. Really think about your questions. You will find that many questions will be answered as you engage in the activities that open each chapter and as you discuss these activities with your classmates. You will also find that many questions will be answered as you continue to read ahead in this book.

We hope that you begin to pose new questions that go beyond what is presented in this book. Share your questions with your fellow students and your instructor. This will make geometry come alive for you. You might even discover something completely new!

## LANGUAGE OF GEOMETRY

In the activities, we assumed that you already know some basic geometric vocabulary. This seems fair enough because you are college students. We hope you recognize most of the geometric terms used so far, even if their meaning is not completely clear in your mind. As you continue this course, it will be increasingly important to use geometric language carefully. It is important to have clear, concise definitions of the terms we are using; thus, you should practice writing these careful definitions.

A *polygon* is a closed figure in the plane that is bounded by line segments. Because it has many (*poly-*) angles (*-gons*), it is called a polygon. A *triangle* could be called a 3-gon, because it has three angles. It could also be called a *trilateral* because it has three sides (*laterals*). A *hexagon* has six angles. The corners of a polygon are called its *vertices* (singular, *vertex*). Does a polygon with *n* vertices always have *n* sides? Why or why not?

In Activities 1, 2, and 9, you worked with several different quadrilaterals. The most general *quadrilateral* is simply a polygon with four sides. The word *quadrilateral* comes from the Latin *quadrilaterus,* meaning four (*quadri-*) sided

(-*laterus*). We could as easily use the Greek roots and call this figure a *tetragon*, which means four (*tetra-*) angled (*-gon*).

While working on Activity 2, you were faced with the question of what sort of figures qualify as quadrilaterals. Certainly there must be four sides. Are these four sides allowed to cross each other? You can easily cause this to happen in your sketch by dragging a vertex. Do you want to call this figure a quadrilateral? Think of the definition stated earlier. A quadrilateral should be a closed figure in the plane that is bounded by line segments (i.e., a polygon), and it should use four segments. Now what do you think? Polygons whose sides intersect are called *self-intersecting* figures.

Even if we avoid self-intersecting quadrilaterals, there is still another issue in Activity 2. Think of the figures you found for which the diagonals did not intersect. Are these legitimate quadrilaterals? Refer again to the definition of *quadrilateral* to help you decide.

Those quadrilaterals for which the diagonals intersected are *convex* quadrilaterals. Quadrilaterals and hexagons can be either convex or not convex. However, every triangle is convex. What about pentagons? More generally, what does it mean for a polygon to be convex? Try to state a clear definition for *convex figure*. Your definition should allow you to decide whether any example of a polygon is convex, no matter what type it is. Experiment again with your sketch for Activity 2, and try to formulate a rule or strategy for determining whether a quadrilateral is convex. Does your rule work only for quadrilaterals, or does it work for other polygons as well?

Within the family of quadrilaterals, there are some special figures—rectangle, square, kite, rhombus, trapezoid, and parallelogram, to name a few. Try to write a one-sentence definition for each. If some of these names are not familiar to you, look up the distinguishing characteristics to help you write the definition.

When you write a definition for a geometric figure, try to use the smallest possible list of requirements to describe it. For example, a square is a quadrilateral with four right angles whose sides are all equal in length. However, we could also define a *square* as a quadrilateral with four equal angles and with one pair of adjacent sides equal in length. This is a considerably shorter set of requirements. Why isn't it necessary to mention that the angles are all right angles? Why is it sufficient to say only that one pair of adjacent sides has the same length?

Every square is a rectangle. Why? Is every rectangle a square? Why or why not? To answer these questions, you must make careful use of the definitions.

A *parallelogram* is a quadrilateral whose opposite sides are parallel. What does it mean for lines to be parallel? In Activity 3, you explored the angles created by a transversal of parallel lines. Using your observations from this activity, can you make any conjectures about the angles of a parallelogram?

What are some other characteristics of a parallelogram? There are many things that can be proven from the simple definition of a parallelogram, and you might remember some from an earlier geometry course. Here is a question you may not have seen: Is a parallelogram a cyclic quadrilateral? To decide this, you must use the definition of *cyclic quadrilateral*, as well as the definition of *parallelogram*.

You will have a chance to answer this and other questions about the properties of parallelograms in the exercises.

Activity 3 asks you to work with parallel lines and to measure angles formed by a *transversal*—that is, a line that lies across (transverses) the two parallel lines. To measure an angle in Sketchpad, select three points that mark the angle, and then choose an option from the Measure menu. What did you learn (or recall) about parallel lines as you worked on Activity 3? Some pairs of the angles in that figure are called *alternate interior angles,* other pairs are called *alternate exterior angles,* and still others are called *interior* (or *exterior*) *angles on the same side of the transversal.* Some of these pairs of angles are congruent—having the same measure—while other pairs are supplementary. What did you observe?

Angles can be classified into three categories: *right angles, acute angles,* and *obtuse angles.* We often say that a right angle has a measure of 90°, but this way of measuring angles is arbitrary. (Some historians have speculated that using 360° in a circle comes from early astronomy and the number of days in a year, but this is not known for certain.) There is nothing special about the size of a degree, and we could develop other systems of measuring angles using other kinds of units. For instance, angles can be measured in *radians,* which you should remember from an earlier course. In highway construction, the angle (incline) of a road is measured by its *grade,* which is stated as a percentage. A road with a 10% grade rises 10 meters for every 100 meters of horizontal distance. How would you express this angle in degrees?

It is not necessary to measure an angle to determine whether it is a right angle, an acute angle, or an obtuse angle. For example, you can take a sheet of paper and very easily fold a right angle. (Try it!) If two lines form right angles (in this case, the lines are represented by two folds in the paper), we say that the lines are *perpendicular.*

Sketchpad has a command on the Construct menu that allows you to construct a line perpendicular to a given line; most of the time, you will be expected to use this tool. But how would you construct a perpendicular line if you were restricted to using only a compass and straightedge—or only **Circle by Center+Point** and **Line** on the Construct menu—and not the full array of commands on the Construct menu? Can you explain why your construction works?

Sketchpad also has a command that allows you to construct a new line on a given point and parallel to a given line. Suppose, however, you are restricted to using only a compass and straightedge (or **Circle by Center+Point** and **Line** on the Construct menu). If you are given a line and a point not on that line, how would you construct a new line that is parallel to the original line? Why does your construction work?

The whole issue of parallel lines is a very interesting story in the history of mathematics. The ancient Greek mathematician Euclid and his contemporaries made some fundamental assumptions about geometry, which were expressed in the form of postulates or axioms. *Postulates,* which are also called *axioms,* are statements to be accepted without proof. They are the beginning of a mathematical theory, and all theorems in that theory are proven—directly or indirectly— from the postulates. In the historical development of geometry, one of Euclid's postulates has generated controversy and a great deal of study and has led to significant developments in mathematical thinking. This is Euclid's Fifth Postulate, the one related to parallel lines.

**Euclid's Fifth Postulate**   If a straight line falling on two straight lines makes the sum of the interior angles on the same side less than the sum of two right angles, then the two straight lines, if produced indefinitely, meet on that side on which the angles are less than two right angles.

If this postulate seems deep (or even convoluted) to you, you are not alone. Mathematicians wrestled with this postulate for nearly 2000 years, attempting to simplify it and to prove it. In general, mathematicians prefer to accept as few postulates as possible. For a long time, mathematicians felt that such a complicated statement surely could be proven from the simpler axioms that preceded it. Euclid himself developed as much of his geometry as possible before using this assumption.

Over the years, other mathematicians formulated different ways of expressing Euclid's Fifth Postulate. Working in the sixteenth century, nearly 1900 years after Euclid, Christopher Clavius formulated an axiom that is almost equivalent to Euclid's Fifth Postulate.

**Clavius' Axiom**   The set of points equidistant from a given line on one side of it forms a straight line (Hartshorne, 2000, 299).

John Playfair, a mathematician working in the late eighteenth century, developed perhaps the best-known alternative to Euclid's Fifth Postulate:

**Playfair's Postulate**   Given any line $\ell$ and any point $P$ not on $\ell$, there is exactly one line through $P$ that is parallel to $\ell$.

Sometimes Playfair's Postulate has been expressed as "two straight lines that intersect one another cannot both be parallel to the same straight line" (Hartshorne, 2000, 300).

Once you have constructed and selected two or more points in Sketchpad, you can use the **Straightedge** tools to construct a line, a line segment, or a ray through any two of the points. In English and other natural languages, we do not always make a careful distinction between a line and a line segment. What exactly is the difference? Try to state this difference as precisely as you can. When you construct a straight object in Sketchpad, the order in which the points $P$ and $Q$ are selected is not important for the line $PQ$, nor is it important for the line segment $PQ$. However, the order of selection makes a critical difference for the two rays, $\overrightarrow{PQ}$ and $\overrightarrow{QP}$. Try these constructions to be sure you understand how the order affects the result.

Because a *line segment* has two endpoints, we can find the middle, or *midpoint,* of the line segment. If a line segment is represented by a piece of string or a fold in a sheet of paper, we could find its midpoint by folding the string or paper in half. Sketchpad has a command that allows you to construct the midpoint of a line segment. How would you construct the midpoint of a segment in Sketchpad without using this shortcut command? (It may be possible to modify the construction you used in Activity 6 to accomplish this.) Why does your construction work?

A *line* is a simple one-dimensional object. When we find the midpoint of a line segment, we are locating the middle point of the segment, and it is clear what this means. The question of middle point is less clear for two-dimensional figures. For example, does a figure like a circle or a triangle have a midpoint? The circle does, of course; what is another name for the midpoint of a circle? How would you find the midpoint of a triangle? There may be several ways to answer this last question. Consider polygons with more sides, such as quadrilaterals, pentagons, hexagons, and so on. How could you find a middle point for these figures? Can you think of more than one way to find a middle point? Does convexity play a role in this?

What conjecture did you make about the diagonal points of the hexagon you constructed in Activity 10? (Before working on this activity, you might have had a mental image of a hexagon as a convex polygon. The hexagon in this activity is certainly not convex.) What is the hypothesis of your conjecture? What is your conclusion? Pappus' Theorem says that the three diagonal points are related in a very special way, which we will discuss in depth in Chapter 10.

If several points lie along a common line, they are said to be *collinear*. Of course, any two points will be collinear; this is one of Euclid's postulates. It is more interesting to consider situations where three or more points are collinear. The corresponding idea for lines is *concurrence*. In the exercises, you will be asked to give a precise definition of concurrence.

## EUCLID'S POSTULATES

In about 300 BC, Euclid and his colleagues gathered the ideas about mathematics known at the time into a book (or series of books) called Euclid's *Elements*. For about 2000 years, this was the most widely circulated book about mathematics, and it served as a textbook for anyone who was considered an educated person. Although it is a geometry text, Euclid's *Elements* is really a catalog of ideas—postulates (or axioms) and theorems—that apply to all areas of mathematics. In Euclid's day, and for many hundreds of years thereafter, arithmetic and algebra were expressed in terms of geometry. (You can see the complete text of Euclid's *Elements*, with interactive diagrams, at http://aleph0.clarku.edu/~djoyce/java/elements/elements.html).

All of the ideas in Euclid's *Elements* are developed from five simple postulates. Well, at least the first four postulates are simple statements—and then there is the fifth postulate!

1. Given two distinct points $P$ and $Q$, there is a line (that is, there is exactly one line) that passes through $P$ and $Q$.

2. Any line segment can be extended indefinitely.

**3.** Given two distinct points $P$ and $Q$, a circle centered at $P$ with radius $PQ$ can be drawn.

**4.** Any two right angles are congruent.

As mentioned earlier, Euclid's Fifth Postulate is a bit more complicated. In modern English, this postulate can be stated as follows:

**5.** If two lines are intersected by a transversal in such a way that the sum of the degree measures of the two interior angles on one side of the transversal is less than the sum of two right angles, then the two lines meet on that side of the transversal.

In the experimenting you did while working on Activity 3, you may have observed that when lines $\ell$ and $m$ are parallel, the sum of the two interior angles on the same side of the transversal, $t$, is equal to two right angles. In Euclidean geometry, we can use this observation as either a condition for parallelism or a consequence of parallelism. That is,

> If two lines are intersected by a transversal in such a way that the sum of the degree measures of the two interior angles on one side of the transversal is equal to the sum of two right angles, then the two lines are parallel; *and* if two lines are parallel, then the sum of the degree measures of the two interior angles formed on one side of a transversal is equal to the sum of two right angles.

This is a long and awkward sentence. We can write it more simply using the abbreviation *iff*, which means "if and only if." In mathematics, iff means that two implication sentences have been joined with the logical connective *and*. In other words, if one statement is true, then the second statement is also true, and if the second statement is true, then the first is true as well. The two statements are equivalent to each other. Thus, we can combine these two implications into the following:

> Two lines are parallel iff the sum of the degree measures of the two interior angles formed on one side of a transversal is equal to the sum of two right angles.

Playfair's Postulate (see page 11) is equivalent to Euclid's Fifth Postulate. This means that if we accept Euclid's Fifth Postulate as an axiom, we can prove Playfair's Postulate—and if we accept Playfair's Postulate as an axiom, we can prove Euclid's Fifth Postulate. Euclid's Fifth Postulate is true iff Playfair's Postulate is true.

We will accept Euclid's first four postulates as axioms—that is, we will accept them as true and will not attempt to prove them. In a certain sense, these postulates will be part of the implicit hypotheses—the unstated assumptions—in all of our conjectures. For the time being, we will also accept Euclid's Fifth Postulate as an axiom. However, this postulate will be a major issue in Chapter 9—even as it was a major issue in the history of mathematics for more than 2000 years!

Euclid's postulates shape the way Sketchpad operates. Let us reflect for a bit on how Euclid's postulates are implemented within Sketchpad's environment.

1. *Given two distinct points P and Q, there is a line that passes through P and Q.* In Sketchpad, you can construct or draw points. If you select two (or more) points, you can construct lines (or segments) through these points.

2. *Any line segment can be extended indefinitely.* If you have constructed a line segment, you can select its endpoints and construct a line through those points.

3. *Given two distinct points P and Q, a circle centered at P with radius PQ can be drawn.* You can construct a circle by selecting two points. The first point selected will be the center of the circle, and the distance between the two points will be the radius. You can also measure a distance or calculate a value and then construct a circle with this radius by choosing a point for the center and this value for the radius.

4. *Any two right angles are congruent.* Given a line or a segment and a point, Sketchpad has a command that allows you to construct a perpendicular line. Also, Sketchpad allows you to move figures around so that you can see whether one figure can be superimposed on another. Any right angle can be moved until it coincides with any other right angle. (We will discuss this idea of moving figures around the plane in Chapter 6.)

5. *Euclid's Fifth Postulate:*
   a. Euclid's statement: *If two lines, ℓ and m, are intersected by a transversal, t, in such a way that the sum of the degree measures of the two interior angles on one side of t is less than the sum of two right angles, then the two lines meet on that side of the transversal.*

      You can draw this configuration in Sketchpad. If you draw ℓ, m, and t as arbitrary lines, Sketchpad will allow you to construct their points of intersection by selecting the lines and using an option on the Construct menu. If you construct ℓ and m as parallel lines, Sketchpad will not let you construct their point of intersection. In other words, Sketchpad is making the same assumption that Euclid made about parallel lines—that they do not intersect.

   b. Playfair's statement: *Given any line ℓ and any point P not on ℓ, there is exactly one line on P that is parallel to ℓ.*

      If, in Sketchpad, you select line ℓ and point P not on ℓ, you can construct a line through P that is parallel to ℓ. However, if you do this multiple times, Sketchpad simply constructs the same line over and over again. You can see this by asking Sketchpad to show the labels. The line through P parallel to ℓ will be labeled only once, because it is just one line. In other words, Sketchpad implements Playfair's Postulate by recognizing only one line through P parallel to ℓ.

## CONGRUENCE

In ordinary English, we sometimes use the words *congruous* or *congruent* to say that two things agree in nature or quality. For example, your instructor's use of Sketchpad as a teaching and learning tool might be congruent with the way she teaches other courses. In mathematics, *congruent* has a more exact or specialized meaning. Two geometric figures are said to be *congruent* if they are exactly the same size and shape, that is, if one could be superimposed exactly on the other to make a perfect fit. If two figures are the same shape but have different sizes, we say they are *similar*. All circles are similar (the same shape); if their radii are equal in length, the circles are also congruent. Two triangles, $\triangle ABC$ and $\triangle DEF$, are similar if there is a one-to-one correspondence between their vertices so that corresponding angles are congruent. Similarity is denoted by $\triangle ABC \sim \triangle DEF$. For two triangles to be congruent, there also must be a correspondence between congruent sides of the triangles. However, it is not necessary to compare all three pairs of angles and all three pairs of sides. There are many ways to verify congruence by checking only a few of these six items. For instance, you can use the famous side-angle-side or side-side-side conditions to prove that two triangles are congruent. You will examine conditions for congruent triangles in the exercises; you will study congruence of triangles in greater depth in Chapter 2.

## IDEAS ABOUT BETWEENNESS

In the two millennia since Euclid, mathematicians have wrestled with his five postulates and have come to understand that there are many issues that Euclid took for granted. One of these issues is *order* of points on a line, which is the notion that given any three collinear points, one of them will be between the other two. This seems pretty obvious, so it is easy to accept this as an axiom. However, it is neither stated nor implied by Euclid's five postulates. In the decades around 1900, many alternative sets of axioms were developed for Euclidean geometry by mathematicians such as Hilbert, Birkhoff, Veblen, and Pasch. Pasch, in particular, developed a careful theory of ordered geometry (Coxeter, 1969, 176–181).

In Activity 4, you saw a situation in which order is a critical concern. Suppose a line $\ell$ enters $\triangle ABC$ by crossing the side $AB$. What you are trying to decide in this activity is how the line $\ell$ leaves the triangle. There are not many choices. Line $\ell$ can intersect a vertex or it can cross another side. Perhaps $\ell$ intersects $\overleftrightarrow{BC}$ between $B$ and $C$—that is, on the segment $BC$, which forms one of the sides of the triangle. Can $\ell$ intersect $\overleftrightarrow{BC}$ at a point exterior to the triangle? Or might $\ell$ miss $\overleftrightarrow{BC}$ entirely? In that case, $\ell$ must intersect $\overleftrightarrow{AC}$, and the same questions arise for this line. Did you find a way to have $\ell$ intersect all three sides without passing through a vertex?

Here are two theorems related to order that seem obvious and that are sometimes useful when you want to show that a line or a ray goes where you expect it to go:

**THEOREM 1.1**  **Pasch's Theorem**  If $A$, $B$, and $C$ are distinct, noncollinear points and $\ell$ is a line that intersects segment $AB$, then $\ell$ also intersects either segment $AC$ or segment $BC$.

-------------------------------------------

Pasch actually used this statement as one of his axioms, as did Hilbert. It is not difficult to prove if we allow another basic assumption.

**Proof of Theorem 1.1**   We will assume that any line divides the plane into two separate pieces. In other words, we accept that whenever point $A$ is on one side of the line and point $B$ is on the other side, the segment $AB$ intersects the line.

Suppose that $\ell$ does not intersect $BC$ or $AC$. Thus, points $B$ and $C$ are on the same side of $\ell$. Furthermore, points $A$ and $C$ are on the same side of $\ell$. Therefore, $A$ and $B$ are on the same side of $\ell$, implying that $AB$ does not intersect $\ell$. We know that $\ell$ does intersect $AB$, so this is impossible.

Pasch's Theorem can be used to prove the following useful theorem. This proof will be left to the exercises.

**THEOREM 1.2**   **Crossbar Theorem**   If $\overrightarrow{AD}$ is between $\overrightarrow{AC}$ and $\overrightarrow{AB}$, then $\overrightarrow{AD}$ intersects segment $BC$.

## CONSTRUCTIONS

We are making a distinction in this course between *drawing* a figure and *constructing* it. If you are careful—that is, if you have a steady hand and a good eye—it is possible to make a very nice drawing that looks like the figure you are trying to construct. For example, you could probably draw a fairly believable equilateral triangle. However, the drawing will not be very robust. If you select one of the vertices and drag it, your equilateral triangle will quickly become scalene. But if you have constructed an equilateral triangle, you can select and drag various points, and although your triangle might change size, it will continue to be equilateral.

In Activity 6, you were asked to construct an equilateral triangle. Because there is no tool or command in Sketchpad to construct an equilateral triangle, you had to develop a strategy using circles and intersection points to make your construction robust. One way to construct a pair of equilateral triangles is to draw a line segment $AB$. Then construct a circle centered at $A$ through the point $B$ and another circle centered at $B$ through the point $A$. These two circles will intersect in two points, $C$ and $D$. Both triangles, $\triangle ABC$ and $\triangle ABD$, are equilateral. Why does this construction guarantee that the sides of these triangles are all the same length? How do you know? Can you use one of Euclid's postulates to prove that this construction is correct?

If you were able to construct an equilateral triangle on the segment $AB$, you should also be able to find the midpoint of the segment $AB$, as well as to construct the perpendicular bisector of the segment (see Figure 1.1). Can you prove that your construction is correct? Look for congruent triangles to help you.

Suppose you are given an arbitrary line $m$, and you want to construct a line perpendicular to $m$ from some point $P$. You can construct a circle, centered at $P$, that intersects $m$ in two points, $R$ and $S$. Now using a construction very similar to constructing an equilateral triangle on $RS$, you should be able to construct the line $\ell$ through $P$ that is perpendicular to $m$. This construction should work whether or not $P$ is on line $m$. If $P$ is not on $m$, the point where $\ell$ intersects $m$ is called the *foot of the perpendicular* from $P$ to $m$.

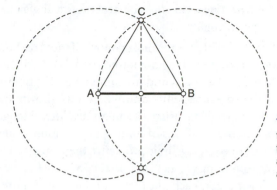

**FIGURE 1.1**
Constructing an Equilateral
Triangle

The constructions to find both
the perpendicular bisector and
the midpoint of a segment
are similar to constructing
an equilateral triangle on segment *AB*.

A line *t* that is tangent to a circle will be perpendicular to the radius of the circle at that point. For a point *A* on the circle, this suggests an easy way to construct a tangent to a circle at that point. From the center of the circle, you can construct a radius (that is, the segment from the center to the point *A*); then you can construct a perpendicular to the radius at point *A*. However, the construction is a little different if point *A* is not on the circle. Can you figure out how to do this? (*Hint:* Use an idea from Activity 8). We will let you ponder this question for a while.

Most geometry textbooks define $\angle PQR$ as two rays, $\overrightarrow{QP}$ and $\overrightarrow{QR}$, that share a common vertex (in this case, the point *Q*). However, Sketchpad requires you to select three points to designate an angle. For example, to designate $\angle PQR$, you must select the points *P*, *Q*, and *R*, in that order. If *P*, *Q*, and *R* are collinear, Sketchpad will allow you to measure the angle, but it will not allow you to bisect it. Of course, bisecting a straight angle is equivalent to erecting a perpendicular at the vertex of the angle; so, in practice, this is not a problem.

Sketchpad provides a command that will bisect an angle for you. But how would you construct the bisector of $\angle PQR$ without using **Angle Bisector** from the Construct menu? The angle bisector is the ray—in this case, $\overrightarrow{QX}$—that divides $\angle PQR$ into two equal angles. So the question arises of how to locate a suitable point *X*. Can you think of a way to do this? The method is in some ways similar to finding the midpoint of a line segment, and we will let you ponder this problem for a while.

From the time of the ancient Greeks up to the nineteenth century, mathematicians were fascinated with the question of whether certain quantities could be constructed using only a compass and an unmarked straightedge (not a ruler). The straightedge is used to construct straight objects—lines, segments, rays—and as such, it represents Euclid's First and Second Postulates. The compass is used to construct circles, and as such, it represents Euclid's Third Postulate. Three construction problems in particular kept mathematicians' attention:

- *Doubling a cube*—constructing a cube with a volume twice that of any given cube

- *Squaring a circle*—constructing a square with the same area as any given circle

- *Trisecting an angle*—constructing an angle one-third the measure of any given angle

This last problem is the most famous of the three because of the many, many people who have claimed to have found a valid construction. These three problems were attempted by Euclid and his contemporaries but resisted solution for 2000 years.

In the nineteenth century, it was shown that problems of constructing geometric figures using a compass and straightedge could be translated into corresponding problems of finding the solution of algebraic equations (Eves, 1976, 410). For some algebraic problems, there is no solution in real numbers. Thus, the corresponding geometric constructions are impossible to realize in the real plane. Using this approach, the three famous problems mentioned above were proven to be impossible constructions. But you should be careful not to leap too quickly to the conclusion that a particular construction is impossible. Some problems, while possible, are simply very difficult!

## PROPERTIES OF TRIANGLES

Triangles may be classified as *equilateral* (all three sides are the same length), *isosceles* (at least two sides are the same length), or *scalene* (all three sides have different lengths). A triangle may also be classified as a *right triangle* if it has a right angle. Some right triangles are scalene, and some are isosceles. Could a right triangle be equilateral? Why or why not?

Triangles will be similar if corresponding pairs of angles have the same measure and corresponding sides of similar triangles have a constant ratio. If two triangles (or any two geometric figures) are congruent, they are also similar. In the congruent case, the constant ratio will be 1. Congruent figures are identical except for their position.

When we look at a triangle, the angles we usually see are the three *interior angles* of the triangle. In Activity 5, you were asked to think about the exterior angles of a triangle. Any *exterior angle* has an obvious relation to its neighboring interior angle (they are supplementary), but in a triangle, an exterior angle is also related to the other interior angles. As you worked on this activity, what did you observe about the exterior angles? Here are two possible conjectures that you might have made:

**Conjecture 1**    An exterior angle of a triangle will have a greater measure than either of the nonadjacent interior angles.

**Conjecture 2**    The measure of an exterior angle of a triangle will be the sum of the measures of the two nonadjacent interior angles.

Conjecture 1 is actually a theorem, which means it has been proven to be true for all triangles. It is called the *Exterior Angle Theorem*. How would you prove this theorem? You will get a chance to do this proof in the exercises.

Conjecture 2 is a little harder to prove. This may surprise you, for a little algebra seems to prove it quite quickly:

$$\text{exterior angle at } A = 180° - \angle A$$
$$= 180° - (180° - (\angle B + \angle C))$$
$$= \angle B + \angle C.$$

What is the big assumption in this proof? It may be difficult to recognize this as an assumption, for it is something you have been told for years. The sum of the (interior) angles in a triangle is, in fact, a more complicated question than you might think. We will examine this question in some detail in Chapter 9.

Once we prove the Exterior Angle Theorem, there is another theorem that is very quick to prove. When one theorem follows very easily from another, we often call it a *corollary* to that theorem.

**COROLLARY 1.3**     **Corollary to the Exterior Angle Theorem**   A perpendicular line from a point to a given line is unique. In other words, from a specified point, there is only one line perpendicular to a given line.

-------------------------------------------------

How would you prove this corollary?

Exterior angles are not limited to triangles. Any polygon has both interior angles and exterior angles. Do you think the Exterior Angle Theorem will work for quadrilaterals? For polygons with even more sides?

You will have an opportunity to prove some properties of triangles in the exercises, and you will study triangles in greater depth in Chapter 2.

## PROPERTIES OF QUADRILATERALS

In Activity 2, did you observe that both diagonals of a convex quadrilateral are interior to the quadrilateral? In fact, a line segment joining any two points of a convex quadrilateral will lie entirely in its interior. This property is one possible definition of *convex*.

There are many special types of quadrilateral. Activity 9 introduces a type that may be new to you. A quadrilateral is *cyclic* if its vertices lie on a common circle. Another way to say this is that a cyclic quadrilateral can be *inscribed* in a circle. Some familiar quadrilaterals, such as squares, are cyclic, but others are not necessarily cyclic. What did you observe about opposite angles in a cyclic quadrilateral? Does your observation still hold if the center of the circle is exterior to the quadrilateral—that is, if the quadrilateral is entirely to one side of the circle? You should be able to use facts from Activity 7 to prove your conjecture. How is your conjecture different if the quadrilateral is self-intersecting?

If the diagonals of a quadrilateral bisect each other, will the quadrilateral be a parallelogram? A rectangle? A cyclic quadrilateral? Can you prove your answers?

## PROPERTIES OF CIRCLES

A *circle* can be defined as a set of points that are equidistant from a fixed center point. Notice in this definition that the circle itself is only the set of points at a fixed distance from the center. This fixed distance from the center is the *radius* of the circle. Points that are closer to the center are not on the circle; rather they are interior to the circle. Points whose distance from the center of the circle is greater than the radius are exterior to the circle.

If *PR* is a fixed chord of a circle and *Q* is any other point on the circle, then ∠*PQR* is *subtended by the chord PR*. This angle is an *inscribed angle* of the circle. The angle *PCR*, where *C* is the center of the circle, is a *central angle* of the circle. As you worked on Activity 7, how did ∠*PQR* change as you moved *Q* around the circle? What did you observe about the relationship between the central angle and the inscribed angles for a fixed chord *PR*? Is your conjecture still true if the central angle is equal to 180°? What if it is greater than 180°? Notice that Sketchpad does not measure angles greater than 180°, though such angles can easily occur in this activity. Does this help you state a conjecture? Isosceles triangles can help you prove your conjecture.

If *PR* is a diameter of the circle, as in Activity 8, the central angle, ∠*PCR*, is formed by two opposite rays, $\overrightarrow{CP}$ and $\overrightarrow{CR}$. In this case, ∠*PCR* is sometimes called a *straight angle*. What is its measure? An angle that subtends a diameter of the circle is said to be an *angle inscribed in a semicircle*. To be consistent with your conjecture from Activity 7—or, if you've proven it, your theorem from that activity—the measure of an angle inscribed in a semicircle should be half the measure of the corresponding central angle. Is this what you observed?

These observations about inscribed and central angles will be useful throughout this course as you work on various problems involving circles.

## EXPLORATION AND CONJECTURE: INDUCTIVE REASONING

As you begin your study of college geometry, you will be invited to look at many examples and to explore many ideas using Sketchpad as a tool for your explorations. You will be asked to make a lot of observations and to formulate conjectures based on your observations. In general, a conjecture is a statement expressed in the form

**If** ... [hypothesis] ..., **then** ... [conclusion] ....

The hypothesis includes the assumptions you are making and the facts or conditions given in the problem. The conclusion is what you claim will always happen if the conditions named in your hypothesis hold. As you grow in mathematical maturity, you may be able to express your conjectures without using *If ... then ...*, but for now we encourage you to use this format. Because the conclusion must follow from the hypothesis, writing your conjectures in this format makes your hypothesis explicit and clear—and this will help you develop clear and robust proofs. Once you have developed a robust proof for a conjecture (and tested it by having your colleagues critique it with you), you can call it a theorem.

The process of making many observations and formulating conjectures based on your observations is called *inductive reasoning*. By examining many examples, you can begin to see patterns and make guesses—conjectures—about what might be true. This process of exploration or experimentation lies at the heart of scientific investigations. Sketchpad is a wonderful tool for examining examples in geometry. Once a figure is constructed, it is easily varied by dragging objects around the picture. By doing this, you get to explore many examples very quickly.

Forming a conjecture is a major step in mathematical investigations. The next step is justifying the conjecture—that is, finding an explanation of why the conjecture is true. This is *proof*, which relies on *deductive reasoning*. Chapter 2 introduces you to rules of logic and deductive reasoning. Throughout this course, we will continue to develop axioms and postulates that are part of the culture of geometry. As the course progresses, you will grow in your skill and confidence at developing valid geometric proofs.

## 1.4 EXERCISES

Give clear and complete answers to the exercises, expressing your explanations in complete sentences. Include diagrams whenever appropriate.

1. What does it mean for two line segments to be congruent? If two line segments were both on the $x$-axis, how would you check whether they were congruent? If two line segments were drawn in the $xy$-plane, how would you check whether they were congruent? What if they were drawn in 3-space? In $n$-space?

2. What does it mean for two circles to be congruent? What is the general form for the equation of a circle? How could you determine from their equations whether two circles are congruent?

3. For two triangles to be congruent, the three sides and the three angles of one triangle must be matched with the sides and angles of the other so that corresponding sides are congruent and corresponding angles are congruent. However, it is not necessary to verify all six items. It is often sufficient to check just three of them. For instance, showing that two sides and the included angle of one triangle are congruent to the corresponding parts of the second triangle is enough, for then the other side and the other

two angles are guaranteed to be congruent as well. This is the side-angle-side (SAS) criterion for congruence of triangles.
   a. List all possible triples of sides and angles (SAS, AAS, etc.).
   b. For each of the triples in your list, either explain why it will guarantee congruence of the two triangles or explain why it will not. (This does not have to be a formal proof; simply explain your answer.) Include diagrams with your explanations.

4. What are the common names of polygons with two, three, four, or more sides? What is the minimum number of sides for a polygon? What is the maximum number of sides?

5. Not all figures in plane geometry are polygons. Find the names of at least three plane figures that are not polygons. Write a definition of each. Make your definitions complete but minimal.

6. What is a rectangle? Rectangles have many properties, but not all of these need to be mentioned in the definition. Write a concise, minimal definition for *rectangle*.

7. What is a parallelogram? What is a minimal set of things you need to show to prove that quadrilateral $RSTU$ is a parallelogram? Can

you develop more than one strategy for doing this?

8. The family of geometric figures we call quadrilaterals includes many types. List the names of as many different quadrilaterals as you can. Write a concise minimal definition of each.

9. Is every square a rectangle? Is every rectangle a square? Are some parallelograms rectangles? Draw a Venn diagram showing the relationships among the various quadrilaterals. Your diagram should make it clear whether you can say "Every *X* is a *Y*," "Some *X*s are *Y*s" or "No *X*s are *Y*s."

10. What is a right angle? Formulate a definition for *right angle* that does not mention the measure of the angle (that is, do not use degrees, radians, or any other system of measurement in your definition). What is an acute angle? An obtuse angle? Again, avoid mentioning the measure of the angle in your definitions.

11. Write a few sentences explaining what is meant by the *interior* of a triangle.

12. Write a clear concise definition of *concurrence*.

13. What does it mean for a polygon to be convex? Write a concise definition of *convexity* for polygons in the plane. Illustrate your definition with both an example and a nonexample. Is there any polygon that is always convex? Is there any polygon that is never convex?

14. What is Viviani's Theorem? Your statement of this theorem should clearly indicate the hypothesis and the conclusion.

15. What is Varignon's Theorem? Your statement of this theorem should clearly indicate the hypothesis and the conclusion.

16. What is Pappus' Theorem? Your statement of this theorem should clearly indicate the hypothesis and the conclusion.

17. a. In $\triangle ABC$, suppose that $\angle A > \angle B$. Prove that $BC > CA$. (*Hint:* Show that the other two cases cannot occur. In other words, show that $BC$ cannot equal $CA$ and that $BC$ cannot be shorter than $CA$. Look for a contradiction in each case.)

b. Now suppose that $BC > CA$. Prove that $\angle A > \angle B$. (This is the *converse* of the implication in part a.)

18. Prove the following basic facts about triangles.
    a. A triangle is isosceles if and only if the base angles are congruent.
    b. The sum of the lengths of any two sides of a triangle is greater than the length of the third side. This is known as the *triangle inequality*.

19. What would you need to know to prove the Crossbar Theorem?

20. In Figure 1.2, *M* is the midpoint of *BC*, and *AM* and *EM* are congruent. Use this to prove that $\angle BCD > \angle B$. This proves part of the Exterior Angle Theorem. How would you prove the other part?

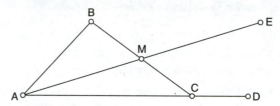

**FIGURE 1.2**
Figure for Exercise 20

21. Suppose that point *P* is not on line $\ell$. Prove that there is only one line through *P* that is perpendicular to $\ell$.

22. Prove these two statements.
    a. If two lines are intersected by a transversal so that a pair of alternate interior angles are congruent, then the lines are parallel.
    b. If two parallel lines are intersected by a transversal, a pair of alternate interior angles will be congruent. (To prove this you can assume Playfair's Postulate.)

23. Prove that a quadrilateral is cyclic if and only if each pair of opposite angles sums to 180°. (Notice that there are two things to prove.)

24. Prove or disprove that a parallelogram is a cyclic quadrilateral.

25. Prove or disprove that the perpendicular bisectors of the sides of a cyclic quadrilateral are concurrent.

26. Prove or disprove the following statements.
    a. The diagonals of a rectangle bisect each other.
    b. The diagonals of a parallelogram bisect each other.
    c. The diagonals of a cyclic quadrilateral bisect each other.
    d. If the diagonals of a quadrilateral bisect each other, then the quadrilateral is cyclic.
    e. If the diagonals of a quadrilateral bisect each other, then the quadrilateral is a rectangle.
    f. If the diagonals of a quadrilateral bisect each other, then the quadrilateral is a parallelogram.

27. Using Sketchpad, draw an arbitrary triangle. Construct the center, or midpoint, $M$, of this triangle. Drag the vertices of the triangle to new positions. Does $M$ continue to be the midpoint of the triangle? Compare your method of constructing the midpoint of a triangle with the method of at least one of your colleagues. Explain why your method of finding the midpoint of a triangle is a robust construction.

28. In Activity 10, you worked with the diagonal points of a hexagon. A hexagon has three diagonal points. How many diagonal points does a tetragon have? (What is a *tetra-gon*?) What about a pentagon? In general, how many diagonal points does an $n$-gon have?

29. Using only a compass and an unmarked straightedge (or only **Circle by Center+Point** and **Line** from the Construct menu), can you find a way to do each of the following constructions? Explain why you think your constructions work.
    a. Construct the midpoint $M$ of line segment $AB$.
    b. Bisect an angle $PQR$.
    c. Given a line $\ell$ and a point $P$ not on $\ell$, construct a line through $P$ *perpendicular* to $\ell$.
    d. Given a line $\ell$ and a point $P$ not on $\ell$, construct a line through $P$ *parallel* to $\ell$.
    e. Construct a tangent line $r$ to a circle $C$ from a point $A$ on the circle.
    f. Construct a tangent line $t$ to a circle $C$ from a point $X$ outside the circle.

The following exercises are more challenging.

30. a. Find a construction to inscribe an equilateral triangle in a circle. Do the same for a square and for a regular hexagon. (*Regular* means that all sides and all angles of the polygon are congruent.)
    b. Here is a construction to inscribe a regular pentagon in a circle: Construct a diameter $AB$ of the circle. At the center, $C$, construct a perpendicular line and let $D$ be one of the line's intersections with the circle. Let $E$ be the midpoint of $CD$. Bisect $\angle AEC$, and let $F$ be the intersection of this bisector with the diameter $AB$. Construct a line $\ell$ through $F$ that is perpendicular to $AB$. The points where $\ell$ intersects the circle, together with $A$, begin the pentagon. Carry out this construction and finish the pentagon.
    c. Prove that the construction in part b actually creates a regular pentagon.

31. The energetic search for solutions to the following problems profoundly influenced the development of geometry for nearly 2000 years before they were proven to be impossible constructions in the nineteenth century. Along the way, a lot of deep and interesting mathematics was developed.
    a. Given an arbitrary angle, $\angle RST$, can you find an approximate construction, using only a compass and an unmarked straightedge (or only **Circle by Center+Point** and **Line** from the Construct menu), to construct two rays, $\overrightarrow{RX}$ and $\overrightarrow{RY}$, that divide $\angle RST$ into three equal parts? How good is your construction? Explain why your construction is approximately correct.
    b. Given an arbitrary circle, $C$, with center at $\mathcal{O}$ and radius $r$, can you find an approximate construction, using only a compass and an unmarked straightedge (or only **Circle by Center+Point** and **Line** from the Construct menu), to construct a square that has the same area as $C$? How good is your construction? Explain why your construction is approximately correct.

Exercises 32 and 33 are especially for future teachers.

32. In the *Principles and Standards for School Mathematics,* the National Council of Teachers of Mathematics (NCTM, 2000, 11) recommends six specific principles for school mathematics that address overarching themes.

   a. Find a copy of the *Principles and Standards,* and study Chapter 2, Principles for School Mathematics (pages 11–27). What are these principles? What do they mean for your future students?

   b. What are specific NCTM recommendations regarding the use of technology in teaching school mathematics?

   c. Find copies of school mathematics texts for the grade levels for which you are seeking teacher certification. How is the NCTM Technology Principle implemented in those textbooks? Cite specific examples.

   d. Write a report in which you present and critique what you learn.

33. Design several classroom activities to introduce a geometric concept using manipulatives or technology that would be appropriate for students in your future classroom. Write a short report explaining how the activities you design reflect both what you have learned in studying this chapter and the recommendations of the NCTM.

Reflect on what you have learned in this chapter.

34. Review the main ideas of this chapter. Describe in your own words the concepts you have studied and what you have learned about them. What are the important ideas? How do they fit together? Which concepts were easy for you? Which were hard?

35. Reflect on the learning environment for this course. Describe aspects of the learning environment that helped you understand the main ideas in this chapter. Which activities did you like? Dislike? Why?

## 1.5 CHAPTER OVERVIEW

In this chapter, you have been asked to do a lot of explorations. You have been asked to observe what is going on, to reflect on what you see, and to make conjectures about your observations. This kind of reasoning, which is grounded in your observations and experiences, is called *inductive reasoning.* In mathematics, as in the sciences, *exploration* leads to *conjecture.* Your conjectures may or may not be valid theorems in geometry. As you work with your colleagues in this course, you will be challenged again and again to justify your conjectures. Those conjectures that can be justified—that is, proven—are called *theorems.* If another result follows very quickly from a particular theorem, it is often called a *corollary* to that theorem.

You also have been reminded of the vocabulary of plane geometry. Much of this geometric language is probably familiar to you, and it has been included here as a review. However, some of the geometric terms mentioned in this chapter may be new for you—or they may be familiar English words used in a precise and technical way in geometry. You may find it valuable to create a personal Geometry Dictionary, in which you add new words as you come across them throughout this course.

How do we know what we know? What do we need to do to verify or prove our conjectures? How can we convince others that our conjectures are correct? We hope you are beginning to see the need for some kind of formal or structured way to construct proofs. We will discuss this in more depth in Chapters 2 and 3.

We spent a lot of time discussing Euclid's postulates and their interpretation in Sketchpad's environment. Each of these postulates can be represented visually, as can the many theorems that derive from the postulates. Sketchpad helps you see, literally, what is going on.

You will be using The Geometer's Sketchpad to *construct* illustrations of many of the ideas you are studying in this course. We make an important distinction between drawing a figure and constructing a figure. It is often relatively easy to sketch or draw a picture to represent an idea; it is quite a different matter to construct the same figure. Geometric constructions, using either a compass and straightedge or using The Geometer's Sketchpad, will challenge you to think more deeply about the underlying geometry of the idea. In this chapter, you have been invited to construct an equilateral triangle and to explain why this robust construction produces a triangle whose sides are all the same length.

One of the powerful features of Sketchpad is that it gives you the ability to define your own tools—that is, if there is a construction you use frequently, you can define and save it as a **Custom** tool. For example, you could define a custom tool to construct an equilateral triangle. Then, whenever you need to construct an equilateral triangle, you could do so by selecting your **ET** tool and clicking on two points.

Although Sketchpad offers tools and commands to do many of the constructions needed for this course, you should be able to do some basic constructions using the standard tools of a compass and straightedge (equivalently, you could use the **Compass** and **Straightedge** tools in Sketchpad). Furthermore, you should be able to explain (prove) why your constructions are correct. The constructions you should master include the following:

- Construct a perpendicular to a line from any point (on or off the line).
- Construct the perpendicular bisector of a given line segment.
- Construct the foot of the perpendicular from a point $P$ to a line $\ell$.
- Construct the tangent line to a circle from a point on the circle.
- Construct the tangent line to a circle from a point not on the circle.
- Construct the bisector of a given angle.

Euclid's Fifth Postulate has been the focus of much attention in the history of mathematics. This postulate is a statement about two lines, $m$ and $n$, with a third line (or *transversal*), $t$, that intersects both $m$ and $n$. You worked with this situation in Activity 3.

If $m$ and $n$ are parallel, then eight angles are formed by the intersection of lines $m$, $n$, and $t$. We can label these angles with letters $a$ through $h$, as in Figure 1.3. These eight angles come in two sizes; that is, there are two sets of four congruent angles: $a$, $c$, $f$, and $h$ are in one set, and $b$, $d$, $e$, and $g$ are in the other set. Angles $a$ and $f$ are *vertical angles,* formed by a pair of intersecting lines. Vertical angles are congruent to each other. Angles $e$ and $d$ are *alternate exterior angles,* while angles $b$ and $g$ are *alternate interior angles.* If lines $m$ and $n$ are parallel, pairs of alternate interior angles will be congruent. The same is true for pairs of alternate exterior angles.

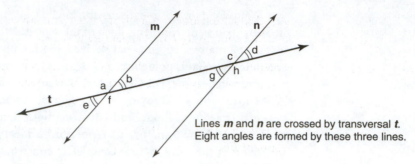

**FIGURE 1.3**
Parallel Lines *m* and *n* Crossed by
Transversal *t*

Lines *m* and *n* are crossed by transversal *t*.
Eight angles are formed by these three lines.

Angles *e* and *g* are *corresponding angles* on the same side of the transversal. Because angles *e* and *b* are vertical angles (thus, congruent to each other) and angles *b* and *g* are alternate interior angles (thus, also congruent to each other), we can conclude that angles *e* and *g* are congruent. In this way, we can prove that any pair of corresponding angles are congruent.

Angles *f* and *g* are *interior angles on the same side of the transversal.* These are the angles referred to in Euclid's Fifth Postulate. If the sum of the measures of these two angles is less than the sum of two right angles, Euclid's Fifth Postulate tells us that lines *m* and *n* will intersect on this side of line *t*. Playfair's Postulate is equivalent to Euclid's Fifth Postulate.

We recalled the notion of similar triangles, which have congruent angles and for which corresponding sides are in the same ratio. We briefly mentioned the tests for congruence of triangles. Congruent triangles are a very useful tool for proofs. After reminding you of various types of quadrilateral, we introduced the *cyclic quadrilateral,* and you explored the properties of its opposite angles. In a circle, inscribed angles are related to central angles in a very strong way.

Because this overview includes a summary of the chapter, it is appropriate for us to list the named theorems that have been introduced in this chapter. Some of these theorems have been given in this chapter and are listed here. Some of the theorems introduced in this chapter, also listed here, were left as exercises for you to write out.

**Viviani's Theorem**   *(Formulation in Exercise 14)*

**Varignon's Theorem**   *(Formulation in Exercise 15)*

**Pappus' Theorem**   *(Formulation in Exercise 16)*

**Euclid's Fifth Postulate**   If a straight line falling on two straight lines makes the sum of the measures of the interior angles on the same side less than the sum of two right angles, then the two straight lines, if produced indefinitely, meet on that side on which the angles are less than two right angles.

**Clavius' Axiom**   The set of points that are equidistant from a given line on one side of it forms a straight line.

**Playfair's Postulate**    Given any line $\ell$ and any point $P$ not on $\ell$, there is exactly one line containing $P$ that is parallel to $\ell$.

**Pasch's Theorem**    If $A$, $B$, and $C$ are distinct noncollinear points and $\ell$ is a line that intersects segment $AB$, then $\ell$ also intersects either segment $AC$ or segment $BC$.

**Crossbar Theorem**    If ray $AD$ is between rays $AC$ and $AB$, then ray $AD$ intersects segment $BC$.

**Exterior Angle Theorem**    An exterior angle of a triangle will have a greater measure than either of the nonadjacent interior angles.

**Corollary to the Exterior Angle Theorem**    A perpendicular line from a point to a given line is unique.

## CHAPTER TWO

In Chapter 1, you were asked to play with a variety of geometric figures—to experiment, to observe, and to make conjectures based on your observations. In doing so, you engaged in a process of *inductive* reasoning. In some problems, you had to provide reasons why your conjectures might be correct. In this chapter, you will begin to develop proofs of your conjectures using clear logical arguments. These mathematical proofs will use *deductive* reasoning. If this is your first course requiring mathematical proofs, you will need to study this chapter carefully—and you will find yourself referring back to this chapter many times throughout this course. If this is your second or third course requiring mathematical proofs, you might find that this chapter is a good review of ideas you have used in the past.

# Mathematical Arguments and Triangle Geometry

Do the following activities, writing your explanations clearly in complete sentences. Include diagrams whenever appropriate. You will be able to answer many of these questions by typing a sentence or two directly into your Sketchpad diagram. Get into the habit of saving your work for each activity, as later work sometimes builds on earlier work. You will find it helpful to read ahead into the chapter as you work on these activities.

1. In a triangle, an *altitude* is a line segment from a vertex perpendicular to the opposite side. Draw a triangle and construct the three altitudes. What do you observe about these segments? What happens for acute triangles, right triangles, and obtuse triangles?

2. In a triangle, a *median* is a line segment from a vertex to the midpoint of the opposite side. Draw a triangle and construct the three medians. What do you observe about these segments? What happens for acute triangles, right triangles, and obtuse triangles?

3. Draw a triangle and construct the three bisectors of the angles at the vertices. (*Note:* You can do this using **Construct | Angle Bisector.**) What do you observe about these segments? What happens for acute triangles, right triangles, and obtuse triangles?

4. Draw a triangle and construct the perpendicular bisectors of the three sides. What do you observe about these segments? What happens for acute triangles, right triangles, and obtuse triangles?

5. Construct a line $\ell$. Then construct a new point $Q$ on $\ell$. (Point $Q$ should not be one of the points you used to construct the line.) Construct a line parallel to $\ell$ with two points $P$ and $R$. Draw $\triangle PQR$, construct its interior, and measure its area. Use a command on the Display menu to animate point $Q$ along line $\ell$, and observe how the area changes. Can you explain what is happening?

6. You are given $\triangle ABC$. Segment $QR$ has been constructed through $B$ so that $QR \parallel AC$. Point $P$ is the intersection of $AR$ and $CQ$. Segment $AR$ intersects side $BC$ at $X$, segment $CQ$ intersects side $AB$ at $Z$. Ray $BP$ intersects side $AC$ at $Y$. Thus, $AX$, $BY$, and $CZ$ are concurrent at $P$. Find all pairs of similar triangles in Figure 2.1.

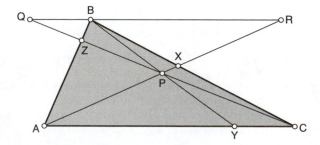

**FIGURE 2.1**
Figure for Activity 6

7. Draw an arbitrary △ABC. Let $X$ be a point on side $BC$ and $Z$ be a point on side $AB$. Lines $AX$ and $CZ$ intersect at $P$. Construct $\overleftrightarrow{BP}$, and find the point $Y$ where $\overleftrightarrow{BP}$ intersects $AC$. Calculate the ratios $\frac{AZ}{ZB}$, $\frac{BX}{XC}$, and $\frac{CY}{YA}$, and find their product. What do you observe about this product? Express your observation as a conjecture. Vary some of the points. Does your conjecture still hold?

8. Draw a triangle and construct the altitudes. The three points where the altitudes meet the sides of the triangle (or the sides of the triangle extended) are called the *feet* of the altitudes.
   a. Construct the circle through these three points. (*Hint:* If $A$ and $B$ are points on a circle, then $AB$ is a chord of the circle, and the center of the circle will lie somewhere on the perpendicular bisector of $AB$.)
   b. What other interesting points does this circle contain?

9. In one triangle, construct the points you discovered in Activities 1, 2, and 4. (You might want to hide some things for clarity.) What do you observe about these three points?

10. Consider an arbitrary triangle, △$ABC$.
   a. Construct point $X$ on side $BC$ and point $Y$ on side $AC$. Construct $\overleftrightarrow{XY}$.
   b. Find the intersection of $\overleftrightarrow{XY}$ and $\overleftrightarrow{AB}$. Label this intersection point $Z$. Try to manipulate the diagram so that $Z$ lies between $A$ and $B$. What is happening?
   c. Calculate the ratios $\frac{AZ}{ZB}$, $\frac{BX}{XC}$, and $\frac{CY}{YA}$, and find their product.
   d. What do you observe? Express your observation as a conjecture.

## 2.2 DISCUSSION

## DEDUCTIVE REASONING

Deductive reasoning is a process of demonstrating that if certain statements are accepted as true, then other statements can be shown to follow logically from those statements. Starting with a set of assumptions—the hypotheses—deductive reasoning builds a chain of implications that lead to a final conclusion. One statement leads to the next statement, which leads to another statement, and so on until the conclusion is reached. As long as each implication in the chain is logically valid, the assumptions at the beginning inevitably lead to the conclusion at the end.

The preceding paragraph suggests several important questions: What sort of statements can be used in deductive reasoning? How are statements combined to form implications? What does it mean for an implication to be logically valid? These questions will be addressed in the following paragraphs.

A mathematical argument, or any instance of deductive reasoning, is built up from statements. A *statement* is a declarative sentence that may be either closed or open. A *closed statement* is a sentence that is either true or false, whereas a statement that has a variable is called an *open statement*. Once the variable is specified, the statement's truth value can be determined. Here are some examples of statements.

- *All houses are red.* We know from experience that some houses are red and some are not. This is a closed statement, and its truth value is false.

- *The house is red.* This is an open statement. To decide whether the statement is true or false, we need to know which house is being discussed. Once we know which house, we will be able to decide the truth value of this statement. In this statement, the word *house* is a variable. Once the variable is specified, we can determine the truth value of the statement.

- *Today is Thursday.* This is an open statement that will be true on Thursdays and false on other days. So *today* can be considered a variable.

- *This figure has four vertices.* This is another example of an open statement. To determine the truth value of the statement, we need to know which figure is being pointed out. So, in this statement, *figure* is a variable.

- *Rectangles have five sides.* We know that rectangles have four sides, so this statement is false.

- *Pentagons have four sides.* A pentagon is a 5-gon, so by definition it has five sides. Because it has five sides, it certainly has four! This makes the statement true. However, the statement "Pentagons have *only* four sides" is false.

If a sentence cannot take on a truth value—that is, if it cannot be evaluated as true or false—then it is a *nonstatement*. The following are some examples of nonstatements:

- *Construct a right angle.* This is a command, or an imperative sentence. It is not a declarative sentence, so it is not a statement.

- *Is the triangle isosceles?* This is a question, or an interrogative sentence. It is not a declarative sentence, so it cannot be a statement.

- *This sentence is false.* Think about this for a minute. Is it true? If so, what would that mean? Is it false? If so, what would that mean? What is going on here?

Notice that in these nonstatements, it does not make sense to ask if the sentence is true or false. The sentence "The angle is right" can be evaluated as true or false, and thus it is a statement. However, it does not make sense to ask if the command "Construct a right angle" is true or false. A simple way to determine whether a sentence is a statement is to ask yourself if it makes sense to talk about the truth value of the sentence.

## RULES OF LOGIC

Many interesting statements are built up from simpler statements. These simpler statements are combined by the logical connectives *and,* and *or.* For example,

- This figure has a right angle, *and* two of its sides are congruent.

- Triangles have three vertices, *and* every rectangle is a square.
- This figure is a rectangle, *or* it is a rhombus.
- The diagonals of a parallelogram bisect each other, *but* the diagonals of a cyclic quadrilateral do not necessarily do so.

To evaluate the truth value of compound statements like these, we first determine the truth value of each component of the statement, then we combine these truth values using rules of logic. For example, the first statement has two components: "This figure has a right angle" is the first component, and "two of its sides are congruent" is the second component. The word *figure* is a variable in each of these components. To evaluate this compound statement, we have to know which figure is being discussed. Then we have to decide whether each part of the statement is true or false for that particular figure. *Both* components of the statement must be true to make the compound statement "This figure has four vertices, *and* two of its sides are congruent" be true. (There are figures that make both parts of this statement true and, hence, make the entire statement true. Can you draw a few examples? Also draw a few nonexamples, that is, figures that make one or both components false.)

In the second statement, we know that triangles indeed have three vertices. But some rectangles are not square. A compound statement of the form *P and Q* is true only if both parts of the statement are true. So the compound statement "Triangles have three vertices, *and* every rectangle is a square" is false.

In the third sentence, we again have a statement with two components, but the components are connected by the word *or*. For this statement to be true, it is only necessary for one of the two components to be true. It can happen that both components are true, which also makes the statement true. Try to draw examples of figures that make the first component true but not the second; figures that make the second component true but not the first; and figures that make both components true. All of your examples will make the third statement true.

The fourth statement is proper English. How do we interpret it logically? There are two components—namely, "the diagonals of a parallelogram bisect each other" and "the diagonals of a cyclic quadrilateral do not necessarily do so." The difficulty is knowing how to treat the word *but* connecting these components. Typically, *but* is used to connect components that contrast in some way, as in the example statement. At its most basic level, however, *but* means *and*. Thus, we are deciding the truth value of "The diagonals of a parallelogram bisect each other, *and* the diagonals of a cyclic quadrilateral do not necessarily do so." Both components are true, so the statement is true.

There is another important logical operator—negation of a statement. The logical connectives *and* and *or* are used to connect two statements, but the word *not* is applied to a single statement. This allows statements such as "This figure is *not* a rectangle." The *not* in this statement reverses the truth value; thus, a rectangle makes the statement false, while a trapezoid makes the statement true.

Let us summarize:

**AND** The compound statement *P and Q* is true only when both components are true. This is called *conjunction*. The mathematical symbol for this is $P \wedge Q$.

**OR** The compound statement *P or Q* is true when one or both components are true. This is called *disjunction*. Its mathematical symbol is $P \vee Q$.

**NOT** The statement *not P* is true when *P* is false, and it is false when *P* is true. This is called *negation*. Its mathematical symbol is $\neg P$. (Some authors use $\sim P$ to represent the negation of *P*. We use $\sim$ to denote similar figures and $\neg$ for negation.)

Is it clear how to simplify $\neg\neg P$? Here is a harder question: How could you simplify $\neg(P \vee Q)$? Be careful; the obvious choice is not correct! Try it for a specific statement: "It is not the case that this figure is a rectangle or it is a rhombus." To make this negated statement true, its components must both be false. How could you say this more naturally?

## CONDITIONAL STATEMENTS: IMPLICATION

It is very common in mathematics to make statements of the form "*If* certain things are assumed to be true, *then* certain other things must also be true." In Activity 5, you may have said something like the following:

*If* a line is chosen parallel to the base *PR* of a triangle and vertex *Q* of the triangle is on this line, *then* the area of $\triangle PQR$ is the same no matter where *Q* is.

This type of statement is called a *conditional statement*, or an *implication*. The pattern is expressed as $P \rightarrow Q$ (read as "*P* implies *Q*"). In a conditional statement, the if-clause (or *P*) is the hypothesis, the condition that is being assumed to be true. The then-clause (or *Q*) is the conclusion, the result that is being deduced from the assumptions.

An implication statement can be true in several different ways. The fundamental idea is that a true hypothesis should always lead to a true conclusion and never to a false conclusion. Here is the beginning of a truth table for implication:

**TABLE 2.1**  **Truth Table for Implication**

| | | *P* implies *Q* |
|---|---|---|
| *P* | *Q* | $P \rightarrow Q$ |
| true | true | true |
| true | false | false |
| false | true | ? |
| false | false | ? |

There remains the question of what to do when the hypothesis is false. If the assumptions in the hypothesis are not true, what can the implication tell us? Actually, when the hypothesis is false, an implication tells us nothing. If vertex $Q$ does not have to stay on the chosen line parallel to the base of the triangle, then the area of $\triangle PQR$ does not have to stay the same. Certain positions for $Q$ will produce equal areas, but other positions will produce vastly different areas. In this situation, we cannot decide whether the conclusion of our statement is true or false. This does not, however, invalidate the implication. After all, when the hypothesis is true, the conclusion is indeed true. From a false hypothesis, it is possible to have a true conclusion or a false conclusion. The implications false → true and false → false are both true. (Now you can complete the truth table.)

In Chapter 1, you worked with Viviani's Theorem. If you expressed your conjecture in the form of a conditional statement, you might have written something like the following statement.

**THEOREM 2.1**    **Viviani's Theorem**    *If $P$ is a point interior to an equilateral triangle, then* the sum of the lengths of the perpendiculars from $P$ to the sides of the triangle is equal to the altitude of the triangle (Wells, 1991, 267).

Suppose the hypothesis of Viviani's Theorem is false. Perhaps $P$ is exterior to the triangle, or perhaps the triangle is not equilateral. It still might happen that the sum of the lengths of the three perpendiculars equals the length of an altitude. You can use Sketchpad to try an example of a nonequilateral triangle if you wish. Can you find a point where the sum equals the length of one of the altitudes and another point where it does not? The theorem is valid when its hypothesis is false, but in this case the theorem does not provide any information about the situation.

Let's look at a conditional statement that is false:

If we construct the diagonals of a trapezoid, then these diagonals bisect each other.

How can we rewrite this as a true statement? The goal is to write the negation of this implication, i.e., to reverse false to true. In formal symbols, we want to write $\neg(P \rightarrow Q)$ in a way that is easy to understand. Be very careful; many people do this incorrectly. Think back to the truth table for $P \rightarrow Q$. There is only one line of this table that is false, namely, the line where $P$ = true and $Q$ = false. This will be the only line that makes the negation true. We can write this symbolically:

$$\neg(P \rightarrow Q) \equiv P \wedge \neg Q,$$

which is read as "The negation of $P$ implies $Q$ is equivalent to $P$ and not $Q$." The negation of our false statement about trapezoids is as follows:

We construct the diagonals of a trapezoid, and these diagonals do not bisect each other.

This can be restated more simply as "The diagonals of a trapezoid do not bisect each other." This new statement is true.

The *converse* of an implication $P \to Q$ is the statement $Q \to P$. Notice that the converse of a statement interchanges the hypothesis and the conclusion. Because the implication and its converse use the same components, it is easy to confuse them. However, these two patterns do not mean the same thing. You should compare truth tables to see that the statement $P \to Q$ does not mean the same thing as its converse. The converse of Viviani's Theorem, for instance, is false:

> If the sum of the lengths of the perpendiculars from $P$ to the sides of a triangle is equal to the altitude of the triangle, then $P$ is a point interior to an equilateral triangle.

It is not difficult to create a counterexample to this converse.

One variation of a conditional statement—the *contrapositive*—can be very useful. The contrapositive of $P \to Q$ is the statement $\neg Q \to \neg P$. The contrapositive of a statement means the same thing as the original statement. Again, it is a valuable exercise to compare their truth tables.

Let us consider a famous example.

**Ceva's Theorem**   Let $\triangle ABC$ have point $X$ on $BC$, point $Y$ on $CA$, and point $Z$ on $AB$. If the lines $AX$, $BY$, and $CZ$ are concurrent, then

$$\frac{AZ}{ZB} \cdot \frac{BX}{XC} \cdot \frac{CY}{YA} = 1.$$

-------------------------------------------

*Concurrence* means that the three lines share a single intersection point. This is what we hope you observed in Activity 7. The contrapositive of this statement is as follows:

> Let $\triangle ABC$ have point $X$ on $BC$, point $Y$ on $CA$, and point $Z$ on $AB$. If
> $$\frac{AZ}{ZB} \cdot \frac{BX}{XC} \cdot \frac{CY}{YA} \neq 1,$$
> then the lines $AX$, $BY$, and $CZ$ are not concurrent.

Is it clear that Ceva's Theorem and its contrapositive mean the same thing? The theorem says that concurrence of the three lines implies that this product is exactly 1. The contrapositive says that failing to have the product equal 1 implies that the lines fail to be concurrent. These are equivalent statements.

Ceva's Theorem is an example of something stronger—a *biconditional statement*. In a biconditional statement, the original implication and its converse are both true. We can express this using the mathematical phrase *if and only if*, which is often abbreviated as *iff*. Thus, we can express Ceva's Theorem as follows:

**THEOREM 2.2**   **Ceva's Theorem and Its Converse**   In $\triangle ABC$ with point $X$ on $BC$, point $Y$ on $CA$, and point $Z$ on $AB$, the lines $AX$, $BY$, and $CZ$ are concurrent if and only if

$$\frac{AZ}{ZB} \cdot \frac{BX}{XC} \cdot \frac{CY}{YA} = 1.$$

-------------------------------------------

For biconditional statements, the two components are interchangeable, that is, they mean the same thing. We can express the pattern for a biconditional statement symbolically as

$$P \leftrightarrow Q \equiv (P \rightarrow Q) \wedge (Q \rightarrow P).$$

To prove an iff statement, it is necessary to prove two separate implications: both $P \rightarrow Q$ and $Q \rightarrow P$. So, a proof of Theorem 2.2 requires two major parts: a proof that concurrence implies the product equals 1, and a proof that a product equal to 1 implies concurrence. We will discuss these proofs later.

Get into the habit of expressing your conjectures clearly in the form of conditional statements. This will help you think clearly about what your hypotheses and your conclusions are. In turn, this will help you develop robust proofs.

## MATHEMATICAL ARGUMENTS

In this course, you will be asked many times to prove geometric statements. The process of developing a robust mathematical proof will challenge you to think very carefully about many things. Is your conjecture stated clearly and correctly? What exactly are your assumptions (hypotheses)? How are these assumptions related to the conclusion? Do the intermediate steps of your proof follow logically from the earlier steps? Have you overlooked anything? These are critical questions.

The first step in developing a robust proof is to write a clear statement of your conjecture. A conjecture is really a conditional statement. If the assumptions you are making are met, then your proof should demonstrate that the conclusion you are claiming will follow. So it is very important to make your assumptions explicit and to express them clearly. Your assumptions—those stated explicitly, as well as those that are implicitly understood (perhaps in the context of the problem)—form your hypothesis.

Sometimes writing a clear statement of your conjecture will show you the connection between the hypothesis and the conclusion. More often, however, finding this connection is the most difficult part of a proof. Try combining various parts of the hypothesis to see what they tell you. Experiment with diagrams to see how the parts of the hypothesis come into play. Rephrase the conjecture, then rephrase it again. Try to prove a special case of your conjecture. Creating proofs is an art that improves with practice, and you will get a lot of practice in this course.

The goal for a robust proof is to develop a *valid* argument, which is an argument that uses the rules of logic correctly. Each step must follow logically from what has come before it, whether from part of the hypothesis or from a statement already proven. Only then will your proof stand up to critical questioning by your colleagues. Your reader (or listener) will either agree with your conclusion or disagree with your assumptions.

Once a conjecture has been proven, we call it a theorem. If the hypothesis of the theorem is accepted as true and the argument is valid, then the conclusion must also be accepted as true.

We have seen that the rules of logic give us a way to determine the truth value of complicated statements. Rules of logic also give us strategies for proving (or disproving) conjectures. These will be discussed in more detail in Chapter 3, but let us mention them here.

**Modus ponens:** *If $P \rightarrow Q$ and $P$ are statements in a proof, then you can conclude $Q$.* If the statement $P$ is part of the hypothesis and you can explain why $P \rightarrow Q$, then you have proven the statement $Q$.

**Syllogism:** *If $P \rightarrow Q$, $Q \rightarrow R$, and $R \rightarrow S$ are statements in a proof, then you can conclude $P \rightarrow S$.* Syllogisms are used to make a *direct proof.* There is a chain of implications leading directly from the hypothesis to the conclusion. *Modus ponens* then completes the proof.

**Modus tollens:** *If $P \rightarrow Q$ and $\neg Q$ are statements in a proof, then you can conclude $\neg P$.* In essence, *modus tollens* suggests that you try to prove $\neg Q \rightarrow \neg P$, which means the same thing as $P \rightarrow Q$. This is called an *indirect proof* and is often an effective proof strategy.

## UNIVERSAL AND EXISTENTIAL QUANTIFIERS

Earlier in this chapter, we talked about open and closed statements. An open statement has a variable. To decide the truth value of an open sentence, we need further information about the variable; that is, we need to close the statement. There are two ways to close an open statement: *substitution* and *quantification.*

We have already seen several examples in which an open statement can be closed by specifying the particular object that is to be substituted for the variable. Here are a couple more examples:

- *The area of a circle is $\pi r^2$.* The radius, $r$, of the circle is a variable. By substituting the value of $r$, we can easily calculate the area of the circle.

- $x + 3 = 5$. This algebraic statement is an open sentence. By substituting a numerical value for the variable $x$, we close the statement and can determine its truth value. Some values for $x$ will make the statement true, while other values for $x$ will make the statement false. Either way, substituting a value for $x$ closes the statement.

Another way to close an open statement is to quantify the variable. In a formal way, an open statement can be viewed as a *predicate,* a function that takes in values for the variable and returns a truth value for the statement. If $S(x)$ is an open statement (a predicate), we must supply input values for $x$ to determine the truth value—the output—of the predicate statement.

Predicates are useful when we want to talk about a set of objects, such as the set of all squares. The predicate can be used to refer to any item from this set. This allows us to say such things as "All squares are rectangles." In this statement, we are talking about an entire set, the set of squares. Every member of that set is supposed to have the property of being a rectangle, that is, every square is supposed to have four sides and four right angles.

Let us be a little more formal about this. Let $S$ denote the set of all squares, and let $x$ represent a member of that set. Let $P(x)$ mean that the variable $x$ has the desired property. Thus, $P(x)$ is the predicate, and $x$ is its variable. Our statement that all squares are rectangles can then be represented by $\forall x \in S,\ P(x)$, which is read as "For all $x$ in the set $S,\ P$ of $x$." For this statement to be true, every member of the set $S$ must have the property $P$.

The symbol $\forall$ is called the *universal quantifier*. (The universal qualifier is an upside-down A—to remember this, think of *All*.) This quantifier says that every member of the universe—in this case, the universe of squares—must have the specified property. To prove a universally quantified statement, you must show that every possible member of the universe has the property stated by the predicate.

Here is another quantified statement: "Some rectangles are not squares." This can be expressed symbolically as $\exists x \in R,\ \neg P(x)$, which is read as "There is an $x$ in the set $R$ that is not $P$ of $x$." This statement uses a different quantifier, a different universe, and a different predicate. The predicate, $P(x)$, is "$x$ is a square." The universe is the set, $R$, of rectangles. The quantifier is the *existential quantifier*. (The existential quantifier symbol is a backward E—to remember this, think of *Exists*.) For the statement $\exists x \in R,\ \neg P(x)$ to be true, there must be at least one rectangle that is not a square. Existential statements can be easy to prove; it is enough to display one object that has the desired property.

*Venn diagrams* are useful tools for understanding quantified statements. Suppose we consider the following sets:

$P$ = the set of parallelograms

$R$ = the set of rectangles

$S$ = the set of squares

Figure 2.2 illustrates the following two statements:

$\forall x \in S,\ x \in R$;

$\forall x \in R,\ x \in P$.

The large rectangle represents the set of all quadrilaterals. Because every square is a rectangle, the set of squares is drawn entirely inside the set of rectangles. Why is set $R$ entirely inside set $P$?

Suppose $R$ was the set of rhombuses instead. How would the Venn diagram change? Now suppose $R$ was the set of kites; how would the Venn diagram change? (*A reminder:* A *kite* is a quadrilateral with two pairs of consecutive congruent sides.) This last one is a bit trickier. You must decide whether all squares are kites, whether only some squares are kites, or whether no squares are kites. Then you must draw the set $S$ accordingly. Are all kites parallelograms? If so, the region for the set of kites should be drawn inside the region for the set of parallelograms. If not, the set of kites might protrude out of the set $P$, or it might be completely outside $P$. You must decide. How should the set of trapezoids be drawn in this Venn diagram?

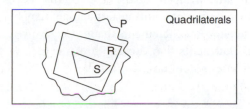

**FIGURE 2.2**
A Venn Diagram

# NEGATING A QUANTIFIED STATEMENT

While developing a proof, it is often necessary to negate a particular statement. For instance, you may be attempting an indirect proof. Using this approach, you would prove the contrapositive $\neg Q \rightarrow \neg P$, which is equivalent to the original implication $P \rightarrow Q$. On the other hand, you may disagree with a conjecture. (Be alert! There are a few false statements in the exercises.) In that case, the goal would be to prove that the given statement is false. In other words, you want to prove that its negation is true.

So, it is important to be able to state negations, of either the entire conjecture or portions of it. We have already seen how to do this for conditional statements:

$$\neg(P \rightarrow Q) \equiv P \wedge \neg Q.$$

Following are the negation patterns for quantified statements.

- *To negate a universally quantified statement:* $\neg(\forall x,\ P(x))$ means the same thing as $\exists x,\ \neg P(x)$. For example, the negation of the statement "All right triangles are isosceles" is "There is (at least one) right triangle that is not isosceles."

- *To negate an existentially quantified statement:* $\neg(\exists x,\ P(x))$ means the same thing as $\forall x,\ \neg P(x)$. For example, the negation of the statement "There is a quadrilateral for which the diagonals do not intersect" is "Every quadrilateral has diagonals that intersect."

(In each of these examples, which statement is true: the original or its negation?)

Sometimes the quantification of a particular statement may not be so obvious or explicit. To negate a statement, we need to think carefully about what is meant. For example, consider the statement "Students like geometry." Does this mean that all students like geometry? In this case, the statement is universally quantified, and its negation is "There is at least one student who doesn't like geometry." On the other hand, if "Students like geometry" is taken to mean that some students like geometry, then it is existentially quantified, and its negation is "All students don't like geometry" (which can be more simply expressed as "No students like geometry").

Notice that the negation of *all* is not *none*. For instance, the statement "All triangles are equilateral" is clearly false. However, the statement "No triangles are equilateral" is also false. Try to write the negations of these two statements. Your negations of each statement will be true, of course, and they will state different things.

Statements that use key words such as *all, any,* or *every*—as well as statements that express an idea about *all, any,* or *every* without actually using these key words—are universally quantified. You may have noticed in the examples above that the negation of a universal statement may be expressed using words like *not all* or *some do not.* Statements that use key words such as *there exists* or *some*—as well as statements that express the same idea without actually using these key words—are existentially quantified.

Suppose you are given the task of negating a statement such as Playfair's Postulate:

> Given any line $\ell$ and any point $P$ not on $\ell$, there is exactly one line $m$ through $P$ that is parallel to $\ell$.

Notice first of all that this statement involves layers (and layers (and layers)) of quantifications. Let us make this very explicit:

> Given any line $\ell$ (and any point $P$ not on $\ell$, (there is exactly one line $m$ through $P$ (that is parallel to $\ell$))).

This is a complicated statement involving three layers of quantification. Let us look just at the structure of the statement:

> $\forall\ell(\forall P(\exists m \text{ (something about } \ell,\ P, \text{ and } m)))$.

To negate a complicated statement of this form, begin at the outermost layer and work toward the center, layer by layer. Thus,

> $\neg(\forall\ell(\forall P(\exists m \text{ (something about } \ell,\ P, \text{ and } m))))$

is equivalent to

> $\exists\ell\ \neg(\forall P(\exists m \text{ (something about } \ell,\ P, \text{ and } m)))$,

which is equivalent to

> $\exists\ell(\exists P\ \neg(\exists m \text{ (something about } \ell,\ P, \text{ and } m)))$.

This, in turn, is equivalent to

> $\exists\ell(\exists P(\forall m\ \neg(\text{something about } \ell,\ P, \text{ and } m)))$;

and then we get the equivalent statement

> $\exists\ell(\exists P(\forall m \text{ (negation of something about } \ell,\ P,\ \text{ and } m)))$.

Finally, we have to decide how to negate the statement "there is exactly one . . . (whatever)." There are, in fact, two possibilities to consider. Either "there are none" or "there is more than one." So, one way to negate Playfair's Postulate would be to say:

> There is a line $\ell$, and there is a point $P$ not on $\ell$, such that *either* no lines through $P$ are parallel to $\ell$ *or* two or more (distinct) lines through $P$ are parallel to $\ell$.

You might express this more clearly as:

> There is a line $\ell$, and there is a point $P$ not on $\ell$, such that either every line through $P$ intersects $\ell$ or at least two distinct lines through $P$ are parallel to $\ell$.

Of course, it will take some practice before you are skilled at negating such a complicated statement!

In the history of mathematical thought, the negation of Playfair's Postulate led to the development of two new geometric worlds—two non-Euclidean geometries: one geometry in which there are no parallel lines, and one geometry in which there are multiple parallels. But we are getting ahead of our story.

# CONGRUENCE CRITERIA FOR TRIANGLES

In Exercise 3 from Chapter 1, you were asked to give informal explanations for the various combinations of criteria (SAS, AAS, and so on) that guarantee the congruence of two triangles. To make these criteria more formal, we will accept one of them as an axiom and develop proofs (or disproofs) of the others.

**Congruence Axiom for Triangles (SAS)**   If two sides and the included angle of one triangle are congruent respectively to two sides and the included angle of another triangle, then the two triangles are congruent.

This is the side-angle-side (SAS) criterion for congruence of triangles, and we accept this axiom without proof.

**THEOREM 2.3**   **ASA Criterion for Triangle Congruence**   If two angles and the included side of one triangle are congruent respectively to two angles and the included side of another triangle, then the two triangles are congruent.

**Proof**   We are given $\triangle ABC$ and $\triangle DEF$ with $\angle A \cong \angle D$, $AC \cong DF$, and $\angle C \cong \angle F$. (See Figure 2.3.) We want to prove that $\triangle ABC \cong \triangle DEF$. Can you justify each step in the following outline of the proof?

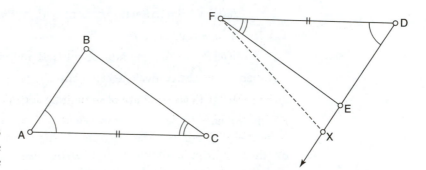

**FIGURE 2.3**
ASA Criterion for Triangle
Congruence

If $AB \cong DE$, these triangles would be congruent. So, assume that $AB$ is not congruent to $DE$. There is a point $X$ on $\overrightarrow{DE}$ so that $AB \cong DX$. By our assumption, the point $X$ is different from the point $E$. We know that $\triangle ABC \cong \triangle DXF$. This implies that $\angle C \cong \angle XFD$. But this is impossible, because $\angle C \cong \angle EFD$. So, $X$ must be the same point as $E$; that is, $AB \cong DX = DE$. (*Note:* We say $DX = DE$ instead of $DX \cong DE$ to emphasize that $DX$ and $DE$ are the same segment.)

Consequently, given a pair of triangles with two angles and the included side of one congruent respectively to two sides and the included angle of the other, we can establish that another pair of sides must be congruent. So, the triangles must be congruent by SAS.

# CONCURRENCE PROPERTIES FOR TRIANGLES

In Activities 1–4, the same sort of thing should have happened. In each activity, three lines were constructed from a triangle, and the three lines intersected at a single point. These lines are said to be *concurrent* at this point. There is nothing particularly special about two lines that intersect. It is much more remarkable, however, for three lines to share a common intersection.

The analogous situation for points is *collinearity*. Two distinct points are of course collinear—this is Euclid's First Postulate. It is more significant when three or more points are collinear. We will see an interesting example of this later in this chapter.

An *altitude* of $\triangle ABC$ is a line segment constructed from a vertex and perpendicular to the line containing the opposite side of the triangle. Because a triangle has three vertices, a triangle has three altitudes. The altitudes of $\triangle ABC$ are each perpendicular to one of the sides of the triangle, so they cannot be parallel to each other and, thus, must intersect. As you worked on Activity 1, you might have made a conjecture something like "If $AX$, $BY$, and $CZ$ are altitudes of $\triangle ABC$, then $AX$, $BY$, and $CZ$ are concurrent." Do you think these segments are really concurrent, or do they just appear so? Can you prove the following theorem?

**THEOREM 2.4**    The three altitudes of a triangle are concurrent. The point where they intersect, called the *orthocenter* of the triangle, is often denoted by $H$.

In Activity 1, you were asked to observe what happens to the orthocenter of $\triangle ABC$ as you varied the triangle. Changing the triangle will test how robust your construction is, for an altitude can be exterior to the triangle. For your diagram to be robust, you should have constructed each altitude as a line through a vertex and perpendicular to the line containing the other two vertices. Then, as you drag one vertex of the triangle around the sketch, you will see that sometimes the orthocenter is interior to the triangle, and sometimes it is exterior to the triangle. What conjecture can you make about when (under what conditions) the orthocenter is *interior to* (or *on* or *exterior to*) the triangle? Can you explain what is happening?

A *median* of $\triangle ABC$ is a line segment from a vertex to the midpoint of the opposite side. As you worked on the activities, you may have noticed that the three medians of $\triangle ABC$ appear to be concurrent. You may have conjectured that if $AM$, $BN$, and $CP$ are medians of $\triangle ABC$, then $AM$, $BN$, and $CP$ are concurrent. But are they really, or do they just appear to be so?

**THEOREM 2.5**    The three medians of a triangle are concurrent. The point where they intersect, called the *centroid*, is often denoted as $G$.

**Proof**    Here is one way to prove this theorem: Draw $\triangle ABC$, and construct two medians, $AD$ and $BE$. Label their point of intersection as $G$.

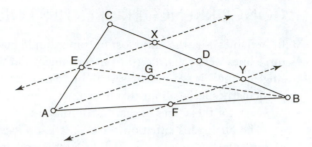

**FIGURE 2.4**
Concurrence of Medians

Look at Figure 2.4. Points $D$, $E$, and $F$ are the midpoints of their respective sides. The new lines, $\overleftrightarrow{EX}$ and $\overleftrightarrow{FY}$, are constructed to be parallel to the median $AD$. This construction creates pairs of similar triangles. It is easy to see that $CX$, $XD$, $DY$, and $YB$ are congruent. Segment $BE$ is divided into congruent segments as well; thus, $|BG| = 2|GE|$. In other words, $G$ is two-thirds of the way along the median $BE$.

Now start again. Let $G'$ be the intersection of $BE$ and $CF$. A similar argument shows that $G'$ is also two-thirds of the way along $BE$. Thus, $G$ and $G'$ are the same point, showing that the three medians intersect at a single point.

Unlike the situation of the orthocenter, the centroid is always interior to its triangle. Why do you think this happens? Is the Crossbar Theorem helpful here?

Consider the three angle bisectors that you worked with in Activity 3. Notice that each angle bisector is a ray, instead of a line segment or a line. (How does Sketchpad know which direction the ray should point?) Did you observe that the three angle bisectors are concurrent?

**THEOREM 2.6**  The three angle bisectors of a triangle are concurrent. The point where they intersect, called the *incenter*, is often denoted as $I$.

Here is an outline of a proof that the three angle bisectors of a triangle are concurrent. First sketch a diagram of $\triangle ABC$. (See Figure 2.5.) Let $\overrightarrow{AD}$ and $\overrightarrow{BE}$ be the bisectors of $\angle A$ and $\angle B$, respectively; mark their point of intersection as $I$. From point $I$, construct perpendiculars to each of the three sides of the triangle. Mark the feet of these perpendiculars as $W$, $X$, and $Y$. What can you say about $IW$, $IX$, and $IY$? (*Hint:* Think about congruent triangles.) Now draw $\overrightarrow{CI}$, and explain why it must bisect $\angle C$. (Again, think about congruent triangles.)

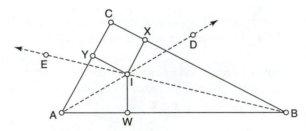

**FIGURE 2.5**
Angle Bisectors of a Triangle

By comparing the congruent triangles in this proof, we can see that the three perpendicular segments are congruent. Thus, we can construct a circle centered at $I$, with $IW$ as a radius. This is the *incircle*, the circle tangent to all three sides of the triangle.

In Activity 4, you observed that the three perpendicular bisectors of the sides of the triangle appear to be concurrent. By this time, you should be expecting the questions "Are these lines really concurrent? How do you know?"

**THEOREM 2.7**  The three perpendicular bisectors of the sides of a triangle are concurrent. The point where they intersect, called the *circumcenter*, is often denoted as $O$.

You will have a chance to prove this theorem in the exercises. The *circumcircle* has its center at $O$ and passes through the three vertices of the triangle. Is the circumcenter always interior to its triangle?

Look back at your work on Exercise 27 in Chapter 1. Which of the points of a triangle could most reasonably be called the "midpoint" of a triangle—the incenter, the circumcenter, the orthocenter, or the centroid? Why do you think so?

You may have noticed as you worked on Activity 9 that the circumcenter, the orthocenter, and the centroid appear to be collinear. These three points are indeed collinear, and the line containing them is called the *Euler line* of the triangle.

One way to prove that three points are collinear is to find the line through two of them and then show that the third point also lies on this line. Consider the line $OG$ through the circumcenter $O$ and the centroid $G$. Locate point $X$ on $\overleftrightarrow{OG}$ so that $G$ is between $O$ and $X$, and $2|OG| = |GX|$. You must show that $X$ lies on all three altitudes of the triangle.

Drop the perpendicular $OD$ from $O$ to side $BC$, draw the median $AD$, and draw the line $AX$. (See Figure 2.6.) Recall that $G$ is two-thirds of the way along $AD$. We have constructed the point $X$ so that $G$ is two-thirds of the way along $XO$. Thus, $\triangle GAX$ and $\triangle GDO$ are similar, which makes $AX$ parallel to $OD$. Because $OD$ is perpendicular to side $BC$, line $AX$ contains the altitude from vertex $A$.

Similar arguments establish that $X$ is also on the altitudes from vertices $B$ and $C$. Because distinct nonparallel lines intersect at a unique point, point $X$ must be the orthocenter ($H$) of the triangle.

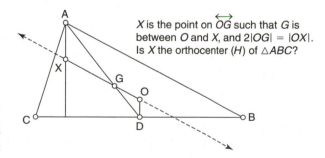

$X$ is the point on $\overleftrightarrow{OG}$ such that $G$ is between $O$ and $X$, and $2|OG| = |OX|$. Is $X$ the orthocenter ($H$) of $\triangle ABC$?

**FIGURE 2.6**
The Euler Line

# BRIEF EXCURSION INTO CIRCLE GEOMETRY

We say that two points determine a line, meaning that given any two points, there is exactly one line that goes through these points. This is one of Euclid's postulates and is implemented in Sketchpad as one of the allowable constructions—from two distinct points you can construct a line. Let us pose another question: How many points determine a circle?

- If we designate one point as the center and a second point to lie on the circle, then there is exactly one circle that can be drawn from that center through the specified point. The radius of the circle will be the distance between the two given points.

- Suppose we stipulate that two points $A$ and $B$ are to lie on the circle. How many different circles can be drawn through $A$ and $B$? Think carefully about where the center of the circle must lie.

- Suppose that we designate three points, $A$, $B$, and $C$, and we want to find the circle through these points. How many distinct circles can be drawn through these three points? If $A$, $B$, and $C$ are not collinear, it is helpful to think about the triangle formed by these three points. How can we construct a circle through the three specified points?

- Now consider four points $A$, $B$, $C$, and $D$. Can we be certain that a circle containing all four of these points is possible?

# THE CIRCUMCIRCLE OF $\triangle ABC$

If you draw a circle and mark three points on it, these points will not be collinear (unless the circle has infinite radius, which would be a very special case!). So these three points will form the vertices of a triangle. It is interesting to ask the question the other way around: If you start with a triangle, can you construct a circle that goes through the three vertices? If so, this would be the circumcircle of the triangle. Can you always construct a circumcircle, or is it only possible for certain special triangles? In other words, are triangles cyclic polygons? Is every triangle a cyclic polygon, or just special triangles?

Let's start by making the problem a bit simpler. Given two points $A$ and $B$, can we construct a circle through these two points? We will have to find a suitable point $X$ for the center of the circle, and $X$ must be the same distance from $A$ as it is from $B$. (Why?) One such point would be the midpoint of $AB$. More generally, if $X$ is any point on the perpendicular bisector of $AB$, then $X$ will be equidistant from the points $A$ and $B$. Can you prove this? (*Hint:* Look for congruent triangles.)

Returning to the question of constructing the circumcircle of $\triangle ABC$, we are looking for a point that is equidistant from the points $A$ and $B$ and simultaneously equidistant from the points $B$ and $C$. Thus, the point we are looking for must be on the perpendicular bisector for $AB$ and on the perpendicular bisector for $BC$. Can we be sure that these two lines—that is, the perpendicular bisectors of $AB$ and $BC$—really do intersect? Either they intersect, or they are parallel. If they were parallel, then $AB$ and $BC$ would be parallel as well. However, we know that $AB$ and

$BC$ intersect at point $B$ and that $AB$ and $BC$ are not the same line. Therefore, the perpendicular bisectors of these segments must intersect. This intersection point $O$ will be the center of the circle, with $OA$ as a radius. (Did you recognize *modus tollens* in this paragraph?)

Hence, we can say that every triangle is a cyclic triangle. The circle that passes through the vertices of a triangle is called its *circumcircle*.

## THE NINE-POINT CIRCLE: A FIRST PASS

For any $\triangle ABC$, the points where the altitudes intersect their opposite sides are called the *feet* of the altitudes. The triangle formed by these three points is sometimes called the *pedal triangle* of $\triangle ABC$, and of course, this triangle has a circumcircle. The circumcircle of the pedal triangle contains quite a few interesting points. As you worked on Activity 8, you might have noticed that not only does this circle pass through the feet of the altitudes, but it also passes through the midpoints of the sides of $\triangle ABC$. We will investigate this circle more closely in the next two chapters and will see that it is sometimes called the *nine-point circle* because it passes through nine interesting points.

Continue experimenting with Sketchpad to compare the length of the diameter of the circumcircle of the pedal triangle with the length of the diameter of the circumcircle for $\triangle ABC$. What do you observe? Does this ratio seem familiar? Can you prove your conjecture?

## CEVA'S THEOREM AND ITS CONVERSE

A *Cevian* is a line segment from a vertex of a triangle to a point on the line containing the opposite side. You have seen several examples of Cevians in this chapter. A median is one example of a Cevian, but not every Cevian is a median. An altitude of a triangle is also a Cevian.

**Ceva's Theorem**   In $\triangle ABC$, if the Cevians $AX$, $BY$, and $CZ$ are concurrent, then

$$\frac{AZ}{ZB} \cdot \frac{BX}{XC} \cdot \frac{CY}{YA} = 1.$$

**Sketch of a Proof**   You saw the beginning of a proof of this in Activity 6. (Look again at Figure 2.1.) In that activity, you found pairs of similar triangles. The pairs needed for the proof are

$$\triangle AZC \sim \triangle BZQ,$$
$$\triangle BXR \sim \triangle CXA,$$
$$\triangle CYP \sim \triangle QBP,$$
$$\triangle YAP \sim \triangle BRP.$$

From each similarity, we get equal ratios. From the first pair, we get

$$\frac{AZ}{ZB} = \frac{AC}{QB},$$

and from the third pair, we get

$$\frac{CY}{QB} = \frac{PY}{PB}.$$

You can derive two more such ratios. With these four equations, it is a simple matter to substitute into the product of the theorem and simplify to 1.

Although the converse of a theorem frequently is not true, the converse of Ceva's Theorem is also a theorem. How would you state the converse of Ceva's Theorem? Once this converse has been stated, we must find a proof for it. In this situation, an indirect proof can be helpful. Think about the converse restated in its contrapositive form. (Try this yourself before reading the next statement.)

In $\triangle ABC$, if the Cevians $AX$, $BY$, and $CZ$ are not concurrent, then $\frac{AZ}{ZB} \cdot \frac{BX}{XC} \cdot \frac{CY}{YA} \neq 1$.

To prove this statement, start by drawing two of the Cevians, and locate their intersection point. Construct a new Cevian that includes this intersection point. For this set of three Cevians, the product will equal 1. The product for the original set of Cevians has two of the same factors, with a third factor that is different. Thus, this product is different from 1. (You will be asked to write a full proof in the exercises.)

Using Ceva's Theorem and its converse, it is easy to prove that the medians of a triangle are concurrent. The proof that the altitudes of a triangle are concurrent (see Activity 1 and Exercise 39) is not much harder; it simply requires a bit of trigonometry. Many other special points, such as the Nagel point and the Gergonne point, can be defined from Ceva's Theorem. Some of these points will appear in the exercises.

## MENELAUS' THEOREM AND ITS CONVERSE

In $\triangle ABC$ of Activity 10, $AX$, $BY$, and $CZ$ are Cevian lines. Because point $Z$ does not lie on side $AB$, but rather on the extension of side $AB$, let us denote the ratio $\frac{AZ}{ZB}$ as negative. We need to do this because $\overrightarrow{AZ}$ and $\overrightarrow{ZB}$ are pointing in opposite directions. We then get Menelaus' Theorem:

**THEOREM 2.8**  **Menelaus' Theorem**  In $\triangle ABC$, suppose that point $X$ is on line $\overleftrightarrow{BC}$, point $Y$ is on $\overleftrightarrow{CA}$, and point $Z$ is on $\overleftrightarrow{AB}$. If the points $X$, $Y$, and $Z$ are collinear, then

$$\frac{AZ}{ZB} \cdot \frac{BX}{XC} \cdot \frac{CY}{YA} = -1.$$

How would you state the converse of Menelaus' Theorem, which is also a theorem?

## 2.3 EXERCISES

Give clear and complete answers to the questions, writing your explanations in complete sentences. Include diagrams whenever appropriate.

1. Identify whether each of the following is a statement or a nonstatement. If the sentence contains a variable, identify the variable. If possible, determine whether the sentence is true or false.
   a. Grizzly bears make affectionate pets.
   b. Some cows are purple.
   c. Do elephants really have good memories?
   d. $X$ is a quadrilateral.
   e. All rectangles are equiangular quadrilaterals.
   f. A kite is a certain kind of quadrilateral.
   g. The first day of spring comes in January.
   h. This class is held in the math lab.
   i. Wow! That is an awesome idea!
   j. All right triangles are equilateral.

2. Consider the sentence, "This sentence is false." Is this sentence true? Is it false? Write a short paragraph explaining what is happening here.

3. Explain how to evaluate (find the truth value) of a statement of the form $P \wedge Q$.

4. Set up a truth table for a statement of the form $P \vee Q$.

5. For each of the following statements, sketch several figures that make the statement true; then sketch several figues that make the statement false. If it is not possible to make some of these sketches, explain what is happening.
   a. This triangle has a right angle, and two of its sides are congruent.
   b. This triangle is equilateral, and it has a right angle.
   c. This quadrilateral is a rectangle, or it is a rhombus.
   d. This quadrilateral has a right angle, and its diagonals bisect each other.
   e. This quadrilateral is a kite, and it is not convex.
   f. This figure is a pentagon, and it is not convex.

6. What is the logical structure of the negation of a statement of the form $\neg(P \wedge Q)$? Express your answer as a rule, and use this rule to negate the statement "This figure is a rectangle, and it is equilateral."

7. What is the logical structure of the negation of a statement of the form $\neg(P \vee Q)$? Express your answer as a rule, and use this rule to negate the statement "This figure is a rectangle, or it is a rhombus."

8. Set up a truth table to compare a statement of the form $P \rightarrow Q$ with its converse and its contrapositive.

9. What is the converse of Ceva's Theorem?

10. What is the converse of Menelaus' Theorem?

11. The converse of Viviani's Theorem is false. Construct a counterexample to prove this.

12. Complete Table 2.1 (page 34), the truth table for implication.

13. A *trapezoid* is a quadrilateral with a pair of opposite sides parallel. (In England, this figure is called a *trapezium*.) Is a parallelogram a trapezoid? Explain your reasoning.

14. Determine whether each of the following statements is true or false.
    a. An octagon has more sides than a dodecagon has.
    b. If $a$ is a quadrilateral, then $a$ is not a square.
    c. Every rectangle has three sides, and all right triangles are equilateral.
    d. Triangle $XYZ$ is isosceles, or a pentagon is a five-sided plane figure.
    e. If $T$ is a right triangle, then $T$ might be an equilateral triangle.
    f. Every right triangle is isosceles.
    g. Every equilateral triangle is isosceles.
    h. Every isosceles triangle is a right triangle.
    i. No rhombus is a square.

15. Some of the statements in Exercise 14 are implications. For each implication, identify the hypothesis and the conclusion. Then write the

converse and the contrapositive of each implication.

16. For each implication from Exercise 14, write the negation.

17. Draw a Venn diagram that illustrates the relationships among various kinds of triangles. Your diagram should include isosceles, equilateral, and right triangles.

18. Draw a Venn diagram that illustrates the relationships among various kinds of quadrilaterals. Your diagram should include kites, parallelograms, rectangles, rhombi, squares, and trapezoids.

19. "All rhombi are squares." "No rhombi are squares." Are these statements negations of each other? Explain your answer.

20. Negate the following statements. In each case, determine which statement is true—the original statement or its negation.
    a. An angle inscribed in a semicircle is a right angle. (*Hint:* What type of quantifier is used in this statement?)
    b. Every triangle has at least three sides.
    c. Every rectangle is a square.
    d. There are exactly three points on every line.
    e. Through any two distinct points there is at least one line.
    f. Every rectangle has three sides, and all right triangles are equilateral.
    g. Triangle *XYZ* is isosceles, or a pentagon is a five-sided plane figure.

21. Negate the following statements.
    a. For every shape *A*, there is a circle *D* such that *D* surrounds *A*.
    b. There is a circle *C* such that every line ℓ intersects *C*.

22. Justify each statement in the proof of the ASA criterion for congruent triangles (see page 42).

23. Prove the SSS criterion for triangle congruence: If three sides of one triangle (△*ABC*) are congruent respectively to three sides of another triangle (△*DEF*), then the two triangles are congruent. (*Hint:* Because $AB \cong DE$, reposition the triangles so vertex *A* coincides with vertex *D*

and vertex *B* coincides with vertex *E*. Now you have a quadrilateral with two pairs of congruent sides. Can you show that any of the pairs of angles are necessarily congruent?)

24. The angle-angle-side (AAS) criterion for congruent triangles says, "If two angles and a nonincluded side of one triangle are congruent respectively to two angles and a nonincluded side of another triangle, then the two triangles are congruent."

    Following is a proof of the AAS criterion for triangle congruence. Justify each step.

    Suppose we are given △*ABC* and △*DEF* with $\angle A \cong \angle D$, $\angle C \cong \angle F$, and $BC \cong EF$. We want to prove that △*ABC* ≅ △*DEF*. If $AC \cong DF$, the triangles would be congruent. (Why?) If *AC* is not congruent to *DF*, then either $AC < DF$ or $AC > DF$.

    Assume $AC < DF$. Then there is a point *X* on *DF* so that $AC \cong XF$. This would make △*ABC* ≅ △*XEF*. So, $\angle A \cong \angle EXF$. But this cannot be, because $\angle A \cong \angle D$. (What theorem is violated?) So, *AC* is not shorter than *DF*.

    In a similar way, we can show that *AC* is not longer than *DF*. Because *AC* is neither shorter than nor longer than *DF*, it must be congruent to *DF*. So, △*ABC* ≅ △*DEF*, by SAS. Thus, if two angles and a nonincluded side of one triangle are congruent respectively to two angles and a nonincluded side of another triangle, then the two triangles are congruent.

25. Given two triangles, △*ABC* and △*DEF*, with right angles at ∠*A* and ∠*D*, with $AB \cong DE$, and with $BC \cong EF$, prove that △*ABC* ≅ △*DEF*. (*Hint:* Because $AB \cong DE$, reposition the triangles so that vertex *A* coincides with vertex *D*, vertex *B* coincides with vertex *E*, and points *C* and *F* lie on opposite sides of $\overleftrightarrow{AB}$. Can you show that points *A* and *D* lie on $\overleftrightarrow{CF}$?)

26. Construct a counterexample to show that SSA is not a criterion for congruent triangles.

27. Construct a counterexample to show that AAA is not a criterion for congruent triangles.

28. Suppose you are given *AB*. Construct two circles—one centered at *A* with radius *AB* and

the other centered at $B$ with radius $BA$. Let $C$ and $D$ denote the intersection points of these two circles. Draw the line segment $CD$. Mark the intersection of segments $AB$ and $CD$ as point $E$.

    a. Prove or disprove that $\triangle ABC$ and $\triangle ABD$ are equilateral.

    b. Prove or disprove that $E$ is the midpoint of $AB$.

    c. Prove or disprove that $E$ is the midpoint of $CD$.

    d. Prove or disprove that $AB$ is perpendicular to $CD$.

    e. Prove or disprove that $AB$ is congruent to $CD$.

    f. Prove or disprove that $ACBD$ is a parallelogram.

    g. Prove or disprove that $ACBD$ is a rhombus.

29. Draw line $m$ and point $P$ not on $m$. Construct the perpendicular from $P$ to $m$, using only a compass and straightedge or only **Circle by Center+Point** and **Line**. Prove that your construction is correct. (Sketchpad has a command to do this automatically, but we want you to figure out how to do it with only these two tools.)

30. Repeat the construction of Exercise 29 with the point $P$ on line $m$. Prove that your construction is correct.

31. Given $\angle PQR$, construct the bisector of this angle, using only a compass and straightedge or only **Circle by Center+Point** and **Line**. Prove that your construction is correct.

32. Prove that any point $X$ on the perpendicular bisector of $AB$ is the center of a circle through points $A$ and $B$.

33. Given three points $A$, $B$, and $C$, prove or disprove that there is exactly one circle through these points. What if $A$, $B$, and $C$ are collinear?

34. Write complete, detailed proofs of Ceva's Theorem and of its converse.

35. Prove that a median divides a triangle into two equal areas.

36. Use the converse of Ceva's Theorem to prove that the three medians of a triangle are concurrent.

37. Use the converse of Ceva's Theorem to prove that the three angle bisectors of a triangle are concurrent. (*Hint:* Use the Law of Sines.)

38. If the sides of a triangle are extended, they form *external angles* of the triangle. These angles can be bisected. Prove that the external angle bisectors at two vertices and the internal angle bisector at the third angle are concurrent. These points are the centers of the *excircles,* which are externally tangent to the extended sides of the triangle.

39. Use the converse of Ceva's Theorem to prove that the altitudes of a triangle are concurrent. (*Hint:* Use trigonometry.)

40. Let $H$ be the orthocenter of $\triangle ABC$. Where is the orthocenter of $\triangle HBC$? Prove your answer.

41. Prove that the Cevians joining the vertices of a triangle with the points of tangency of the incircle are concurrent. This is the *Gergonne point* of the triangle.

42. The three excircles of $\triangle ABC$ will be tangent to the (nonextended) sides at three points. Prove that the Cevians joining the vertices of the triangle with these points of tangency are concurrent.

43. Prove that the perpendicular bisectors of the sides of a triangle are concurrent. (*Hint:* Let $O$ be the intersection of two of the perpendicular bisectors. By finding congruent triangles, prove that the line through $O$ perpendicular to the third side is also a bisector.)

44. Prove that the circumcenter of a triangle is the center of a circle that contains all three vertices.

45. In this chapter, you learned about six of the special points of the nine-point circle. Make a diagram of an arbitrary triangle, and construct these six special points.

The following exercises are more challenging.

46. For $\triangle ABC$, suppose that point $X$ is halfway around the perimeter from $A$, that $Y$ is halfway around the perimeter from $B$, and that $Z$ is halfway around the perimeter from $C$. Prove that the Cevians $AX$, $BY$, and $CZ$ are concurrent.

This point of concurrence is called the *Nagel point* of the triangle.

**47.** Let $P$ be a point on the circumcircle of $\triangle ABC$. Drop perpendiculars from $P$ to the (possibly extended) sides of the triangle. Prove that the feet of these perpendiculars are collinear. This is the *Simson line* for point $P$. (*Hint*: Use the converse of Menelaus' Theorem, some trigonometry, and facts about subtended angles.)

**48.** Construct an example of Menelaus' Theorem in which all three of the ratios are negative. Also, give both a geometric reason and an arithmetic reason for why there cannot be an example with exactly two of the ratios negative.

**49.** Two triangles, $\triangle ABC$ and $\triangle A'B'C'$, are said to be *perspective from a point P* if the lines formed by corresponding points ($\overleftrightarrow{AA'}$, $\overleftrightarrow{BB'}$, and $\overleftrightarrow{CC'}$) are concurrent at $P$. Two triangles are said to be *perspective from a line m* if the intersection points of corresponding sides ($\overleftrightarrow{AB} \cap \overleftrightarrow{A'B'}$, $\overleftrightarrow{BC} \cap \overleftrightarrow{B'C'}$, and $\overleftrightarrow{CA} \cap \overleftrightarrow{C'A'}$) are collinear on $m$ (see Figure 2.7).

*Part of Desargues' Theorem*: If two triangles are perspective from a point, then they are perspective from a line.

   a. Create a diagram of Desargues' Theorem in which the point of perspectivity is between the two triangles.

   b. Create another diagram in which the point of perspectivity is interior to both triangles.

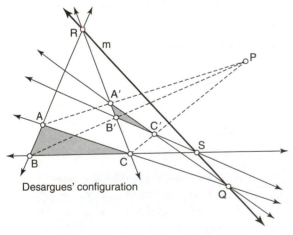

**FIGURE 2.7**
Figure for Exercises 49 and 50

**50.** Use Menelaus' Theorem to prove the part of Desargues' Theorem stated in Exercise 49. (*Hint*: In Figure 2.7, look at $\triangle ABP$ and the line $RA'B'$. Then consider $\triangle BCP$ and the line $B'C'S$, plus $\triangle ACP$ and the line $QA'C'$. Combine the Menelaus products from these three situations to get three points on the sides of $\triangle ABC$.)

   (*Note*: The converse of this statement is also true but is a bit more difficult to prove.)

**51.** For $\triangle ABC$, the angle bisector at $A$ will meet the opposite side $BC$ at point $D$. Use the Law of Sines to prove that

$$\frac{|BD|}{|DC|} = \frac{|AB|}{|AC|}.$$

Exercises 52–55 are especially for future teachers.

**52.** In the *Principles and Standards for School Mathematics* (NCTM, 2000, 56), the National Council of Teachers of Mathematics (NCTM) recommends that

> Instructional programs from prekindergarten through grade 12 should enable all students to
>
> • recognize reasoning and proof as fundamental aspects of mathematics;
> • make and investigate mathematical conjectures;
> • develop and evaluate mathematical arguments;
> • select and use various types of reasoning and methods of proof.

What does this mean for your future students?

   a. Find a copy of the *Principles and Standards,* and study the discussion of the Reasoning and Proof Standard (NCTM, 2000, 56–59). What are the specific NCTM recommendations with regard to reasoning and proof in school mathematics? What are some expectations that the NCTM has about the growth in mathematical reasoning of children and teenagers?

   b. In using this textbook to study college geometry, you are engaging in exploration and conjecture, and you are being asked to explain your observations and prove your conjectures. You are engaging in both inductive and deductive reasoning. How is the work you are doing in this course related to the Reasoning and Proof Standard developed by the NCTM? Cite specific examples.

c. Write a report in which you present and critique what you learn in studying the NCTM Reasoning and Proof Standard in light of your experiences in this course. Your report should include your answers to the questions in parts a and b.

53. Find copies of school mathematics texts for one of the grade levels for which you are seeking teacher certification.
   a. How is the NCTM Reasoning and Proof Standard reflected in the presentation of mathematics given in those school mathematics texts?
   b. Design several classroom activities involving reasoning and proof that would be appropriate for students in your future classroom.
   c. Write a short report explaining how the activities you design reflect both what you are learning in this class (in Chapters 1 and 2) and the NCTM's recommendations on Reasoning and Proof in school mathematics.

Reflect on what you have learned in this chapter.

54. Review the main ideas of this chapter. Describe, in your own words, the concepts you have studied, and what you have learned about them. What are the important ideas? How do they fit together? Which concepts were easy for you? Which were hard?

55. Reflect on the learning environment for this course. Describe aspects of the learning environment that helped you understand the main ideas in this chapter. Which activities did you like? Dislike? Why?

## 2.4  CHAPTER OVERVIEW

This chapter introduced some basic rules of mathematical reasoning, as well as a lot of ideas about triangles. Throughout this course, you will be experimenting with diagrams constructed in Sketchpad, and you will be making conjectures based on what you observe. You will use the rules of logic—the rules of mathematical inference—as you learn to develop mathematical proofs of your conjectures. Truth tables and Venn diagrams can also be helpful tools for logical reasoning.

A statement that uses the logical connective *and* is called a *conjunction,* while a statement that uses the connective *or* is called a *disjunction.* The negation of a conjunction is a disjunction, and the negation of a disjunction is a conjunction.

$$\neg(A \wedge B) \equiv \neg A \vee \neg B$$
$$\neg(A \vee B) \equiv \neg A \wedge \neg B$$

You should be able to negate a given statement, as well as to write both the converse and the contrapositive of a given conditional statement. The converse of a conditional statement simply reverses the order of the implication; that is, the *converse* of $A \to B$ is $B \to A$. The *contrapositive* of a conditional statement negates each part of the statement and reverses the order so that the contrapositive of $A \to B$ is $\neg B \to \neg A$. The contrapositive of a conditional statement means the same thing as the original statement, whereas the converse does not.

You should be able to recognize a statement that has *universal* or *existential* *quantifiers,* even if the quantifiers are not explicitly stated. You should be able to write a given statement using the symbolic language of logic—the existential and universal quantifiers $\exists x$ and $\forall y$, as well as the logical connectives *and* $(a \wedge b)$, *or* $(p \vee q)$, and *not* $(\neg s)$.

The process of negating a quantified statement is essentially a two-step process: You must first change the quantifier and then negate the statement:

$$\neg(\exists x\, P(x)) \equiv \forall x \neg P(x)$$
$$\neg(\forall x\, P(x)) \equiv \exists x \neg P(x).$$

A sentence such as the one in Exercise 2 ("This sentence is false.") is an example of a *paradox.* This sentence appears, on the surface, to be a simple statement; yet when we try to assign it a truth value, it gets quite complicated. No matter what truth value—true or false—we try to assign, it is not quite right. Paradoxes such as this are reminders that human thought is rich and complex, and not always reducible to simple true/false logic.

The area of a triangle can be measured by calculating one-half the product of the base and the height; that is,

$$\text{area}(\triangle ABC) = \frac{1}{2}(\text{base})(\text{height}).$$

The length of any side of the triangle can be taken as the base, while the height is measured as the length of the altitude perpendicular to that base. Because a triangle has three sides, there are three different calculations that we can use to find the area—and all three give the same result.

This chapter focused on language and theorems related to the geometry of triangles, including

- the altitudes and orthocenter of a triangle,
- the medians and centroid of a triangle,
- the incenter and incircle of a triangle,
- the excenters and excircles of a triangle,
- the Euler line of a triangle, and
- the nine-point circle of a triangle.

Any three noncollinear points determine a unique circle. For any $\triangle ABC$, the circle through the points $A$, $B$, and $C$ is called the *circumcircle* of $\triangle ABC$. The feet of the three altitudes of $\triangle ABC$ are also noncollinear; thus, they form another triangle, called the *pedal triangle,* of $\triangle ABC$. The pedal triangle has its own circumcircle, which passes through quite a few interesting points. As you worked on Activity 8, you might have noticed that this circle passes through the midpoints of the sides of $\triangle ABC$. The circumcircle of the pedal triangle actually passes through at least nine special points of $\triangle ABC$. For this reason, it is commonly called the "*nine-point circle.*" We will continue our investigations of this special circle in Chapters 3 and 4.

In this chapter's activities, you saw that the three altitudes of a triangle are concurrent, as are the three medians, the three angle bisectors, and the three perpendicular bisectors of the sides. The intersection point of the altitudes is called the *orthocenter,* often denoted by $H$. The medians intersect at the *centroid* $G$, and the perpendicular bisectors of the sides intersect at the *circumcenter O*. The circumcenter of a triangle is the center of the circle through the three vertices of the triangle. The three angle bisectors of a triangle are concurrent at a point $I$, called the *incenter* of the triangle. Point $I$ is also the center of the *incircle,* a circle tangent to all three sides of the triangle.

Three points—the orthocenter ($H$), the centroid ($G$), and the circumcenter ($O$)—lie on the *Euler line* of the triangle.

The theorems of Ceva and Menelaus are closely related. The first involves concurrence of three lines, while the second involves collinearity of three points.

**Ceva's Theorem**    In $\triangle ABC$, if the Cevians $AX$, $BY$, and $CZ$ are concurrent, then

$$\frac{AZ}{ZB} \cdot \frac{BX}{XC} \cdot \frac{CY}{YA} = 1.$$

**Converse of Ceva's Theorem**    *(Left as Exercise 9.)*

**Menelaus' Theorem**    In $\triangle ABC$, suppose that point $X$ is on $\overleftrightarrow{BC}$, point $Y$ is on $\overleftrightarrow{CA}$, and point $Z$ is on $\overleftrightarrow{AB}$. If the points $X$, $Y$, and $Z$ are collinear, then

$$\frac{AZ}{ZB} \cdot \frac{BX}{XC} \cdot \frac{CY}{YA} = -1.$$

(To make this work, a ratio must be considered negative if the two segments are in opposite directions.)

**Converse of Menelaus' Theorem**    *(Left as Exercise 10.)*

An important concept in geometry is the congruence of triangles. Two triangles are considered to be congruent if they have the same shape and the same size. Beginning with Exercise 3 of Chapter 1, you have been asked to think about conditions that are necessary for a given pair of triangles to be congruent. In this chapter, we accepted side-angle-side (SAS) as a congruence axiom for triangles.

**Congruence Axiom for Triangles (SAS)**    If two sides and the included angle of one triangle are congruent respectively to two sides and the included angle of another triangle, then the two triangles are congruent.

Using this axiom, we are able to prove several congruence theorems. The first of these is the ASA criterion for triangle congruence.

**ASA Criterion for Triangle Congruence**    If two angles and the included side of one triangle are congruent respectively to two angles and the included side of another triangle, then the two triangles are congruent.

You investigated several additional criteria for triangle congruence in Exercises 22–27.

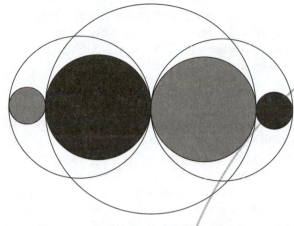

The Geometer's Sketchpad sees things through Euclidean eyes—that is, the default worldview of Sketchpad is the Euclidean plane. Each construction you are allowed to perform on a Sketchpad worksheet is related to Euclid's five postulates. As you work on the following activities, think about Euclid's postulates. Which is being applied as you use Sketchpad to construct each diagram? If the diagrams you construct are robust, you will be shaping your geometric reasoning according to a Euclidean perspective. After constructing a diagram in Sketchpad, reflect on the steps you needed to take to produce the diagram. This will give you a way to begin to set up your proofs.

# Circle Geometry, Robust Constructions, and Proofs

Do the following activities, expressing your explanations clearly in complete sentences. Include diagrams whenever appropriate. Save your work for each activity, as later work sometimes builds on earlier work. You will find it helpful to read ahead into the chapter as you work on these activities.

1.  Draw an arbitrary triangle, $\triangle ABC$.
    a.  Construct an equilateral triangle on each leg of $\triangle ABC$. The constructed triangles should remain equilateral when you drag the vertices of $\triangle ABC$ to new positions. Calculate the areas of each equilateral triangle. Drag the vertices $A$, $B$, and $C$ to see what happens when $\triangle ABC$ is acute, right, or obtuse. What do you observe? Make a conjecture.
    b.  Repeat part a, using circles on each side instead of equilateral triangles. Construct a circle on each leg so that the leg of the triangle is a diameter of the circle. Calculate the area of each circle. What do you observe? What happens when $\triangle ABC$ is acute, right, or obtuse? Make a conjecture.
    c.  Prove your conjectures from parts a and b.

2.  Construct a cyclic quadrilateral $ABCD$. Extend side $AB$, creating an exterior angle at $B$. Construct the diagonal $AC$. Measure the exterior angle at $B$ and each of the interior angles of the quadrilateral.
    a.  What is the relationship between the sizes of $\angle ABC$ and $\angle ADC$? Does your observation still hold if you move the vertices of the quadrilateral or if you change the size of the circle?
    b.  What is the relationship between the measure of the exterior angle at $B$ and that of each of the interior angles of the quadrilateral? Does your observation still hold if you move the vertices of the quadrilateral or if you change the size of the circle?

3.  Construct a cyclic quadrilateral $WXYZ$. Extend sides $WX$ and $YZ$ to lines, and label the intersection of these lines $P$.
    a.  Identify all pairs of similar triangles in this figure.
    b.  Calculate the products $PW \cdot PX$ and $PZ \cdot PY$. Make a conjecture.
    c.  Does your conjecture still hold if point $P$ is inside or on the circle?
    d.  Prove your conjecture.

4.  Draw a circle.
    a.  Draw a point $A$ outside the circle, and construct a tangent to the circle from $A$. (*Hint:* Use an idea from Activity 8, Chapter 1.)
    b.  Mark point $B$ on the circle, and construct a tangent to the circle through $B$.
    c.  Draw a point $C$ interior to the circle, and construct a tangent to the circle from $C$.

5.  Create a new sketch of the same figure you constructed for Activity 3. Construct tangent $PT$ to the circle.

a. If the center of the circle is labeled $O$, what is the measure of $\angle PTO$?

b. Find an expression for $PT$ (or for $PT^2$) in terms of $PO$ and $TO$.

c. Find an expression for $PT$ (or for $PT^2$) in terms of $PW$ and $PX$.

d. Construct a new point, $Q$, on the circle. Construct $\overleftrightarrow{PQ}$, and label the other point where this line intersects the circle as point $R$. Calculate the product $PQ \cdot PR$. Animate point $Q$ around the circle. What do you observe? Can you explain what is happening here?

6. Given a circle $C$ and a point $P$, the *power of $P$ with respect to $C$* is a function that depends only on the radius, $r$, of $C$ and the distance, $d$, of $P$ from the center of $C$. In the previous activity, you saw two ways to calculate the power of a particular point $P$ with respect to a given circle.

a. Draw two overlapping circles, $C_1$ and $C_2$, and mark their common chord $AB$.

b. Construct a third circle, $C_3$, that also has $AB$ as a chord. What can you say about the centers of the family of circles that have $AB$ as a common chord?

c. Construct a point $P$ on $\overleftrightarrow{AB}$. Calculate the powers of $P$ with respect to each of the circles, $C_1, C_2,$ and $C_3$. Move $P$ along $\overleftrightarrow{AB}$. Make a conjecture about the values of the power of $P$ with respect to $C_k$, where $k$ can be either 1 or 2 or 3. Prove your conjecture.

7. Draw a circle, $C$, with diameter $AB$. (Construct $AB$ so that it is certain to be a diameter of $C$.)

a. Construct a point, $P$, on the diameter $AB$, and construct two additional circles with diameters $AP$ and $PB$.

b. The region bounded by the three semicircular arcs on one side of diameter $AB$ is called an *arbelos*. Calculate the area of the arbelos.

c. Construct a perpendicular to $AB$ at point $P$. Label the intersections of this perpendicular with the circle $C$ as $R$ and $S$. Construct a circle with diameter $PR$, and calculate its area.

d. Make a conjecture about these areas. Prove your conjecture.

8. Draw a circle with center $A$ and radius $AB$.

a. Construct a tangent to this circle at $B$. Call this tangent line $t$.

b. Two circles are said to be *orthogonal* if their tangents are perpendicular at their points of intersection. Construct a point $Q$ on $t$. Construct a circle with center $Q$ that is orthogonal to the first circle.

c. Animate point $Q$ along line $t$. Describe the family of circles with centers on line $t$ that are orthogonal to the circle centered at $A$.

d. If point $Q$ could move off to infinity, what could you say about the circle?

9. Draw a triangle, $\triangle XYZ$.

a. Construct the incenter, $I$, of $\triangle XYZ$. Construct the incircle of this triangle.

b. Extend sides $XY$ and $XZ$ to lines. Construct the angle bisectors of the exterior angles at $Y$ and $Z$, and mark their intersection as point $P$. Using $P$ as the center, construct a circle tangent to side $YZ$.

c. Can you prove that the circle you constructed in part b is also tangent to $\overleftrightarrow{XZ}$ and $\overleftrightarrow{YZ}$? (This circle is called an *excircle* of $\triangle XYZ$. A triangle has three excircles.)

**10.** Draw circle $C$ centered at $O$ with radius $r$. Draw a line, $\ell$, that does not go through the center of $C$.

SKETCHPAD TIP    Once you have drawn the circle $C$, you can measure its radius by measuring the distance from the center $O$ to a point on $C$. Then with this measurement selected, choose **Properties** from the Edit menu, and change the label to $r$.

a. Choose point $P$ on $\ell$, and construct $\overrightarrow{OP}$. Construct point $P'$ on $\overrightarrow{OP}$ so that

$$|OP'| = \frac{r^2}{|OP|}.$$

(*Hint:* You have to think about how to construct a point that is on $\overrightarrow{OP}$ and that is a certain distance from point $O$.) The point $P'$ is called the *inversion of P with respect to circle C*.

b. Arrange your diagram so that $\ell$ intersects $C$. Drag $P$ along $\ell$, and observe the location of $P'$. When $P$ is interior to $C$, where is $P'$? Where is $P'$ when $P$ is exterior to $C$? What if $P$ is on circle $C$? Can you explain why this happens?

c. Construct the locus of $P'$ as $P$ moves along line $\ell$. Experiment with different positions of line $\ell$. What happens when $\ell$ intersects $C$? What happens when the entire line $\ell$ is exterior to $C$? What happens when $\ell$ goes through the center of $C$?

SKETCHPAD TIP    Select the points $P$ and $P'$, then choose **Locus** from the Construct menu.

d. Sketchpad allows $P$ to move along only the portion of $\ell$ that appears on screen. What do you think would happen if $P$ were allowed to move along the entire (infinite) line $\ell$?

## 3.2 DISCUSSION

### AXIOM SYSTEMS: ANCIENT AND MODERN APPROACHES

Around 300 BC, Euclid and his colleagues set down five postulates, as well as hundreds of propositions (theorems) that follow from those postulates, in the thirteen books of Euclid's *Elements*. Book I of the *Elements* opens with a list of definitions:

A *point* is that which has no part. A *line* is breadthless length. The extremities of a line are points. A *straight line* is a line which lies evenly with the points on itself.

When a straight line set up on a straight line makes the adjacent angles equal to one another, each of the equal angles is *right,* and the straight line standing on the other is called a *perpendicular* to that on which it stands. An *obtuse angle* is an angle greater than a right angle. An *acute angle* is an angle less than a right angle.

*Parallel* straight lines are straight lines which, being in the same plane and being produced indefinitely in both directions, do not meet one another in either direction (Heath, 1956, 153–154).

There are actually more definitions in this list. We included just a few here to give you a bit of the flavor of Euclid.

At the end of the nineteenth century, David Hilbert, doing research in the foundations of geometry, refined and clarified Euclid's postulates and definitions. He was attempting to clean up weaknesses and ambiguities that had been noticed in Euclid's *Elements* and to set out very clearly and exactly what Euclid's postulates meant. In building the language for geometry, Hilbert identified certain terms as the basic building blocks of this language. The basic geometric objects are point, line, and plane. Unlike Euclid, Hilbert called these *undefined terms.* He wanted to avoid the kind of circular reasoning in which the definition of one term depends on the definition of another, which in turn depends on the definition of the first. In using these undefined terms to build the language and theory of geometry, Hilbert was not simply saying, "Oh well, we know what points, lines, and planes are, so let's just agree to use these terms without defining them." Rather, starting from these undefined terms, Hilbert developed sets of axioms to specify the properties of these objects and the relationships that exist among them. In his *Grundlagen der Geometrie,* published in 1899, Hilbert identified five sets of properties for points, lines, planes, and the possible relationships among them. He organized his axioms into five groups: incidence, betweenness, congruence, continuity, and parallelism. Hilbert's *axioms of incidence* specify exactly what we mean when we say that a point is "on a line," a line "goes through a point," or a line "lies in a plane." His *axioms of betweenness* address how we know when a point is between two other points or when a ray is between two other rays. The *axioms of congruence* specify exactly what conditions must be met for one object to be congruent to another.

From a purely axiomatic perspective, we should be able to substitute any words—even nonsensical words such as *abba, dabba,* and *jobba*—for the undefined terms *point, line,* and *plane.* For example, Euclid's first postulate says that

Given two distinct *points* P and Q, there is exactly one *line* ℓ that passes through P and Q.

To underscore the idea that *point* and *line* are undefined terms, we might express this postulate as

Given two distinct *abbas* a and b, there is exactly one *dabba* d that passes through a and b.

Because our focus in this course is more constructive than axiomatic, we will leave this discussion of axiom systems for now. In closing, we observe that many contemporary high school and college geometry texts use Hilbert's axioms. In fact, your own high school geometry course might have used axioms of incidence, betweenness, and congruence that were based directly on Hilbert's formulation of these axioms (Eves, 1976; Greenberg, 1980).

## ROBUST CONSTRUCTIONS: DEVELOPING A VISUAL PROOF

We are making a distinction in this course between *drawing* a figure and *constructing* it. If your construction is robust, you will be able to select any of the vertices or edges of the figure and move them around the sketch without losing the properties that you constructed. If the figure you have constructed is not robust, it might fall apart when you drag some of the vertices or edges.

In Sketchpad, the allowable constructions are based on Euclid's postulates (see pages 12–13). Constructions involving points and lines have their justification in Euclid's first two postulates. Constructions involving circles have their justification in Euclid's Third Postulate. Constructions involving perpendicular lines are justified by Euclid's Fourth Postulate, while constructions involving parallel lines are justified by Euclid's Fifth Postulate. If you can construct a diagram in Sketchpad, you are using Euclid's postulates implicitly.

As you construct a robust diagram in Sketchpad to illustrate a geometric theorem, you are developing a visual proof of that theorem. We now turn our attention to how to develop a proof from a diagram. You have been using Sketchpad to construct figures. By this time, you might have noticed how using Sketchpad really requires you to make deliberate choices about what you are constructing. For example, when you use Sketchpad to construct a line, $\ell$, Sketchpad starts with two points and constructs $\ell$ using these points. If you animate one of these points, it is free to wander all over the plane. If you construct a new point $Q$ on $\ell$ and animate it, $Q$ will be constrained to move only along line $\ell$. In teaching this class, we have noticed that having to make these deliberate choices about how to construct a Sketchpad diagram helps our students think more carefully about important ideas as they work to develop proofs. Retracing the steps that you take in developing a robust construction can lead to a robust proof.

## STEP-BY-STEP PROOFS

Let's see how we can use Sketchpad to help construct robust proofs. We start with carefully constructed step-by-step proofs, which give us a clear way to begin writing correct proofs. As you grow confident in writing mathematical proofs, you may gradually shift from writing step-by-step proofs to writing your proofs in a flowing paragraph form. In a step-by-step proof, each line of the proof presents one new idea or concept, which together with previous steps, produces a new result (Maher, 1994).

Write out each line of your proof as a complete sentence, clearly justifying the step. In developing a geometric argument, you may use the following types of justifications for each step:

- Your justification may be based on the conditions that are given when the problem is posed: "We are given . . . ", or "By hypothesis . . . ".

- The justification may be based on the definitions, postulates, and axioms of the geometric system we are using: "By definition . . . ", "By postulate . . . ", or "By axiom . . . ".

- The constructions you can do with Sketchpad are implicitly linked to axioms or postulates of geometry. You can make this link explicit by saying, "This construction is allowed by . . . ".

- Any previously proved theorem can be used as a justification in a proof: "By theorem . . . ". We don't expect you to memorize the theorems, but it can be helpful to refer to the theorem by its name or number (if it has one).

- As you develop a proof, one step of your proof may depend directly on a previous step in the argument. The justification can be given simply as "By step . . . ".

- In the *Elements,* Euclid listed some "common notions" as allowable justifications in a proof. These common notions include properties of equality and congruence, arithmetic and algebraic computations, and rules of logic. In giving your justification, you may name the property or rule you are using.

Many of the diagrams you have been constructing with Sketchpad—assuming they are robust diagrams—can be viewed as visual proofs or demonstrations of geometric theorems. Consider, for example, the figure you constructed for Activity 5 in Chapter 2. Recall that you constructed $\triangle PQR$ with point $Q$ on line $\ell$, which was parallel to the line containing points $P$ and $R$. The base of $\triangle PQR$ was the fixed line segment $PR$, and the altitude was the perpendicular distance between the parallel lines $\ell$ and $PR$. If your diagram was a robust construction, the distance between the parallel lines did not change as $Q$ moved along line $\ell$. Your diagram demonstrated very clearly that the area of $\triangle PQR$ remained constant no matter how the shape of the triangle changed.

Let's use a similar diagram to prove that the interior angles of a triangle add up to a straight angle. We develop this proof in a step-by-step format.

**THEOREM 3.1**   The interior angles of a triangle add up to a straight angle.

**Proof**

1. We are given an arbitrary triangle. Let's call it $\triangle PQR$ (see Figure 3.1, next page). We are interested in the sum of the angles of $\triangle PQR$; that is, we are interested in the sum

   $$m\angle RPQ + m\angle PQR + m\angle QRP.$$

2. Construct a line through $Q$ parallel to side $PR$, making use of Playfair's Postulate, which is equivalent to Euclid's Fifth Postulate. Call this line $j$.

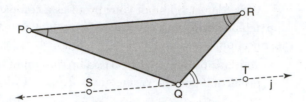

FIGURE 3.1
Angle Sum of a Triangle

3. Because $\overleftrightarrow{PQ}$ and $\overleftrightarrow{RQ}$ are transversals to parallel lines $j$ and $PR$, we know that pairs of alternate interior angles are congruent by Euclid's Fifth Postulate. Construct two additional points, $S$ and $T$, on line $j$ so that

$$\angle RPQ \cong \angle SQP \text{ and } \angle QRP \cong \angle RQT.$$

4. Substituting equal quantities for equal quantities in the expression given in step 1, we get

$$m\angle SQP + m\angle PQR + m\angle RQT.$$

5. Because $j$ is a straight line, we know that

$$m\angle SQP + m\angle PQR + m\angle RQT = \text{straight angle} = 180°.$$

6. Thus, by again substituting equal quantities, we can write

$$m\angle RPQ + m\angle PQR + m\angle QRP = 180°.$$

So, we have proven that the interior angles of a triangle add up to a straight angle.

---------------------------------------------

Notice that we used Euclid's Fifth Postulate (in steps 2 and 3) to prove this theorem. In a non-Euclidean world, it is *not* the case that there is *exactly* one line through $P$ parallel to line $j$. There might be more than one line, or there might not be any line at all through $P$ that is parallel to $j$. However, changing Euclid's Fifth Postulate would change this proof—and might even change the theorem. We will investigate this situation in greater depth in Chapter 9.

In Chapter 2, we suggested a strategy for developing a proof of the theorem that the angle bisectors of a triangle are concurrent (see Theorem 2.6). We now offer an outline of that proof in a step-by-step format. (In the exercises, you will be asked to provide the justifications for each step.)

**THEOREM 3.2**  The three angle bisectors of a triangle are concurrent. (See Figure 3.2.)

**Proof**

1. We are given $\triangle ABC$ (see Figure 3.2).

2. Construct $\overrightarrow{AD}$ so that it bisects $\angle A$ and $\overrightarrow{BE}$ so that it bisects $\angle B$.

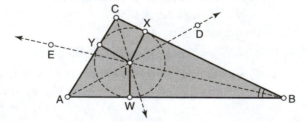

FIGURE 3.2
Angle Bisectors of a Triangle Are
Concurrent

3. Rays $AD$ and $BE$ intersect at point $I$ interior to $\triangle ABC$. We must show that $\overrightarrow{CI}$ bisects $\angle C$.

4. Construct perpendiculars from point $I$ to each side of $\triangle ABC$. Let $W$, $X$, and $Y$ be the feet of the perpendiculars from $I$ to sides $AB$, $BC$, and $AC$, respectively. Thus, we have

$$IW \perp AB, \quad IX \perp BC, \quad \text{and} \quad IY \perp AC.$$

5. Triangles $AIY$ and $AIW$ are congruent.

6. $IY \cong IW$.

7. Triangles $BIW$ and $BIX$ are congruent.

8. $IW \cong IX$.

9. Triangles $CIY$ and $CIX$ are congruent.

10. $\angle YCI \cong \angle XCI$. In other words, $\overrightarrow{CI}$ bisects $\angle C$.

So, we have proven that the three angle bisectors of a triangle are concurrent (once you have provided the justification for each step of this proof).

------------------------------------------

## INCIRCLES AND EXCIRCLES

We have just seen that the three angle bisectors of the interior angles of a triangle are concurrent at a point, which is commonly denoted $I$. In the proof, we saw that the distance from $I$ to a side of the triangle is the same for all three sides. Therefore, the incenter $I$ is the center of the incircle, a circle interior to the triangle that is tangent to all three sides of the triangle.

We can also consider the angle bisector of an exterior angle of a triangle $XYZ$. An exterior angle is formed at $Y$ when either side $XY$ or $YZ$ is extended beyond the point $Y$. (The two exterior angles at $Y$ are congruent because they are vertical angles to each other.) The bisectors of the exterior angles at $Y$ and $Z$ will be concurrent with the bisector of the interior angle at $X$. In other words, if the bisectors of the exterior angles at $Y$ and $Z$ intersect at point $P$, the bisector of the interior angle at $X$ also goes through $P$. Can you prove this? You will have a chance to do this in the exercises. (*Hint:* Look for congruent triangles.)

The point $P$ will be the center of an excircle of the triangle, a circle exterior to the triangle and tangent to one side and to extensions of the other two sides (see Figure 3.3). One way to construct the circle in Activity 9b is to drop a perpendicular

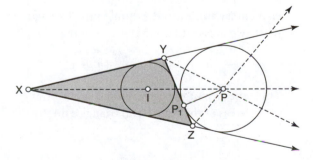

**FIGURE 3.3**
The Incircle and One of the
Excircles of a Triangle

from $P$ to side $YZ$. Call the foot of this perpendicular $P_1$. The segment $PP_1$ will be the radius of the excircle centered at $P$. When you construct the circle centered at $P$ with radius $PP_1$, it certainly appears to be tangent to $\overleftrightarrow{XY}$ and $\overleftrightarrow{XZ}$. But how can we prove that this is actually the case? Because this circle is tangent to side $YZ$ by construction, we only need to show that it must be tangent to the extended sides $\overleftrightarrow{XY}$ and $\overleftrightarrow{XZ}$ of our triangle. Drop perpendiculars from $P$ to $\overleftrightarrow{XY}$ and $\overleftrightarrow{XZ}$, calling the feet of these perpendiculars $P_2$ and $P_3$, respectively. Using a method similar to that used in the proof of Theorem 3.2, which says that the three angle bisectors of a triangle are concurrent, we can prove that $PP_1$, $PP_2$, and $PP_3$ are congruent segments. Thus, $P_1$, $P_2$, and $P_3$ must be points on the circle centered at $P$. Therefore, the excircle is tangent to side $YZ$ and to the extensions of sides $XY$ and $YZ$ (Baragar, 2001, 36; Coxeter, 1969, 11).

## THE PYTHAGOREAN THEOREM

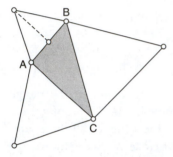

**FIGURE 3.4**
Pythagorean Theorem

Perhaps the most famous theorem of the ancient Greeks is the Pythagorean Theorem. For any right triangle, the area of a square constructed on the hypotenuse will be equal to the sum of the areas of the squares constructed on the other two sides. It is interesting to ask if there is anything special about the squares in this theorem (see Figure 3.4). For example, in Activity 1, you were asked to construct equilateral triangles on each side of $\triangle ABC$ and to calculate the areas of the three equilateral triangles. Perhaps you observed that when one of the angles of $\triangle ABC$ was pretty close to being a right angle, the areas of two of the equilateral triangles appeared to add up to the area of the third equilateral triangle. Is the area of the equilateral triangle on the hypotenuse really equal to the sum of the areas of the equilateral triangles on the other two sides?

**CONJECTURE 3.3**

We are given $\triangle ABC$ with a right angle at $A$. The area of the equilateral triangle on the hypotenuse (side $BC$) is equal to the sum of the areas of the equilateral triangles on the other two sides (sides $AB$ and $AC$).

**Proof** Let $\triangle ABC$ be a right triangle with right angle at $A$. Let $|AC| = b$, $|AB| = c$, and $|BC| = a$. Then, by the Pythagorean Theorem, we know that $b^2 + c^2 = a^2$.

Now construct an equilateral triangle on each side of $\triangle ABC$. The triangle on side $AC$ will have sides of length $b$. The area of a triangle is calculated as $\frac{1}{2}(\text{base} \times \text{height})$. So the area of the triangle on side $AC$ is $\frac{1}{2}(b \times \frac{\sqrt{3}}{2}b) = \frac{\sqrt{3}}{4}b^2$. Similarly, the area of the triangle on side $AB$ is $\frac{1}{2}(c \times \frac{\sqrt{3}}{2}c) = \frac{\sqrt{3}}{4}c^2$. Using elementary algebra and assuming the Pythagorean Theorem, we can see that these two areas add up to the area of the triangle on side $BC$:

$$\frac{\sqrt{3}}{4}b^2 + \frac{\sqrt{3}}{4}c^2 = \frac{\sqrt{3}}{4}(b^2 + c^2) = \frac{\sqrt{3}}{4}a^2.$$

Thus, the area of the equilateral triangle on the hypotenuse is equal to the sum of the areas of the equilateral triangles on the other two sides.

In Activity 1b, you were asked to repeat this experiment with circles on each side of $\triangle ABC$. Again, you might have noticed that the areas of two of the circles came pretty close to adding up to the area of the third circle when one of the angles was pretty close to being a right angle. You should be able to prove this conjecture using a method similar to that used in the proof of Conjecture 3.3.

Suppose we construct semicircles on each side of a right triangle. Is the area of the semicircle on the hypotenuse equal to the sum of the areas of the semicircles on the other two sides? What if we construct rectangles of constant height $h$ and widths equal to the lengths of each side of the right triangle? Will the area of the rectangle on the hypotenuse equal the sum of the areas of the rectangles on each of the other two sides? You will have a chance to prove or disprove these questions in the exercises.

## LANGUAGE OF CIRCLES

A *circle* is the set of points at a fixed distance, $r$, from a fixed point, $O$. The point $O$ is called the *center* of the circle, and the distance $r$ is called the *radius* of the circle. Points whose distance from the center point $O$ is less than $r$—that is, the set of points $\{P : d(P, O) < r\}$—are *interior to* the circle, whereas points whose distance from $P$ is greater than $r$—that is, the set of points $\{P : d(P, O) > r\}$—are *exterior to* the circle.

A *chord* of a circle is a line segment joining two points on the circle. A chord that passes through the center of the circle is called a *diameter*. Using Sketchpad, we can construct a diameter of a circle by constructing a line through the center of the circle and then finding the line segment between the points where this line intersects the circle.

Whereas a chord intersects a circle in two points, a *tangent* is a line that intersects a circle in exactly one point. The point where a tangent line touches the circle is called the *point of tangency*. It is not difficult to prove that a tangent line is perpendicular to a radius at the point of tangency.

The *circumference* of a circle is the length of its perimeter. An *arc* of a circle is a piece of the circle. A *sector* of a circle is a pie-shaped portion of the interior of the circle, bounded by an arc of the circle and two radii. A *segment* of a circle is the region bounded by an arc and a chord of the circle. If $P$, $Q$, and $R$ are three points on a circle with center at $O$, then angle $POR$ is called a *central angle* of the circle, and angle $PQR$ is called an *inscribed angle*. The angle $PQR$ may also be called an angle *subtended by the chord PR*.

As you worked on Activities 7 and 8 in Chapter 1, you should have noticed a fixed relationship between a central angle and an inscribed angle that share the same chord. An angle inscribed in a circle subtended by a fixed chord of the circle is one-half the size of the corresponding central angle. You needed to use this idea as you worked on Activity 2 in this chapter. You can prove this relationship by working with several isosceles triangles that appear in a diagram of this situation. A consequence of this relationship between inscribed and central angles is that any angle inscribed in a semicircle will be a right angle. Also, because the diagonal of a cyclic quadrilateral is a chord of the circle, opposite interior angles of a convex cyclic quadrilateral are supplementary. Moreover, an exterior angle of a cyclic quadrilateral will be congruent to the opposite interior angle. Recognizing pairs of such angles is the key to seeing the similar triangles in Activity 3 in this chapter and to proving that $PW \cdot PX = PZ \cdot PY$.

Any three noncollinear points form the vertices of a triangle. As you worked on some of the activities for Chapter 2 (particularly Activities 4 and 8), you should have observed that a circle can always be constructed so that it goes through the vertices of a triangle. The perpendicular bisectors of the three sides of the triangle will be concurrent at the point $O$, which is the center of this circumscribed circle.

Quadrilaterals, on the other hand, do not always have all four vertices lying on a circle. Those special quadrilaterals that do are called *cyclic quadrilaterals.* Look again at the different quadrilaterals you worked with in Exercises 8 and 9, Chapter 1, and Exercise 18, Chapter 2. Where do cyclic quadrilaterals fit into this scheme of quadrilaterals? Every square is a cyclic quadrilateral, because all the angles are right angles. What about parallelograms? Trapezoids? Kites? Rectangles?

Because a tangent to a circle, $C$, is a line perpendicular to a radius of $C$ at the point of tangency, it is very easy to construct a tangent to circle $C$ at a point, $B$, that is on $C$. To construct a tangent to $C$ from a point $A$ that is exterior to the circle, use the idea that an angle inscribed in a semicircle is a right angle. Construct a line segment from point $A$ to the center of $C$, then construct the segment's midpoint, $M$. A circle centered at $M$ with radius $AM$ will intersect $C$ in two points, $P_1$ and $P_2$. Lines $AP_1$ and $AP_2$ will both be tangent to the circle $C$. Why is it impossible to construct a tangent to the circle $C$ from a point interior to $C$?

## SOME INTERESTING FAMILIES OF CIRCLES

In Activities 6 and 8, you worked with two different families of circles. One is a family of circles that share a common chord, and the second is a family of circles orthogonal to a given circle.

All the circles in Activity 6 share the common chord $AB$. Did you notice that the centers of $C_1$, $C_2$, and $C_3$ are collinear? Perhaps you used this idea to construct circle $C_3$. You can choose any point $X$ on the line that passes through the centers of $C_1$ and $C_2$, and $X$ will be the center of a circle that has $AB$ as a chord.

Suppose, however, that you start with segment $AB$, and you want to construct a circle that has $AB$ as a chord. In other words, suppose you don't have circles $C_1$ and $C_2$—so that you can't use the line through their centers. How will you find the line where the centers of the family of circles through $AB$ lie? The centers of this family of circles will lie on the perpendicular bisector of $AB$. To prove this, choose any point $X$ on the perpendicular bisector of $AB$ (see Figure 3.5). Triangle $XAB$ is an isosceles triangle with $XA \cong XB$. So, the circle centered at $X$ that passes through $A$ will also pass through $B$.

Suppose you are given a circle (or an arc of the circle, or even just a few points on the circle), and you want to find the center of the circle. Or suppose you are given a triangle and you want to construct the circumcircle of that triangle. How

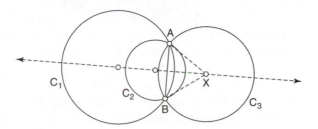

**FIGURE 3.5**
Constructing Circles That Share
a Common Chord

can you use the fact that the center of the circle must lie on the perpendicular bisector of any chord of the circle to solve these problems? You will have a chance to do this in the exercises.

In Activity 8, you investigated a family of circles orthogonal to the circle centered at $A$ with radius $AB$.

> *Remember:* Two circles are said to be *orthogonal* if their tangents are perpendicular at their points of intersection.

Starting with circle $C_1$ centered at $A$ with radius $AB$, it is easy to construct a tangent line $t$ to $C_1$ at point $B$ (see Figure 3.6). Choose any point $Q$ on $t$, and construct the circle centered at $Q$ with radius $QB$. Call this second circle $C_2$. Then $\overleftrightarrow{AB}$ is tangent to circle $C_2$, and $\overleftrightarrow{QB}$ is tangent to circle $C_1$. Moreover, $\overleftrightarrow{AB} \perp \overleftrightarrow{QB}$. So, circles $C_1$ and $C_2$ are orthogonal circles.

Observe also that circles $C_1$ and $C_2$ intersect at a second point $R$. Because $AR$ is a radius of $C_1$, it will also be a tangent to circle $C_2$. (How can you be sure that $AR$ and $QR$ really are perpendicular?) In other words, both tangent lines to circle $C_1$ from point $Q$ are perpendicular to the corresponding tangent lines to circle $C_2$ from point $A$. This makes $ABQR$ a cyclic quadrilateral. (Can you prove this?)

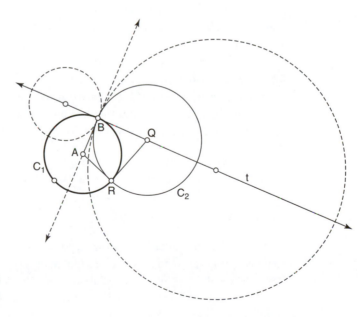

**FIGURE 3.6**
Orthogonal Circles

When you animated point $Q$ in Activity 8, you saw that the circle with center $Q$ grows larger as $Q$ moves away from $B$, and it becomes smaller as $Q$ moves closer to $B$. In fact, as $Q$ moves across the point $B$, the radius of $C_2$ shrinks momentarily to 0, so that circle $C_2$ shrinks (momentarily) to a single point.

As $Q$ moves along line $t$, it is constrained to stay within the viewing window. However, we can imagine line $t$ extending far beyond the edges of this window. What happens to the circle $C_2$ as $Q$ moves off toward $\infty$ (that is, toward "infinity")? As point $Q$ moves farther and farther away from point $B$, the radius of circle $C_2$ gets larger and larger. If $Q$ were allowed to move off to infinity, the circle would have infinite radius, and its curvature would become very "straight"; in other words, the circle would become a straight line.

## POWER OF A POINT

The idea of *function* is an important mathematical idea. A function is a process that takes one or more inputs from a specified set and turns them into an output. You studied functions in your high school or college algebra course. For example, you might have worked with expressions that looked something like this:

$$y = f(x) = 5x^3 + 4x.$$

In this expression, $x$ is the input, $y$ is the output, and the expression $5x^3 + 4x$ is a rule that tells you how to convert the input to an output of the function. For this example, $x$ can be any real number, so the set of all real numbers is the domain of the function $f$. You might also have encountered some functions that required more than one input. For example,

$$g(x, y) = 5x^2 + 2xy + 3y^3$$

is a function that requires two inputs. Here you have to supply two numbers, $x$ and $y$, then use the expression $5x^2 + 2xy + 3y^3$ to calculate the output. The inputs, $x$ and $y$, can be any real numbers, and the output of the function $g$ will again be a real number.

In Activities 5 and 6, you worked with a different kind of function. This function has two inputs—point $P$ and circle $C$. The process of this function requires you (or Sketchpad) to figure out the radius, $r$, of $C$ and the distance, $d$, of $P$ from the center of $C$, and then to calculate the value $d^2 - r^2$. So, the output of this function is a number. As noted in the activities:

**DEFINITION**   Given a circle $C$ and a point $P$, *the power of P with respect to* $C$, denoted Power($P,C$), is a function that depends only on the radius, $r$, of $C$ and the distance, $d$, of $P$ from the center of $C$:

$$\text{Power}(P,C) = d^2 - r^2.$$

So, Power is a function that takes two geometric objects (a point and a circle) and returns a number. Notice that if $P$ is interior to $C$, $d$ will be smaller than $r$, and Power($P,C$) will be a negative number.

As you worked on Activities 3, 5, and 6, you probably observed that if $Q$ and $R$ are points on the circle $C$, and $P$ is collinear with $Q$ and $R$, then the product

$PQ \cdot PR$ gives another way of calculating the Power($P,\mathcal{C}$). But this observation needs to be proved. We can use the similar triangles in Activity 3 to show that $PQ \cdot PR$ is constant, no matter where $Q$ and $R$ are on the circle (as long as $P$, $Q$, and $R$ are collinear). But we still need to prove that $PQ \cdot PR = d^2 - r^2$. You will have a chance to do this in the exercises.

So, you can calculate the power of point $P$ with respect to circle $\mathcal{C}$ using either of these methods. If you know the radius of the circle, $r$, and the distance of $P$ from the center of the circle, $d$, you can calculate Power($P,\mathcal{C}$) using the defining formula $d^2 - r^2$. On the other hand, if you have a line through $P$ that intersects the circle at points $Q$ and $R$, you can calculate the power of $P$ with respect to $\mathcal{C}$ using $PQ \cdot PR$. If $P$ is interior to the circle, we have seen that Power($P,\mathcal{C}$) $< 0$ (because $d^2 < r^2$). If this happens, then $Q$ and $R$ will be on opposite sides of point $P$. In other words, $\overrightarrow{PQ}$ and $\overrightarrow{PR}$ will be pointing in opposite directions. In this case, we take $PQ \cdot PR$ to be negative (Coxeter, 1991, 81; Sved, 1991, 15).

## INVERSION IN A CIRCLE

As you saw in the discussion about Power($P,\mathcal{C}$), the inputs to a function are not always numbers. Power is a function that takes two geometric objects—a point, $P$, and a circle, $\mathcal{C}$—and returns a number. Inversion in a circle gives another example of a function whose inputs are not real numbers. Moreover, the output of inversion is a new *point*, not a number.

**DEFINITION**  We are given a circle, $\mathcal{C}$, centered at $O$ with radius $r$. For any point $P$ in the plane (*except* $O$, the center of $\mathcal{C}$), the *inversion of P with respect to $\mathcal{C}$* is the point $P'$ on $\overrightarrow{OP}$ such that

$$|OP| \cdot |OP'| = r^2.$$

To underscore the fact that inversion is a function that takes two inputs (point $P$ and circle $\mathcal{C}$) and returns one output (point $P'$), we can write Inversion($P,\mathcal{C}$) $= P'$.

Suppose you are given (as you were in Activity 10) a circle, $\mathcal{C}$, centered at $O$ with radius $r$. Then for any point $P$ in the plane, you can construct $\overrightarrow{OP}$. If you measure the length of segment $OP$, you can construct a circle centered at $O$ with radius $r^2/|OP|$. The intersection of this new circle with $\overrightarrow{OP}$ gives point $P'$. Observe that if $P$ is exterior to $\mathcal{C}$, then $r^2/|OP| < r$; so, $P'$ will be interior to $\mathcal{C}$. If $P$ is interior to $\mathcal{C}$, then $r^2/|OP| > r$; so, $P'$ will be exterior to $\mathcal{C}$. What happens when $P$ lies on the circle $\mathcal{C}$?

Inversion is an example of a function that takes points in the plane and returns points in the plane. There is a difficulty with the origin. Do you see why? Can you explain why the origin has no image under inversion? In Chapter 6, we will study other functions that take points in the plane and return points in the plane. For now, let's investigate what effect inversion has on lines and circles.

How does inversion transform line $\ell$? For example, what happens when $\ell$ is completely outside circle $\mathcal{C}$? Did you observe that the inversion of $\ell$ appears to be a circular arc? In your Sketchpad worksheet, point $P$ was only able to move along a portion of line $\ell$, that is, along a line segment. In fact, the inversion of the entire set of points lying on line $\ell$ is a circle that goes through point $O$.

**THEOREM 3.4**  We are given a circle, $\mathcal{C}$, with center $O$ and radius $r$. Let $\ell$ be a line that does not go through $O$. The inversion of $\ell$ in $\mathcal{C}$ will be a circle through point $O$.

**Proof**  There are two cases: (1) $\ell$ does not intersect $\mathcal{C}$, and (2) $\ell$ intersects $\mathcal{C}$. Here, we consider the first case, where $\ell$ does not intersect $\mathcal{C}$. (You will have an opportunity to consider the second case, where $\ell$ intersects $\mathcal{C}$, in the exercises.) We want to show that for any point $P$ on $\ell$, point $P'$ lies on some circle that goes through point $O$.

Construct a perpendicular from $O$ to $\ell$, and call the foot of this perpendicular $T$. See Figure 3.7. Construct $T' = \text{Inversion}(T, \mathcal{C})$. Construct the circle with diameter $OT'$. For any point $P$ on $\ell$, we will show that $P'$ lies on this circle.

Let $P^*$ be the point where $\overrightarrow{OP}$ intersects the circle with diameter $OT'$. We claim that $P^* = P'$. To see this, observe that $\triangle OTP$ and $\triangle OP^*T'$ are similar. (Why?) This gives us

$$\frac{OT}{OP^*} = \frac{OP}{OT'} \quad \Rightarrow \quad |OT| \cdot |OT'| = |OP^*| \cdot |OP| = r^2$$
$$\Rightarrow \quad OP^* = OP'.$$

To complete this proof, you need to consider the case where $\ell$ intersects $\mathcal{C}$.

------------------------------------------

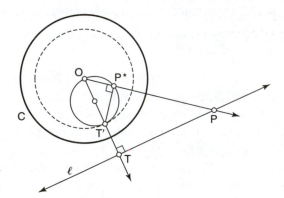

**FIGURE 3.7**
The Inversion of Line $\ell$

Observe that the inversion of a point on circle $\mathcal{C}$ will be the same point—that is, for any point $P$ that lies on $\mathcal{C}$, the $\text{Inversion}(P, \mathcal{C}) = P$. Such a point is called a *fixed point* of the inversion. Thus, the inversion of circle $\mathcal{C}$ will be $\mathcal{C}$ itself. (In Chapter 6, we will investigate other functions of the plane. Some of those functions will have fixed points and some will not.)

It is relatively easy to show that the inversion of the inversion of a point, $(P')'$, will be $P$. We leave this to the exercises.

Consider a line, $m$, that passes through $O$, the center of circle $\mathcal{C}$. For any point $P$ on $m$, point $P' = \text{Inversion}(P, \mathcal{C})$ will lie on $\overrightarrow{OP}$, which is a subset of line $m$.

Thus, the inversion of line $m$ is line $m$ itself. (However, most of the points of $m$ will not be fixed points of this inversion. Only two points of $m$ are fixed points. Which two?)

What happens to point $O$ in inversion? Given circle $C$ with center $O$ and radius $r$, what is Inversion$(O,C)$? To answer this question, we need to find the point $O'$ such that $|OO| \cdot |OO'| = r^2$. But the length of $\overrightarrow{OO}$ (that is, the ray from point $O$ to the same point $O$) is 0. So, $\overrightarrow{OO'}$ would have to be infinitely long for this product to be equal to $r^2$. In fact, that is exactly how mathematicians like to think of this situation. We can imagine that there is a point beyond the plane that is called the *point at infinity*. This point is often denoted by the Greek letter $\Omega$. (This is the letter at the end of the Greek alphabet, pronounced "omega".) So, Inversion$(O,C)$, which we call $O'$ for short, is $\Omega$, the point at infinity. We will consider this idea more deeply in Chapter 9 (Baragar, 2001; Kimberling, 2003; Sved, 1991).

## THE ARBELOS AND THE SALINON

Archimedes, a significant figure among the Greek mathematicians who followed Euclid, left us a number of interesting problems in his *Book of Lemmas* (or *Liber Assumptorum*). Among these are several problems involving the arbelos (also called a "shoemaker's knife") and the salinon (or "salt cellar"). Perhaps part of the fascination of these figures is that they are apparently very simple—being bounded by semicircular arcs—yet they provide a rich source of problems.

You worked with the arbelos in Activity 7. If you constructed the arbelos in Sketchpad, you had an opportunity to observe that the area of the arbelos appears to be equal to the area of the circle with diameter $PR$ (see Figure 3.8). How can you prove this?

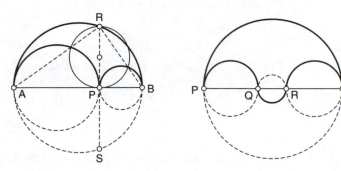

**FIGURE 3.8**
The Arbelos (Left) and the Salinon (Right)

To prove that the areas are equal, we can use an algebraic approach; that is, we can express the areas we are looking for using algebraic expressions. We can find the area of the arbelos by calculating the area of the large semicircle and subtracting the areas of the two small semicircles. Because point $P$ is on diameter $AB$, we know that the diameters of the two smaller semicircles must add up to the diameter of the largest semicircle. So, we can find an expression for the area of the arbelos in terms of the lengths of the diameters (or the radii) of the three semicircles.

Let $r_1$ be the radius of the largest circle, and let $r_2$ and $r_3$ be the radii of the two smaller circles. Note that $r_1 = r_2 + r_3$. So, the area of the arbelos is given by

$$\frac{1}{2}\left(\pi r_1^2 - \left(\pi r_2^2 + \pi r_3^2\right)\right).$$

This can be simplified to

$$\frac{\pi}{2}\left(r_1^2 - r_2^2 - r_3^2\right).$$

Because $P$ can be anywhere on $AB$, how can we find the area of the circle with diameter $PR$? Observe that $\angle ARB$ is a right angle. (How do we know this?) Using this fact and the Pythagorean Theorem, we can find an expression for the area of the circle with diameter $PR$ in terms of the radii of the other three circles.

Let $d_1$ be the diameter of the largest circle, and let $d_2$ and $d_3$ be the diameters of the two smaller circles. Let $d_4$ denote the diameter of the circle on $PR$. Using the Pythagorean Theorem three times, we have

$$d_1^2 = |AR|^2 + |RB|^2$$
$$|AR|^2 = d_2^2 + d_4^2$$
$$|RB|^2 = d_3^2 + d_4^2.$$

Substituting for equal quantities, we can write

$$d_1^2 = \left(d_2^2 + d_4^2\right) + \left(d_3^2 + d_4^2\right).$$

Solving for $d_4^2$ and simplifying, we get

$$d_4^2 = \frac{1}{2}\left(d_1^2 - d_2^2 - d_3^2\right).$$

Because we used radii instead of diameters to calculate the area of the arbelos, we express $d_4^2$ in terms of $r_i$ instead of $d_i$:

$$d_4^2 = \frac{1}{2}\left((2r_1)^2 - (2r_2)^2 - (2r_3)^2\right) = \frac{4}{2}\left(r_1^2 - r_2^2 - r_3^2\right).$$

Finally, the area of the circle on $PR$ is calculated by

$$\pi\left(\frac{d_4}{2}\right)^2 = \frac{\pi}{4}d_4^2 = \frac{\pi}{4}\frac{4}{2}\left(r_1^2 - r_2^2 - r_3^2\right)$$
$$= \frac{\pi}{2}\left(r_1^2 - r_2^2 - r_3^2\right),$$

which is exactly the result we got when we calculated the area of the arbelos.

The salinon is another interesting figure that we can construct using semicircular arcs. Given a circle with diameter $PS$, construct circles with diameters $PQ$, $QR$, and $RS$. The salinon is formed by taking three semicircular arcs on one side of $PS$ and one arc on the other side of $PS$, as in Figure 3.8 (Eves, 1976, 156; Knight, 2000; Nelsen, 2002; Wells, 1991, 5). You will have a chance to work with the salinon in the exercises.

# THE NINE-POINT CIRCLE: A SECOND PASS

The arbelos and the salinon were figures studied by the Greeks in the third century BC. Now we turn our attention to an interesting geometric problem of the nineteenth century AD. In Chapter 2, you began to experiment with the so-called *nine-point circle.* (See Activity 8, from Chapter 2.)

The circle that passes through the feet of the three altitudes of a triangle contains quite a few interesting points. In 1821, the French mathematicians Charles Julien Brianchon and Jean Victor Poncelet published a theorem that for any triangle, the *midpoints of the sides,* the *feet of the altitudes,* and the *midpoints of the lines joining the vertices of the triangle to the orthocenter* all lie on a circle (see Figure 3.9). The English mathematician Benjamin Bevan had posed a very similar problem 17 years earlier.

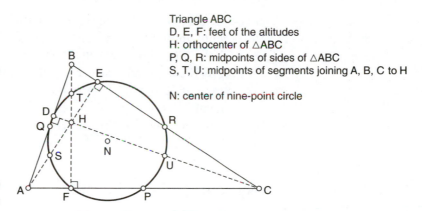

Triangle ABC
D, E, F: feet of the altitudes
H: orthocenter of △ABC
P, Q, R: midpoints of sides of △ABC
S, T, U: midpoints of segments joining A, B, C to H

N: center of nine-point circle

**FIGURE 3.9**
The Nine-Point Circle

In Germany, Karl Wilhelm Feuerbach, proved—by algebraically calculating the circles' radii and the distances between their centers—that the nine-point circle touches the incircle and each of the excircles of the triangle. His proof added another four special points to the nine-point circle (which is now sometimes called the Feuerbach circle). We will continue our investigations of this special circle in Chapter 4, where you will have an opportunity to use methods of analytic geometry to develop some of the proofs (Thomas, 2002, 69; Wells, 1991, 76, 159).

## METHODS OF PROOF

Let's say you have been thinking about a particular geometric idea—perhaps a theorem you're trying to prove. Having experimented with a Sketchpad diagram, you think you understand the construction. Maybe you've even had one of those "Aha!" moments, and you feel ready to develop a proof. How do you get started?

Actually, you've already taken the first, perhaps most important, step—you have a sense that you understand what is going on. Now you feel ready to try to write out a proof, which is a certain kind of explanation of what is going on. Your task is to set out the steps you took to develop your sketch in a clear way so that others can follow your reasoning. Your argument must flow from the geometric ideas, with justifications for what you claim you can do at each step. You may include diagrams and tables to help communicate your ideas, but your proof needs to be carried forward in the logical progression of ideas.

Start by being clear about your assumptions. For the present, assume Euclid's postulates (or axioms), upon which the allowable constructions in a Sketchpad worksheet are based. Thus, Euclid's postulates are implicit assumptions for all of your conjectures.

Clearly state the conjecture or theorem you are proposing to prove. What are the givens—that is, what is the hypothesis of your conjecture? What things result from the givens—that is, what is the conclusion of your conjecture? In general, a conjecture or a theorem is a statement of the form $P \rightarrow Q$. Express the conjecture or theorem you are trying to prove in the form

*If* hypothesis, *then* conclusion.

If you have been able to construct a robust Sketchpad diagram, the steps in your proof might follow in the same order as the steps in your construction. (If your construction is robust, it will not fall apart as you drag individual points.) By reflecting on the steps you took to construct the diagram, you can develop the steps you need in the proof. If your proof is robust, it will stand up to the scrutiny of your colleagues. Once you have developed a proof, it is a good idea to talk through it with someone else. Let your colleagues question your justification of each step. A sound proof—like a robust construction—will not fall apart under such questioning.

## Constructing a Direct Proof

Just as most of the steps in your construction must be done in a certain order, so too do most of the steps in a proof follow from earlier steps. We start with the objects and assumptions given in the hypothesis ($P$), and we work step-by-step toward the conclusion ($Q$). In a direct proof, we work logically forward, one step at a time, toward the desired conclusion. This mimics the development of a construction in Sketchpad—starting with a line, a point, a circle, whatever, and building step by step toward the conclusion.

*Modus ponens* (see Chapter 2) is the rule of logic that allows us to move a proof forward in a direct way. If we are able to assert that $A \rightarrow B$ at one step of the proof, and at another step of the proof we show that we have the conditions required for $A$, then we can use *modus ponens* to conclude $B$. The assertion $A \rightarrow B$ has the form of a theorem, and we usually have this by invoking a previously proved theorem (or a postulate).

For example, in step 2 of the proof of Theorem 3.1, we invoked Playfair's Postulate, which says that if we have a line $\ell$ and a point not on $\ell$, there will be exactly one line through the point parallel to $\ell$. We had a line ($\overleftrightarrow{PR}$) and a point ($Q$) that was not on $\overleftrightarrow{PR}$; so, we used the logic rule of *modus ponens* to conclude that there was a line $j$ through $Q$ parallel to $\overleftrightarrow{PR}$.

## Constructing an Indirect Proof

Sometimes a direct proof doesn't work. In this case, we can back into the proof using the logic rule of *modus tollens* (see Chapter 2), which is based on the fact that the contrapositive of a statement ($\neg Q \rightarrow \neg P$) means the same thing as the original statement ($P \rightarrow Q$) (see page 36). The idea here is that we want to prove $P \rightarrow Q$, and we begin by saying "Suppose *not Q*;" that is, we consider what the situation would be if $Q$ did not occur. Assuming $\neg Q$, we work through the steps as before and reach the *negation of P*. If we can show that $\neg Q$ leads to $\neg P$ (that is, $\neg Q \rightarrow \neg P$), then by contraposition, we can conclude $P \rightarrow Q$.

A subtly different approach is based on the idea that the negation of $P \rightarrow Q$ is $P$ and $\neg Q$, which can be written symbolically as

$$\neg(P \rightarrow Q) \equiv P \wedge \neg Q.$$

Using this approach, we again begin by assuming $\neg Q$, but this time we assume that $P$ is true; that is, we say, "Suppose $P$ and *not Q*." We then use logical reasoning to look for a contradiction to something that is known to be true. If $P \wedge \neg Q$ leads to a contradiction, then $\neg(P \wedge \neg Q)$—that is, $P \rightarrow Q$—must be true.

We used this strategy in our discussion of the construction of the circumcircle of $\triangle ABC$ in Chapter 2. We worked with $\triangle ABC$, and we wanted to show that the perpendicular bisectors of sides $AB$ and $BC$ intersect. Let's call these two perpendicular bisectors $\ell$ and $m$. Either $\ell$ and $m$ intersect, or they are parallel. We want to prove that $\ell$ and $m$ intersect, but what happens if we assume that $\ell \parallel m$? (Assume $\neg Q$.) If $\ell \parallel m$, then $AB \parallel BC$ (that is, $\neg Q \rightarrow \neg P$). But we know that $AB$ and $BC$ are two different sides of $\triangle ABC$ and have point $B$ in common. So, $AB$ cannot be parallel to $BC$. (We have a contradiction.) Thus, $\ell$ and $m$ must intersect in a point, and this point turns out to be the center of the circumcircle of $\triangle ABC$.

In summary, an indirect proof follows one of these forms:

- **Assume $\neg Q$.** Show by a direct proof that $\neg Q$ leads to $\neg P$, that is, $\neg Q \rightarrow \neg P$. This is equivalent to $P \rightarrow Q$. We have $P$. Therefore, $Q$.

- **Assume $P \wedge \neg Q$.** Show by a direct proof that this leads to a contradiction. This proves that $P \wedge \neg Q$ must be false, so that $\neg(P \wedge \neg Q)$, which is equivalent to $P \rightarrow Q$, must be true.

## Using a Counterexample in a Proof

Not every conjecture is a theorem. By this time, you have probably made some conjectures that you have found to be invalid. In fact, not every statement you are asked to prove in the exercises is a valid theorem. You must learn to be critical of any statement you are trying to prove. Sometimes as you think about a statement or develop a Sketchpad diagram to illustrate a statement, you will find an example for which all the requirements in the hypothesis hold, but the conclusion does not. Such an example is called a *counterexample*. A counterexample can serve as a *disproof* of a conjecture.

Consider, for example, the conjecture "All right triangles are isosceles." It is not hard to produce an example of a right triangle that is not isosceles, thus showing that this conjecture is not correct.

## Proving a State ent of the For $P \leftrightarrow Q$

Both a statement and its converse may be theorems. This is not always the case, but it does happen. In this special situation, the theorems can be expressed as a single statement in the form

$$P \rightarrow Q \quad \text{and} \quad Q \rightarrow P.$$

Such a theorem may be expressed more compactly as

$P$ if and only if $Q$.

The expression *if and only if* is used so often in mathematics that we have a special abbreviation (iff) to express it. *If and only if* can also be written in symbolic form as a double-headed arrow ($\leftrightarrow$).

$P$ if and only if $Q$ can be written as $P$ iff $Q$ or even as $P \leftrightarrow Q$.

For example, Ceva's Theorem and its converse can be expressed as a single statement:

**Ceva's Theorem** Suppose *ABC* is a triangle. The lines containing the Cevians *AX*, *BY*, and *CZ* are concurrent if and only if

$$\frac{AZ}{ZB} \cdot \frac{BX}{XC} \cdot \frac{CY}{YA} = 1.$$

Ceva's Theorem and its converse are two theorems. To prove Ceva's Theorem and its converse requires two separate proofs. To prove any statement of the form $P \leftrightarrow Q$, two separate proofs are needed; that is, you must prove *both* (1) $P \rightarrow Q$, *and* (2) $Q \rightarrow P$. These two proofs are independent of each other and may even use different proof strategies (Fenton and Dubinsky, 1996; Smith, Eggen, and St. Andre, 2001).

---

## 3.3 EXERCISES

Give clear and complete answers to the exercises, expressing your explanations in complete sentences. Include diagrams whenever appropriate.

1. The following statements have all appeared in the activities or exercises of this or the previous chapters. You have already constructed diagrams in Sketchpad illustrating these statements, and you may already have written informal proofs of some of these statements. Write out a careful step-by-step proof for each statement.

   a. Let $O$ be the center of a circle, and let $P$, $Q$, and $R$ be points on the circle. Prove that the measure of the central angle, $\angle POR$, is twice the measure of the inscribed angle, $\angle PQR$.

   b. Let $O$ be the center of a circle, and let $PR$ be a diameter of this circle. If $Q$ is a point on the circle, prove that $\angle PQR$ is a right angle.

   c. A median divides its triangle into two equal areas.

   d. The three medians of a triangle are concurrent at a point called the *centroid*, often denoted as $G$.

   e. The three altitudes of a triangle are concurrent at a point called the *orthocenter*, often denoted as $H$.

   f. The perpendicular bisectors of the three sides of a triangle are concurrent at a point called the *circumcenter*, often denoted as $O$.

g. The circumcenter, $O$, of a triangle is the center of a circle that passes through the three vertices of the triangle.

h. The opposite interior angles of a convex cyclic quadrilateral are supplementary.

2. The circumcenter, centroid, and orthocenter lie on a common line, known as the *Euler line,* of the triangle. (*Hint:* See page 45, where the Euler line is first discussed.)

3. Prove or disprove the following:
   a. A rectangle is a cyclic quadrilateral.
   b. A parallelogram is a cyclic quadrilateral.

4. Draw a Venn diagram illustrating the property of being a cyclic quadrilateral. Your diagram should make it clear which quadrilaterals—kites, parallelograms, rectangles, rhombi, squares, and trapezoids—are always cyclic, sometimes cyclic, or never cyclic.

5. Prove that a tangent line to a circle is perpendicular to a radius of that circle at the point of tangency.

6. Write a short essay explaining why it is impossible to construct a tangent to a circle from a point interior to the circle.

7. Suppose $ABC$ is a triangle. Let $B'$ and $C'$ be points on sides $AB$ and $AC$, respectively. Prove that $B'C' \parallel BC$ if and only if
$$\frac{|AB'|}{|AB|} = \frac{|AC'|}{|AC|}.$$
   (*Note:* There are two things to prove.)

8. Let $ABCD$ be a convex cyclic quadrilateral.
   a. Prove that the interior angles at $A$ and $C$ are supplementary.
   b. Prove that the exterior angle at $B$ is congruent to the interior angle at $D$.
   c. How does this situation change if $ABCD$ is not convex?

9. Let $WXYZ$ be a cyclic quadrilateral. (*Note:* $WXYZ$ is not necessarily convex.) Let $P$ be the point of intersection of sides $WX$ and $YZ$ (possibly extended).
   a. If $P$ is exterior to the circle, prove that $PW \cdot PX = PZ \cdot PY$.

   b. Under what conditions will $P$ be interior to or on the circle?
   c. Prove or disprove that
$$PW \cdot PX = PZ \cdot PY$$
   when $P$ is on or interior to the circle.

10. Complete the proof of Theorem 3.2 by supplying justifications for each given step.

11. Given a right triangle, $\triangle ABC$, with right angle at $A$, prove that the area of the circle whose diameter is on the hypotenuse is equal to the sum of the areas of the circles whose diameters are on the other two sides.

12. Given a right triangle, $\triangle ABC$, with right angle at $A$, prove or disprove that the area of the semicircle on the hypotenuse is equal to the sum of the areas of the semicircles on the other two sides.

13. Let $\triangle ABC$ be a right triangle with right angle at $A$. Rectangles, each of height 3 units, are constructed along each side of this triangle. Prove or disprove that the area of the rectangle on the hypotenuse is equal to the sum of the areas of the rectangles on each of the other two sides.

14. Exercises 11–13 use the Pythagorean Theorem. You were allowed to assume the Pythagorean Theorem and to use it as a justification for one of the steps in your proof.
   a. Develop a proof of the Pythagorean Theorem itself.
   b. Find out how Pythagoras (or one of the other ancient Greeks) proved this theorem.
   c. Find at least one additional proof of the Pythagorean Theorem.

15. Let $\triangle XYZ$ be an arbitrary triangle. Extend sides $XY$ and $XZ$ to lines. Construct the angle bisectors of the exterior angles at $Y$ and $Z$, and call their intersection $P$. Construct the circle centered at $P$ tangent to side $YZ$ of the triangle. Prove that this circle is also tangent to $\overleftrightarrow{XY}$ and $\overleftrightarrow{XZ}$.

16. Given a triangle, $\triangle XYZ$, extend sides $XY$ and $XZ$ creating exterior angles at $Y$ and $Z$.
   a. Prove that the bisectors of the exterior angles at $Y$ and $Z$ are concurrent with the

bisector of the internal angle at $X$. Call this point of intersection $R$.

b. Construct a circle, $\mathcal{C}$, centered at $R$, that is tangent to $\overleftrightarrow{XZ}$.

c. Prove that $\mathcal{C}$ is an excircle of $\triangle XYZ$. (See Exercise 15.)

17. Prove that if two circles are orthogonal, then they intersect at exactly two points and their tangents are perpendicular at both of those points.

18. Prove that $ABQR$ in Figure 3.6 is a cyclic quadrilateral.

19. You are given $\triangle ABC$.

a. Construct the center of the circumcircle of $\triangle ABC$. Prove that your construction is correct.

b. Explain how you can use the construction from part a to find an entire circle given just an arc of the circle.

c. What is the smallest number, $n$, of noncollinear points that determine a unique circle? Explain how you would construct a circle given just $n$ points.

20. Consider circle $\mathcal{C}$ with center $O$ and radius $r$. Let $A_1 A_2$ be a diameter of $\mathcal{C}$. Let $P$ be a point on $\overleftrightarrow{A_1 A_2}$. Let $d$ denote the distance from $P$ to $O$.

a. Assume that $P$ lies outside of $\mathcal{C}$. Let $\overleftrightarrow{PT}$ be tangent to $\mathcal{C}$ at $T$. Show that $PT^2 = d^2 - r^2$.

b. Still assuming that $P$ lies outside of $\mathcal{C}$, show that $PA_1 \cdot PA_2 = d^2 - r^2$.

c. If $P$ lies inside of $\mathcal{C}$, it is not possible to construct the tangent from $P$ to $\mathcal{C}$. However, it may still be possible to calculate $PA_1 \cdot PA_2$. Assume that $P$ lies inside of $\mathcal{C}$. Show that the product

$$PA_1 \cdot PA_2 = d^2 - r^2.$$

d. Explain how the power of $P$ with respect to a fixed circle $\mathcal{C}$, $\mathrm{Power}(P, \mathcal{C})$, can always be calculated by $d^2 - r^2$.

e. What is the significance of $\mathrm{Power}(P, \mathcal{C})$ being positive, zero, or negative? In other words, if you know that $\mathrm{Power}(P, \mathcal{C})$ is

$$> 0 \quad \text{or} \quad = 0 \quad \text{or} \quad < 0,$$

what do you know about the point $P$?

21. Let $X$ and $Y$ be any points on a circle of radius $r$. Let $P$ be a point exterior to the circle and collinear with $X$ and $Y$. If the distance from $P$ to the center of the circle is $d$, prove that $PX \cdot PY = d^2 - r^2$. Does this relationship still hold if $P$ is interior to the circle?

22. You are given a circle $\mathcal{C}$ with center $O$ and radius $r$. Prove that for any point $P$ in the plane (except the point $O$), the inversion of the inversion of $P$ with respect to $\mathcal{C}$ is the point $P$—that is, $(P')' = P$.

23. Complete the proof of Theorem 3.4 by considering the case where $\ell$ intersects $\mathcal{C}$.

24. **The Arbelos**  Refer to Figure 3.8. Let $T$ and $U$ be the points where $\overleftrightarrow{RA}$ and $\overleftrightarrow{RB}$ intersect the smaller arcs of the arbelos.

a. Prove that $PR$ and $TU$ are congruent line segments.

b. Prove that $PTRU$ is a parallelogram.

c. Prove that $\overleftrightarrow{TU}$ is tangent to the two smaller circular arcs of the arbelos.

25. **The Salinon**  Figure 3.8 gives a diagram of a salinon. Points $P$, $Q$, $R$, and $S$ are collinear (and in that order), with $PQ \cong RS$. Semicircles with diameters $PQ$, $RS$, and $PS$ all lie on the same side of $\overleftrightarrow{PS}$, while the semicircle with diameter $QR$ lies on the opposite side of $\overleftrightarrow{PS}$.

a. Construct a diagram of a salinon in Sketchpad. Make your diagram robust enough that $PQ$ stays congruent to $RS$ even if you move the points.

b. Calculate the area of the salinon.

c. The perpendicular bisector of $PS$ is the axis of symmetry of the salinon. Construct this axis of symmetry, and let $M$ and $N$ be the points where this line intersects the semicircles on diameters $PS$ and $QR$, respectively. Construct a circle with diameter $MN$.

d. Prove that the area of the salinon is equal to the area of the circle with diameter $MN$.

The following problems are more challenging.

26. Prove that the nine-point circle for $\triangle ABC$ is tangent to the incircle of $\triangle ABC$.

27. Prove that the nine-point circle of $\triangle ABC$ is tangent to one of the excircles of $\triangle ABC$.

28. You are given a circle $C$ centered at $O$ with radius $r$. Consider a second circle $C^*$, which is different from $C$. Choose a point $Q$ on $C^*$, and construct $Q' = \text{Inversion}(Q,C)$. Use the methods from Activity 10 to construct the locus of $Q'$. Describe the inversion of $C^*$ with respect to $C$. Prove that your observation is correct.

29. For any $\triangle ABC$, it is claimed that the diameter of the nine-point circle is half the length of the diameter of the circumcircle of $\triangle ABC$.
    a. If this claim can be proved, how will the radius of the nine-point circle compare with the radius of the circumcircle? Explain.
    b. What impact will this have on the relative areas of the nine-point circle and the circumcircle?
    c. Prove (or disprove) the claim.

30. Prove that for any $\triangle ABC$, the center of the nine-point circle lies on the Euler line of the triangle, midway between the circumcenter and the orthocenter.

31. We have a method for constructing a tangent to a circle $C$ from a point $A$ that may be on or exterior to $C$ (see page 68). Suppose you have two circles, $C_1$ and $C_2$. Develop a strategy for constructing a line that is tangent to both $C_1$ and $C_2$. (*Hint:* If $r_1$ and $r_2$ are the radii of $C_1$ and $C_2$, respectively, it will be helpful to construct a circle with radius $|r_1 - r_2|$.)

32. For $\triangle ABC$, let $a = |BC|$, $b = |CA|$, and $c = |AB|$. (So, $a$ is the length of the side opposite $\angle A$, etc.) For convenience, let $s$ be the *semiperimeter* $(a + b + c)/2$.
    a. The incircle of the triangle is tangent to the three sides. Label these points of tangency as $D$, $E$, and $F$ on sides $AB$, $BC$, and $CA$, respectively. Find expressions for the lengths of the segments $AD$, $DB$, $BE$, $EC$, $CF$, and $FA$ in terms of $a$, $b$, $c$, and $s$.
    b. Consider one of the excircles for $\triangle ABC$. This circle is also tangent to the three sides, at points $X$, $Y$, and $Z$ on sides $AB$, $BC$, and $CA$, respectively. Find expressions for the lengths of the segments $AX$, $XB$, $BY$, $YC$, $CZ$, and $ZA$ in terms of $a$, $b$, $c$, and $s$.

33. Using the notation of Exercise 32, part a, prove that the Cevians $AE$, $BF$, and $CD$ are concurrent. This is the *Gergonne point* of the triangle.

Exercises 34–35 are especially for future teachers.

34. In the *Principles and Standards for School Mathematics* (NCTM, 2000, 56), the National Council of Teachers of Mathematics (NCTM) recommends that

    Instructional programs from prekindergarten through grade 12 should enable all students to

    - recognize reasoning and proof as fundamental aspects of mathematics;
    - make and investigate mathematical conjectures;
    - develop and evaluate mathematical arguments;
    - select and use various types of reasoning and methods of proof.

    What does this mean for your future students?
    a. The NCTM has developed specific instructional recommendations for each of four different grade bands: Pre-K–2, 3–5, 6–8, and 9–12. Find a copy of the *Principles and Standards,* and study the discussion of the Reasoning and Proof Standard (NCTM, 2000, 56–59) for one of these grade bands. Choose one of the grade levels for which you are seeking certification. What are the specific NCTM recommendations with regard to reasoning and proof for your chosen grade band?
    b. Find some mathematics textbooks for these same grade levels. How are the NCTM recommendations implemented in these textbooks? Cite specific examples.
    c. Write a report in which you present and critique what you learn in studying the NCTM Reasoning and Proof Standard in light of your experiences in this course. Your report should include your answers to parts a and b.

**35.** Design several classroom activities involving reasoning and proof that would be appropriate for students in your future classroom. Write a short report explaining how the activities you design reflect both the NCTM recommendations and what you are learning about reasoning and proof in this class (in Chapters 1–3).

Reflect on what you have learned in this chapter.

**36.** Review the main ideas of this chapter. Describe, in your own words, the concepts you have studied and what you have learned about them. What are the important ideas? How do they fit together? Which concepts were easy for you? Which were hard?

**37.** Reflect on the learning environment for this course. Describe aspects of the learning environment that helped you understand the main ideas in this chapter. Which activities did you like? Dislike? Why?

## 3.4 CHAPTER OVERVIEW

This chapter highlighted two important themes of this course: the concept of a function and the development of sound mathematical proofs.

The concept of a function is foundational in mathematics. In your experiences prior to this course, functions may have been all about numbers—numbers as inputs and numbers as output. In this chapter, we investigated two new kinds of functions: *power* and *inversion*.

- Power is a function that takes two inputs—a point and a circle—and returns a number. The number returned by Power($P,C$) is a measure of how far $P$ is from $C$, where both the *distance* ($d$) of $P$ from the center of $C$ and the *radius* ($r$) of $C$ are used in computing this measure: Power($P,C$) $= d^2 - r^2$. If Power($P,C$) is positive, we know that $d > r$ and, thus, $P$ lies outside of $C$. If Power($P,C$) is negative, we know that $d < r$ and, thus, $P$ lies interior to $C$. If $P$ lies on $C$, then $d = r$; in this case, Power($P,C$) $= 0$.

- Inversion is a function that also accepts a point and a circle and returns a point. Thus, Inversion($P,C$) maps points of the plane to points of the plane. We will investigate this idea of function of the plane more deeply in Chapter 6, with special attention to the transformations of the plane.

Writing good mathematical proofs is an important theme in this course. Developing a sound proof is a complex skill. First, you must understand the *ideas* in the conjecture or theorem you are attempting to prove. Then you must use *logical reasoning* to clearly lay out the steps of your proof. This chapter identified several common strategies for developing mathematical proofs. We will refer back to these strategies throughout the rest of this course.

The *scientific method*, an important investigative strategy used in the natural sciences, involves experimentation followed by reflection on the results of the experiments. This is an *inductive process*. In this course, you have been encouraged to do a lot of experimentation of geometric ideas using Sketchpad diagrams and to make conjectures based on your experimentation. This inductive process can be very effective for learning, partly because working with geometric constructions in a Sketchpad diagram is a hands-on activity. The constructions are concrete objects (in the pedagogical sense), and the way we humans learn tends to move from concrete experiences toward abstract thought.

In general, the *mathematical method* approaches problems in the opposite way. Developing a sound mathematical proof requires you to move from axioms to theorems, from assumptions to results. Mathematicians tend to start with assumptions in the form of *axioms* (or *postulates*) and deduce results (*theorems*) from the axioms. This is a *deductive process*. Getting an idea for a conjecture requires some experimentation and inductive thinking, while developing a proof requires logical reasoning and deductive thinking.

In these first three chapters, we have been working with Euclid's axioms. In a later chapter, we will change one of these axioms, giving us an opportunity to see how much our assumptions shape our worldview. (In other words, what we see depends on our axioms.)

Step-by-step proofs are similar to the two-column proofs you may have used in your high school geometry course. Setting up your proofs in a step-by-step format is a way to begin organizing your thoughts. This organization will help you ensure that you are providing a justification for each step in your proof. As you become more comfortable with writing proofs, you may begin to write proofs in a flowing paragraph form, as you may have seen in many mathematics texts.

If you were able to develop a robust construction using Sketchpad, you may be able to translate the steps you took in constructing the diagram into steps in your proof. At the very least, your experimentation in constructing the diagram will have challenged you to think about the ideas you will need as you develop a proof.

The geometric content of this chapter focused on properties of circles. You were reminded of a lot of the terminology that is used in talking about circles. You probably encountered some new and challenging ideas in the geometry of circles. There are a number of circles associated with a triangle, and this provides a bridge back to the work of Chapter 2. Each triangle has one *incircle* and three *excircles;* these circles are associated with the bisectors of the triangle's internal and external angles. The *circumcircle* passes through the triangle's three vertices, and its center can be found by constructing the perpendicular bisectors of the sides. The *nine-point circle* of a triangle passes through many interesting points. We will continue our investigations of the nine-point circle in Chapter 4, where we will be able to use methods of analytic geometry to develop some of the proofs.

The *Pythagorean Theorem* says that the area of the square on the hypotenuse is equal to the sum of the areas of the squares on the other two legs of a right triangle. Our investigations in this chapter extended this theorem to include several new possibilities. If $\triangle ABC$ is a right triangle, then the area of the circle on the hypotenuse is equal to the sum of the areas of the circles on the other two sides. Squares, circles, and equilateral triangles work in this situation because they are families of similar figures. All circles are similar to each other; that is, they are the same shape, even when they have different sizes. Because of this similarity, their areas are proportional. This is why the Pythagorean Theorem can be extended to include circles but does not work for all situations involving rectangles.

We can construct various families of circles—circles that share particular characteristics. For example, there is a family of circles that all have the same center point, $O$; another family of circles that all pass through a point, $A$; and a family of circles with the same radius. We investigated two interesting families of circles in this chapter: circles that share a common chord and circles that are orthogonal to a given fixed circle. Although these two families of circles are interesting to study for themselves, they will also be important in investigations that we will take up in later chapters.

Mathematics is a living study. At every age of history, people have been investigating interesting problems. Our investigations of the arbelos and the salinon gave us an opportunity to step back into the early history of mathematics. Although these figures are constructed using simple arcs of circles, many interesting problems about them can be posed. In this chapter, we investigated just a few of these problems. Another interesting collection of problems involves the nine-point circle. We will continue our investigations of the nine-point circle (which comes from the late nineteenth century) in Chapter 4, where we will be able to use methods of analytic geometry—coordinates and equations—as tools to investigate these problems and to prove our conjectures.

At the end of Chapter 1, we listed the following constructions that you should master:

- Construct a perpendicular to a line from any point (on or off the line).
- Construct the perpendicular bisector of a given line segment.
- Construct the foot of the perpendicular from a point $P$ to a line $\ell$.
- Construct the tangent line to a circle from a point on the circle.
- Construct the tangent line to a circle from a point not on the circle.
- Construct the bisector of a given angle.

Now, as you are coming to the end of Chapter 3, you should recognize that each of these constructions depends in a fundamental way on some properties of circles. Each of these constructions can be done using the intersection points of one, two, or three circles. The geometric idea of a circle is very simple, and yet circles have amazingly many applications.

This chapter concluded with a discussion of various methods for developing mathematical proofs. The first step in developing a proof is to be clear in your own mind about what it is you are attempting to prove. Once you are able to formulate a clear statement of the conjecture you are trying to prove, you might use *modus ponens* to develop a direct proof.

When direct proof doesn't work, you can try an indirect approach. We discussed two slightly different strategies for developing an indirect proof. Suppose you are trying to prove a statement of the form $P \rightarrow Q$. One strategy is to prove the contrapositive, $\neg Q \rightarrow \neg P$. Because a statement is logically equivalent to its contrapositive (*modus tollens*), proving the contrapositive proves the original statement. The other indirect strategy is to assume the negation of what you are trying to prove, that is, assume $P$ and $\neg Q$, and then find a contradiction to something that is known to be true. This method is sometimes called *proof by contradiction.*

If a conjecture is not true, you might discover a counterexample to disprove it.

Sometimes a statement, $P \to Q$, and its converse, $Q \to P$, are both true; such a theorem can be compactly expressed using the expression "$P$ if and only if $Q$" (or $P$ iff $Q$, for short). In this case, the proof must have two parts: one part proving $P \to Q$, and another part proving $Q \to P$. Ceva's Theorem was cited as an example of this kind of theorem:

**Ceva's Theorem and Its Converse**    We are given $\triangle ABC$. The Cevians $AX$, $BY$, and $CZ$ are concurrent if and only if

$$\frac{AZ}{ZB} \cdot \frac{BX}{XC} \cdot \frac{CY}{YA} = 1.$$

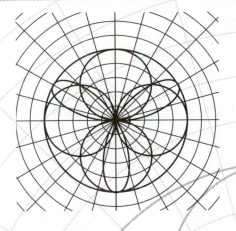

E arly seventeenth-century mathematicians established a correspondence between points of the plane and ordered pairs of real numbers, or coordinates. The significance of this development is that geometric lines and curves can be described using equations in two variables, and many problems of geometry can be studied using methods of algebra and analysis. This *analytic geometry* gives us a new tool for solving geometric problems (Eves, 1976, 278).

# Analytic Geometry

Do the following activities, writing your explanations clearly in complete sentences. Include diagrams whenever appropriate. Save your work for each activity, as later work sometimes builds on earlier work. You will find it helpful to read ahead into the chapter as you work on these activities.

1. Draw a line, and create two new points on it. Label the new points 0 and 1. Then *construct* (don't draw) points to represent 2, $\frac{1}{2}$, $-2$, and $\sqrt{2}$. Label your points accordingly.

   Now drag the point labeled 1. What happens to the other points as you do this?

   *A challenge:* Can you also construct the points for $\frac{1}{3}$ and $\sqrt{3}$?

2. Figure 4.1 shows an unusual coordinate system. As usual, however, you plot a point $(x, y)$ by starting at the origin, $O$, then moving $x$ units in the $x$-direction and $y$ units in the $y$-direction.

   a. On a sheet of paper, make a sketch of this coordinate system, plot the following points, and label them.

   $\quad$ **A** $(5, 0)$
   $\quad$ **B** $(0, 2)$
   $\quad$ **C** $(1, \sqrt{2})$
   $\quad$ **D** $(0, -4)$
   $\quad$ **E** $(-\sqrt{3}, \sqrt{3})$

   b. Find the distance between points $A$ and $B$. (*Hint:* This is equivalent to finding the length of side $AB$ of $\triangle OAB$.)

   c. Find the distance between points $A$ and $E$.

3. Using the Graph menu, create a set of axes, and hide the grid. Draw a line that intersects both axes, and construct two new points, $A$ and $B$, on this line. Measure the *abscissas* (the $x$-coordinates) and the *ordinates* (the $y$-coordinates) of both points. Then calculate the ratio

$$\frac{y_B - y_A}{x_B - x_A}.$$

Animate point $A$ (using an option on the Display menu), and observe the value of this ratio as $A$ moves. Now animate point $B$ also, perhaps at a different speed, and again observe the ratio. Explain what you see happening. Now change your line. Is your observation still valid?

4. Create a set of coordinate axes, and hide the grid.

   a. Draw a line that intersects the negative $x$-axis and the positive $y$-axis. Find these points of intersection. Measure the individual coordinates of the intersection points (the abscissas and the ordinates), and use these values to calculate the *slope* of your line. (You can choose **Properties** from the Edit menu to label your calculation as the slope.)

   Measure the angle between the $x$-axis and your line. This is the *angle of inclination* for this line. You should be able to vary the line and have the calculations update automatically.

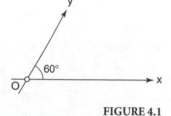

**FIGURE 4.1**
Coordinate System for Activity 2

b. How are the slope and the angle of inclination related? (*Hint:* Consider the triangle formed by your line and the two axes.)

c. Construct another line parallel to your line. Measure both the slope and the angle of inclination for this new line. How are these values related to those values of your original line? Explain why.

d. Construct a line perpendicular to your original line, and measure its slope and its angle of inclination. How are these values related to the slope and angle of inclination of the original line? Explain why.

5. Suppose you are given a point, $P$, and a line, $\ell$. Sketchpad can measure the distance between $P$ and $\ell$. Precisely what do you think this measurement means? Explain why your idea makes sense as a way to measure the distance from a point to a line.

6. Create a set of coordinate axes. Select **Square Grid** from the **Grid Form** choices, then hide the grid.

a. From the Graph menu, select **New Function,** and define the function $f(x) = 3 + \sqrt{25 - (x - 2)^2}$. Again, select **New Function,** and define the function $g(x) = 3 - \sqrt{25 - (x - 2)^2}$.

b. Select both functions, $f(x)$ and $g(x)$, and choose **Plot Functions.**

c. What is the shape of this figure?

d. Check your answer to part c by trying to find the center and radius of the figure.

7. Construct a parallelogram, $ABCD$, with $A$ at the origin and $B$ on the positive $x$-axis. Draw the diagonal $AC$, and construct point $P$ on this diagonal. Then draw segments $PB$ and $PD$. This creates four small triangles. Construct the interior of each triangle (select the three vertices and go to the Construct menu), and measure the area of each triangle. Vary the location of $P$, and form a conjecture about these areas. Then use coordinates to prove your conjecture in general.

8. In navigation, it is common to give directions in terms of *distance* and *compass point*.

a. On a sheet of paper, draw the following points.
   - the origin
   - 2 km east
   - 5 km north
   - 3 km southwest
   - 4 km east-southeast

b. How should we interpret −3 km west? Add this point to your drawing.

c. If east $= 0°$ and angles are measured counterclockwise, describe each of the points from parts a and b as (distance, angle°).

d. Choose **Grid Form** and then **Polar Grid** from the Graph menu. Select **Plot Points** from the Graph menu to graph each description from part c. Then use Sketchpad to measure the coordinates of these points. Explain any differences between your descriptions and the measured coordinates.

9. In the grid of a Cartesian coordinate system, the basic objects are vertical lines and horizontal lines, which can be described by equations such as $x = a$ and $y = b$. What are the basic objects of the grid for a polar coordinate system? How would you describe them by equations?

10. On a polar grid, plot the function $f(\theta) = \sqrt{3 \sin(2\theta)}$. Use radians for your plot. How does the $\sqrt{3}$ appear in the graph? Explain why this plot does not go into quadrants II or IV.

11. a. On a polar grid, plot the following functions, one at a time. These functions are called *rose curves*.
     - $r = \sin(3\theta)$
     - $r = \sin(4\theta)$
     - $r = \sin(5\theta)$
     - $r = \sin(6\theta)$

    b. For $r = \sin(n\theta)$, how many petals will the rose have?

    c. Create a rose with 16 petals. Create a rose with 17 petals. Create a rose with 18 petals. (*Hint:* For 18 petals, the coefficient of $\theta$ does not have to be an integer.)

# 4.2 DISCUSSION

Analytic geometry unites geometry and algebra in a powerful way. Coordinate systems allow us to use algebra to answer geometric questions. Because we have a large collection of algebraic tools, this is a valuable connection that works both ways. In the past century or so, much research has been done on what geometry can tell us about algebra. However, this discussion goes well beyond what we plan to cover in this chapter.

To use analytic geometry in proofs, we need ways to describe basic geometric concepts in algebraic terms. Let's look at points, lines, and distance.

## POINTS

Activity 1 asks you to create something familiar, a *number line.* By labeling one point as 0 and another point as 1, you have located the *origin* for your number line and have shown how large the *unit length* will be. All other coordinates follow from these two. If you actually constructed each point, then as you varied the location of point 1, the other points should have adjusted accordingly.

The critical idea of a number line is that there is a correspondence between the set of real numbers and the set of points on the line. Each number describes the location of a point, and each point has a unique number describing it. More precisely, there is a *one-to-one correspondence* between these two sets, meaning each real number corresponds to one and only one point on the line, and the point is labeled by this number.

By repeating the unit measure, we can locate the natural numbers on the number line. You might have done this by constructing a sequence of circles and finding their intersections with your number line.

Measuring in one direction from the origin along the number line gives the positive integers, and measuring in the opposite direction from the origin allows us to locate the negative integers. Although it is usual for the negative numbers to go to the left on a number line and the positive numbers to the right, there is no mathematical reason for this; it is merely a convention. The constructions for locating rational numbers are not difficult. Even finding $\sqrt{2}$ and $\sqrt{3}$ is pretty straightforward—though $\sqrt{3}$ takes a bit of thought. A number such as $\pi$, however, is a different matter: $\pi$ is not a *constructible number*. This means we cannot construct the precise location for $\pi$. Even so, $\pi$ has its unique point on the line (Jacobson, 1974, 263–277).

With only a single line, we are limited in how much geometry we can do. One important concept is the *distance between two points* on the number line, which is calculated as follows:

$$d(x_1, x_2) = |x_1 - x_2| = \sqrt{(x_1 - x_2)^2}.$$

The second formula may look more complicated than necessary, but bear with us! This pattern will be helpful later.

Things get more interesting when we go to two dimensions. This requires two axes, that is, two number lines set at an angle to each other. Because of the two axes, the coordinate plane is often denoted by $\mathfrak{R}^2$, the *Cartesian product* $\mathfrak{R} \times \mathfrak{R}$ of two number lines. In this situation, the *origin*, which is the point where the two axes intersect, is the zero point for both axes. The origin is the starting point for locating other points in the system.

Activity 2 presents a pair of coordinate axes. These axes are somewhat unusual because they are not perpendicular. However, in this skew coordinate system, positive $x$-values still go to the right and positive $y$-values still go upward. To plot the point $(1, \sqrt{2})$, for instance, we would begin at the origin and move a distance 1 to the right and then move upward, parallel to the $y$-axis, a distance $\sqrt{2}$. The other points in Activity 2 can be located similarly.

In Activity 2, we deliberately did not show what scale to use on these axes. As in the case of a number line, the unit length is arbitrary. With two axes, there can even be two different unit lengths, one for each axis. The **Square Grid** coordinate system in Sketchpad uses the same scale on both the horizontal and vertical axes, but the **Rectangular Grid** coordinate system does not. Other graphing tools, such as a calculator or mathematical software, will frequently use different scales on the vertical and horizontal axes. The scale will affect the image on the screen, so you must be aware of scale in any coordinate system.

The distance between point $A(5, 0)$ and point $B(0, 2)$ in Activity 2 can be viewed as the length of one side of $\triangle OAB$ (see Figure 4.2, next page). Knowing two sides and the included angle, we can use the *Law of Cosines* to find the third side, $AB$:

$$|AB|^2 = d^2 = |OA|^2 + |OB|^2 - 2 \cdot |OA| \cdot |OB| \cdot \cos(60°)$$

$$|AB| = d = \sqrt{5^2 + 2^2 - 2 \cdot 5 \cdot 2 \cdot \frac{1}{2}} = \sqrt{19}.$$

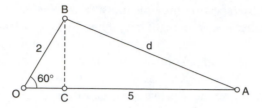

**FIGURE 4.2**
Distance in a Skew Coordinate
System

We could have used the Pythagorean Theorem (twice) to get the same result. In Figure 4.2, how do we know that the coordinates of point $C$ are $(1, 0)$? We find the length of side $BC$, then we use this result to find the length of side $AB$, which is the distance $d$ that we want:

$$BC = \sqrt{2^2 - 1^2} = \sqrt{3},$$

$$d = \sqrt{(\sqrt{3})^2 + 4^2} = \sqrt{19}.$$

A similar calculation will find the distance between points $A$ and $E$ in Activity 2. One approach is to use $\triangle AOE$, with an angle of $120°$ at the origin. Another method is to keep the $60°$ angle and use a longer base, which will keep two sides of the triangle parallel to the two axes (see Figure 4.3).

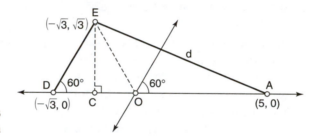

**FIGURE 4.3**
Another Distance Calculation

The general formula for the distance between two points, $P_1(x_1, y_1)$ and $P_2(x_2, y_2)$, in this $60°$-coordinate system is

$$d = \sqrt{(x_1 - x_2)^2 + (y_1 - y_2)^2 - 2(x_1 - x_2)(y_1 - y_2)\cos(60°)}.$$

Using this formula, we can calculate the distance between points $A(5, 0)$ and $E(-\sqrt{3}, \sqrt{3})$ as

$$|AE| = d = \sqrt{(5 + \sqrt{3})^2 + (\sqrt{3})^2 - 2 \cdot (5 + \sqrt{3}) \cdot \sqrt{3} \cdot \frac{1}{2}}$$

$$= \sqrt{25 + 10\sqrt{3} + 3 + 3 - 5\sqrt{3} - 3}$$

$$= \sqrt{28 + 5\sqrt{3}}.$$

Again, we could have calculated this distance using two applications of the Pythagorean Theorem. (Having two different strategies gives us a way to check our work.) Notice in Figure 4.3 the location of point $E$ with respect to the origin. Point $C$ on the $x$-axis directly below point $E$ has coordinates $(-\frac{\sqrt{3}}{2}, 0)$, and $EC$

has length $\frac{3}{2}$. This gives

$$|AE| = \sqrt{\left(5 + \frac{\sqrt{3}}{2}\right)^2 + \left(\frac{3}{2}\right)^2} = \sqrt{25 + 5\sqrt{3} + \frac{3}{4} + \frac{9}{4}}$$

$$= \sqrt{28 + 5\sqrt{3}},$$

which is the same result we got before.

The presence of the $\cos(60°)$ in the general distance formula for the $60°$-coordinate system makes the calculation a bit messy. When we put the coordinate axes at right angles to each other, the distance formula is the same except that the angle is $90°$. This makes the formula much nicer because $\cos(90°)$ is simply 0. This is one reason perpendicular axes are preferred.

Having a formula for distance opens many possible questions. The following theorem presents a useful result.

**THEOREM 4.1**   The midpoint of the segment between points $P(x_p, y_p)$ and $Q(x_q, y_q)$ is the point

$$\left(\frac{x_p + x_q}{2}, \frac{y_p + y_q}{2}\right).$$

Theorem 4.1 is straightforward to prove, especially in the case of perpendicular axes. The proof will be left to the exercises.

## LINES

A point is a zero-dimensional object that has only location. A point's location is described by its coordinate on a number line. A line, however, is a one-dimensional object that has both location and direction. An algebraic description of a line must include both location and direction. Here are three common ways to describe a line, as well as one less common way:

$$y - y_0 = m(x - x_0) \quad \textit{point-slope form}$$
$$y = mx + b \quad \textit{slope-intercept form}$$
$$\frac{x}{a} + \frac{y}{b} = 1 \quad \textit{intercept form}$$
$$Ax + By = C \quad \textit{general form}$$

In point-slope form, the point $(x_0, y_0)$ tells the line's location and the slope $m$ tells its direction. (It is traditional to refer to the slope as $m$, and we will continue the tradition.) In slope-intercept form, the point $(0, b)$ is the intercept point on the vertical axis. This point tells the line's location, and once again, $m$ tells the direction. In *intercept form*, the points $(a, 0)$ and $(0, b)$ tell the line's location and indirectly give the direction.

The general form is the most powerful of these four ways to describe a line, for it can be used to describe horizontal and vertical lines, as well as any other line. Each of the other forms has limitations, which will be examined in the exercises.

Most of this information should be familiar to you. What we want to do in this section is carefully examine why we can use these forms to describe a line. To do so, we first need a clear understanding of slope.

In Activity 3, you should have found that the ratio $\dfrac{y_B - y_A}{x_B - x_A}$ remained constant, no matter where $A$ and $B$ were on the line. This formula calculates the *slope* of the line. We often say that the slope is the rise over the run, because slope is a measure of the vertical change (the rise) from point $A$ to point $B$ divided by the horizontal change (the run) from $A$ to $B$. The vertical change can be denoted by $\Delta y$ and the horizontal change by $\Delta x$. Thus, the calculation for slope can be written as

$$\frac{\text{vertical change}}{\text{horizontal change}} = \frac{\Delta y}{\Delta x}$$

and can be read as "the change in $y$ divided by the change in $x$."

Figure 4.4 illustrates one situation that might occur. In this figure, the segments $\Delta x_i$ and $\Delta y_i$ are parallel to the $x$- and $y$-axes, respectively. Why is the slope from $A_1$ to $B_1$ equal to the slope from $A_2$ to $B_2$? (*Hint:* Think of similar triangles!) Notice that it really doesn't matter which points we use to calculate the slope of the line, because the triangle for points $A_1$ and $B_1$ is similar to the triangle for points $A_2$ and $B_2$. Both of these triangles are also similar to the triangle for $A_1$ and $B_2$. Because the triangles are similar, ratios of the lengths of corresponding sides will be equal.

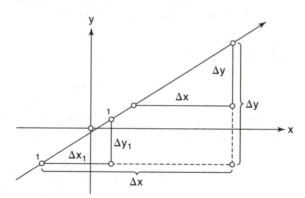

**FIGURE 4.4**
Slope Is Well-Defined

Does all this continue to work if either $\Delta x$ or $\Delta y$ is 0? We also must consider two special cases, the horizontal lines and the vertical lines. For a horizontal line, we cannot draw triangles with vertical and horizontal sides. Instead, think about the coordinates of the points along this line. Each of these points is the same distance from the horizontal axis, so the $y$-coordinate is always the same. This means that the numerator $\Delta y$ will always be 0, making the slope equal to 0.

**THEORE  4.2**  For a nonvertical line, the slope is well-defined. In other words, no matter which two points, $A(x_A, y_A)$ and $B(x_B, y_B)$, are used to calculate the slope, the value

$$\frac{y_B - y_A}{x_B - x_A} = \frac{\Delta y}{\Delta x}$$

will be the same.

------------------------------------

There is still the awkward case of the vertical lines. Points along any particular vertical line will all have the same $x$-coordinate. Thus, we encounter the difficult situation of the denominator $\Delta x = 0$, which makes the fraction undefined. This is our answer: The slope of a vertical line is undefined.

It is not uncommon to hear people say that the slope of a vertical line is infinity. This is not correct for at least two reasons. First, $\infty$ is not really a number. It is not part of the real number system, and arithmetic with $\infty$ does not work in the usual way. Second, there are two contradictory ways to interpret a vertical line. Such a line can be viewed as rising infinitely quickly, and some might call this a slope of $+\infty$. However, the line can also be viewed as falling infinitely quickly, leading to a slope of $-\infty$. If the slope concept is to be well-defined, there cannot be one line with two different slopes.

Now suppose we know a point, $(x_0, y_0)$, on a particular line and we also know its slope, $m$. If we let $(x, y)$ stand for any other point on this line, we can write

$$\frac{y - y_0}{x - x_0} = m$$

and simplify to get

$$y - y_0 = m(x - x_0),$$

which is the *point-slope form* of the line. This form of the equation of a line is useful for converting information about a line into an equation for the line. It takes only a little more algebra to rewrite this as

$$y = mx + (y_0 - mx_0) = mx + b,$$

the familiar *slope-intercept form* of the line. The slope-intercept form of the equation of a line is useful for converting an equation of a line into a picture (or graph) of the line.

The *general form* of a linear equation is $Ax + By = C$. If $A = 0$, we have the equation $y = \frac{C}{B}$, a horizontal line. If $B = 0$, we have the equation $x = \frac{C}{A}$, a vertical line. If $A$ and $B$ are both nonzero, the general form can be rearranged into slope-intercept form. This shows that we can use the general form to represent any line.

**THEORE  4.3**   A line can be described by a linear equation, and a linear equation describes a line.

----

Activity 4 introduces an alternative way to specify the direction of a line, namely, the angle of inclination, $\alpha$. This angle is measured at the point where the line intersects the horizontal axis, and it should have a value $0 \leq \alpha < \pi$. There are, of course, lines that do not intersect the horizontal axis. For convenience later, let us define $\alpha = 0$ for all horizontal lines, whether or not they intersect the horizontal axis. Figure 4.5, next page, shows a relationship between the slope of a line and the angle of inclination. The important relationship is $m = \tan \alpha$. Notice that this relationship holds even for the special cases. For horizontal lines, $m = 0 = \tan 0$. For vertical lines, $m$ is undefined and $\tan(\frac{\pi}{2})$ is undefined. This relationship helps us prove some important facts.

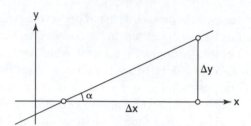

**FIGURE 4.5**
Slope and Angle of Inclination

**THEORE** **4.4**   Two lines are parallel if and only if the two lines have equal slopes.

**Outline of a Proof**   It is necessary to deal with the horizontal case separately. However, we will first prove the general case. Figure 4.6 gives a generic picture of the situation.

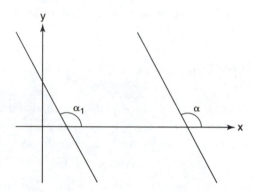

**FIGURE 4.6**
Parallelism and Angles of Inclination

Suppose two lines that both intersect the $x$-axis are parallel. Then the $x$-axis is a transversal of the two lines, and the two angles of inclination are equal. Thus, the tangents of the two angles are equal, so the slopes are equal.

The reverse argument also works. Suppose the slopes are equal. Then the tangents of the two angles are equal, and so the angles of inclination are equal. (Recall that these angles are between 0 and $\pi$.) Using the $x$-axis as a transversal, the two lines are parallel.

The argument just given will not work for horizontal lines, because most horizontal lines do not intersect the $x$-axis. However, all horizontal lines are parallel, and all horizontal lines have slope $= 0 = \tan 0$. So, the theorem is true in this case also.

There are several ways to prove the next theorem. In the proof shown here, we illustrate an approach based on the Pythagorean Theorem.

**THEORE** **4.5**   Let $\ell_1$ and $\ell_2$ be two lines, neither one vertical, with slopes $m_1$ and $m_2$, respectively. The two lines are perpendicular, written $\ell_1 \perp \ell_2$, if and only if $m_1 m_2 = -1$.

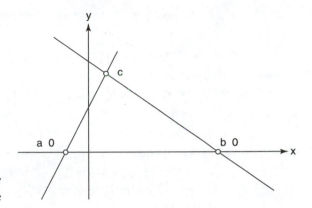

**FIGURE 4.7**
Perpendicularity and Slope

**Proof**   For the lines in Figure 4.7, $\angle ACB$ is $90°$ if and only if the Pythagorean Theorem holds for $\triangle ACB$. So, we want $|AB|^2 = |AC|^2 + |CB|^2$. This holds iff

$$(b - a)^2 = ((c - a)^2 + (d - 0)^2) + ((b - c)^2 + (0 - d)^2)$$

iff   $-ab + ac + bc - c^2 = d^2$

iff   $(c - a)(b - c) = d^2$

iff   $-1 = \left( \dfrac{d - 0}{c - a} \right) \left( \dfrac{0 - d}{b - c} \right) = (\text{slope of } \overleftrightarrow{AC})(\text{slope of } \overleftrightarrow{CB}).$

(You should check our calculations.) If $\overleftrightarrow{AC}$ is $\ell_1$ and $\overleftrightarrow{CB}$ is $\ell_2$, these calculations say that the product of the two slopes, $m_1 m_2$, is $-1$.

-----

A consequence of $m_1 m_2 = -1$ is that $m_1 = \dfrac{-1}{m_2}$. This means $m_1$ is the opposite of $m_2$ in both arithmetic senses: $m_1$ is both *negative* to and *reciprocal* to $m_2$.

Here is an example that combines two of the concepts we have just seen: Given a point $A = (a, 0)$ on the horizontal axis and another point $B = (0, b)$ on the vertical axis, find an equation for the perpendicular bisector of segment $AB$. To do this, we need to know a location for and the direction of this line. One possible location is the midpoint of $AB$, which is the point $\left( \dfrac{a}{2}, \dfrac{b}{2} \right)$. The slope of $AB$ is $\dfrac{b}{-a} = -\dfrac{b}{a}$; so, the slope we need for a perpendicular line is $\dfrac{a}{b}$. Using point-slope form, an equation for the perpendicular bisector is

$$y - \frac{b}{2} = \frac{a}{b} \left( x - \frac{a}{2} \right).$$

## DISTANCE

A circle is a set of points, all of which are the same distance from a fixed center point. So, a circle can be described by its center and its radius—these two pieces of data are enough to let us write an equation for a circle. Let us be more specific: Suppose the center of our circle is the point $(a, b)$ and the radius is $r$. Any point

$(x, y)$ on the circle must be a distance $r$ from the center, and the distance formula gives

$$\sqrt{(x - a)^2 + (y - b)^2} = r.$$

Equations with radical symbols can be awkward to use. Fortunately, we can improve this particular equation by squaring both sides to get

$$(x - a)^2 + (y - b)^2 = r^2.$$

Suppose, for example, that you wanted to use an equation to represent a circle with radius 5 units and center at $(2, 3)$. Using the equation form developed above, this circle would be represented by $(x - 2)^2 + (y - 3)^2 = 25$. Solving this for $y$, we get two equations (because of the square root):

$$y = 3 \pm \sqrt{25 - (x - 2)^2}.$$

These are the two equations you plotted in Activity 6.

We used the formula for the distance between two points to find the equation of a circle because we wanted an equation for the set of all points $(x, y)$ at a fixed distance $r$ from the point $(a, b)$. A harder question is to find the distance between a point and a line. It may not even be clear what we mean by the distance between point $P$ and line $\ell$, for there are many points on $\ell$, and many distances could be calculated between $P$ and one of the many points on $\ell$. How do we know which of these distances to choose? This is the issue that you had to think about in Activity 5. How did Sketchpad find the distance? By choosing points farther and farther away from $P$, the distance can be made arbitrarily large. However, there is a limit to how small this distance can be, and that minimum distance is what is meant by the distance from $P$ to $\ell$.

So, the problem is to decide which point on $\ell$ is closest to point $P$. One way to solve this problem is to find the circle centered at $P$ that is tangent to line $\ell$. This solution is the reverse of some of the problems posed in Chapter 3, where you were trying to find the line through a fixed point that was tangent to a given circle. (See Activity 4, part a, from Chapter 3.) Here we want to find the radius of the smallest circle centered at $P$ and tangent to line $\ell$. It should be clear that the shortest route from $P$ to $\ell$ is along a line perpendicular to $\ell$. This needs a proof, however, and you will be asked to supply one in the exercises.

To find our desired formula, we want the distance from $P$ to the intersection point (labeled $Q$ in Figure 4.8). First, we need some algebraic descriptions. Let $P$ be the point $(a, b)$, and let $\ell$ be the line $y = mx + c$. (Because we are using $b$ as a coordinate for $P$, we cannot use $b$ in the equation for the line.) The line through $P$ and perpendicular to $\ell$ is

$$y - b = -\frac{1}{m}(x - a).$$

The point $Q$ is the intersection of this new line with $\ell$. To find $Q$, solve the following system of equations:

$$\begin{cases} y = mx + c, \\ y = -\frac{1}{m}(x - a) + b. \end{cases}$$

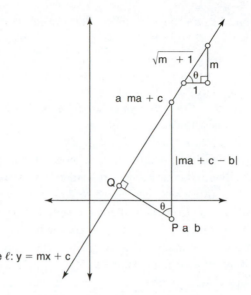

FIGURE 4.8

Distance from a Point to a Line

line $\ell$: y = mx + c

After solving this system, which you should check, we see that $Q$ is the point

$$\left( \frac{a + bm - cm}{m^2 + 1}, \frac{am + bm^2 + c}{m^2 + 1} \right).$$

Using the distance formula, we can now calculate the distance between points $P$ and $Q$. The algebra is tedious, so we will show only portions of the work. You should fill in the missing steps. The distance from $P$ to $Q$ is

$$= \sqrt{\left( \frac{a + bm - cm}{m^2 + 1} - a \right)^2 + \left( \frac{am + bm^2 + c}{m^2 + 1} - b \right)^2}$$

$$= \frac{1}{m^2 + 1} \sqrt{(bm - cm - am^2)^2 + (am + c - b)^2}$$

$$= \frac{1}{m^2 + 1} \sqrt{m^2(am - b + c)^2 + (am - b + c)^2}$$

$$= \frac{1}{m^2 + 1} |am - b + c| \sqrt{m^2 + 1}$$

$$= \frac{|am + c - b|}{\sqrt{m^2 + 1}}.$$

This formula looks somewhat nicer if we use the general form of the line. For the point $P(a, b)$ and the line $Ax + By - C = 0$, we can write the distance from $P$ to this line as

$$\frac{|Aa + Bb - C|}{\sqrt{A^2 + B^2}}.$$

The preceding discussion relied on a great deal of algebra, but does not really explain why the formula works. Here is an alternative explanation that may be more illuminating.

In Figure 4.8, the smaller triangle is constructed specifically to show the slope of the line. How do we know that the two angles marked $\theta$ are truly equal and that the two triangles are similar? We get

$$\frac{d}{1} = \frac{|am + c - b|}{\sqrt{m^2 + 1}}.$$

There are some issues in this equation that need attention. Why is the absolute value necessary? Recall that the picture in Figure 4.8 is only one of the possible ways that this problem could arise, and $P$ could be above the line. Neither of the preceding explanations deals with the possibility that line $\ell$ is vertical. If $\ell$ is vertical, does the formula still give a correct answer? Obviously, the version with $m$ will not, but what about the general form version?

In mathematics, we are sometimes interested in verifying that a particular calculation gives the desired result. At other times, we are interested in using a calculation to explain what is happening in a particular situation. In the preceding discussion, we used both of these methods to present the idea of calculating the distance from a point to a line. The first method verifies that the calculation works, while the second explains why it works. Both of these strategies could be called a "proof." Only one of these proofs makes sense for a horizontal line; which one? These issues will appear in the exercises.

## USING COORDINATES IN PROOFS

A great benefit of using coordinates is the connection it gives us between geometric and algebraic ideas. Lines and linear equations are a classic example. The geometric ideas of position and direction of a line correspond to the algebraic ideas of the constant term and the leading coefficient of a linear equation in the slope-intercept form. We can now use the wide variety of familiar algebraic tools, such as solving equations and solving systems of equations, to prove geometric facts.

Activity 7 presents a theorem to be proved. Although not a particularly important theorem in itself, the activity gave you a chance to try some of the algebraic techniques we have developed so far. To develop this proof, you need to use things like slope, linear equations, distance between points, and distance between a point and a line. One possible picture for Activity 7 is shown in Figure 4.9.

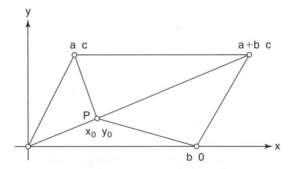

**FIGURE 4.9**
Graph for Activity 7

First let us discuss this picture. Does it fit the description in the problem? Points $A$ and $B$ are in the correct places, and point $D$ is chosen arbitrarily. So, the coordinate values $a$, $b$, and $c$ can be anything. They could even be negative values!

However, point $C$ must be chosen carefully so that we actually get a parallelogram. The coordinates of point $C$ make $DC$ both parallel to and congruent to $AB$.

Notice also that there are three 0s among the coordinates. It is usually a good idea to place the coordinate system to get as many 0s as possible. In Figure 4.9, the coordinate system was placed so that the origin and one of the axes were parts of the parallelogram. Remember that we will be doing algebra with these coordinate values, and having lots of 0s will usually make the algebra easier.

We hope you noticed that no matter where point $P$ is, the four small triangles fall into pairs with equal area. Specifically, $\triangle APB$ and $\triangle APD$ appear to have equal areas, as do $\triangle CPB$ and $\triangle CPD$. How can we prove this? Notice that $\triangle ACB$ is congruent to $\triangle CAD$. (Why?) Thus, it is only necessary to verify that one of the pairs has equal areas, for subtracting areas would then prove that the other pair also has equal areas.

We will focus on $\triangle APB$ and $\triangle APD$. One of these areas is easy to calculate:

$$\text{area of } \triangle APB = \frac{1}{2}by_0.$$

The area of $\triangle APD$ is more difficult. Think of segment $AD$ as the base. Then, the height of this triangle is the distance from $P$ to the line $AD$. The formula for this line is

$$y = \frac{c}{a}x.$$

Now we can use formulas for distance to get

$$\text{area of } \triangle APD = \frac{1}{2}\sqrt{a^2 + c^2}\frac{|\frac{c}{a}x_0 - y_0|}{\sqrt{\left(\frac{c}{a}\right)^2 + 1}},$$

which simplifies to $\frac{1}{2}|cx_0 - ay_0|$. To decide whether these two algebraic expressions are equal, recall that $x_0$ and $y_0$ are related through $\overleftrightarrow{AC}$, which is

$$y = \left(\frac{c}{a+b}\right)x, \quad \text{so} \quad y_0 = \left(\frac{c}{a+b}\right)x_0.$$

With this substitution, we get

$$\text{area of } \triangle APB = \frac{1}{2} \cdot b \cdot \frac{c}{a+b} \cdot x_0 = \frac{bc}{2(a+b)} \cdot x_0$$

and

$$\text{area of } \triangle APD = \frac{1}{2}\left|c \cdot x_0 - a\frac{c}{a+b} \cdot x_0\right| = \frac{1}{2}\left|\frac{bc}{(a+b)} \cdot x_0\right|.$$

Because of the way we set up our coordinates, $a$, $b$, and $c$ are positive values. Therefore, the two expressions are equal, making the two areas equal.

When setting up a coordinate system for a proof, it is always tempting to use the origin as one of the important points. The "double zeros" at the origin are often helpful when writing the equations. In some situations, however, there is an advantage to using points other than the origin. For example, let us prove algebraically that the perpendiculars from the vertices of a triangle to the opposite sides are concurrent (at the orthocenter, remember?). Figure 4.10, next page, shows one way to draw coordinates for the triangle.

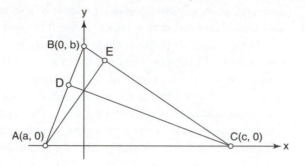

**FIGURE 4.10**
Proving Concurrence of Altitudes

By putting $B$ on the vertical axis, we have made the vertical axis one of the perpendiculars of the triangle. Our desired concurrency point must lie on the vertical axis, where $x = 0$.

The line $AB$ is $y = -\frac{b}{a}x$. The slope of its perpendicular is then $\frac{a}{b}$, and $\overleftrightarrow{CD}$ is

$$y - 0 = \frac{a}{b}(x - c)$$

$$y = \frac{a}{b}x - \frac{ac}{b}.$$

Now we do a similar thing with side $BC$. The equation of the line through points $B$ and $C$ is $y = -\frac{b}{c}x + b$. So the equation of the line perpendicular to $\overleftrightarrow{BC}$ through the point $A$ is

$$y - 0 = \frac{c}{b}(x - a)$$

$$y = \frac{c}{b}x - \frac{ac}{b}.$$

Thus, all three perpendiculars pass through point $(0, -\frac{ac}{b})$. This point, which lies on the vertical axis, is the orthocenter of the triangle.

If the minus sign looks out of place in the coordinates of the orthocenter, recall that Figure 4.10 shows $a$ as a negative number, while $b$ and $c$ are positive. However, nothing prevents $a$ from being positive. Does the proof still make sense if point $A$ is to the right of the origin? Where will the orthocenter be in that situation? What happens if $A$ is at the origin?

## POLAR COORDINATES

The $xy$-coordinate system is a comfortable one for people accustomed to city living. With our experiences at navigating in terms of streets and intersections, the notion of moving along two perpendicular directions is quite natural. Indeed, many cities—such as Indianapolis, Indiana; St. Petersburg, Florida; and Milwaukee, Wisconsin—have a rectangular grid for a large portion of their street system. Navigating on the ocean, however, is quite different from navigating in an urban area. On the sea, it is more natural to specify a direction and a distance to travel in that direction. This is the idea underlying the polar coordinate system.

The *polar coordinate system* uses an origin point and a single axis. This axis is actually just a ray, not a line. Typically, this ray is drawn to the right, as though it were the positive portion of the $x$-axis. We can describe then a point, $P$, in the plane by giving two numbers: its distance from the origin $O$ and its angle, measured between the polar axis and the ray $OP$. The convention in mathematics is that counterclockwise angles are considered to be positive and clockwise angles are negative; that convention applies here.

It is common to refer to polar coordinates as $(r, \theta)$, where $r$ is the distance from the origin—the radius of the circle containing this point—and $\theta$ is the angle measured counterclockwise from the polar axis. These variables can take on any real number values: positive, zero, or negative. For example, we can plot the points $(3, \pi)$, $(-3, \pi)$, $(-3, -\pi)$, and $(3, -\pi)$. Unlike Cartesian coordinates, however, this list of four ordered pairs actually represents only two distinct points. There are a couple of reasons for this. One difficulty is that $\pi$ and $-\pi$ are different angles that correspond to the same ray, namely, the negative $x$-axis. So, $(3, \pi)$ and $(3, -\pi)$ are different names for the same point. Similarly, $(-3, \pi)$ and $(-3, -\pi)$ are the same point. However, locating a point whose first coordinate is negative when expressed in polar coordinates can be a little confusing. In Activity 8b, how did you interpret $-3$ km west? If we think of this as instructions for moving, we should face to the west, then move $-3$ km—in other words, back up 3 km. On the coordinate system, this point lies on the positive $x$-axis, that is, on the polar ray.

The fact that points do not have unique coordinates in the polar system can cause trouble when analyzing problems algebraically. For example, suppose we want to find the intersection points of the curves described by

$$\begin{cases} r = \sin\theta, \\ r = \cos\theta. \end{cases}$$

The standard approach is to set the two expressions for $r$ equal, and solve for the variable $\theta$:

$$\sin\theta = \cos\theta$$
$$\tan\theta = 1$$
$$\theta = \left\{ \frac{\pi}{4}, \frac{5\pi}{4} \right\}.$$

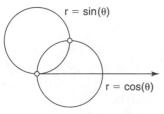

r = sin(θ)

r = cos(θ)

**FIGURE 4.11**
Intersecting Circles

Substituting these values into the two functions gives two intersection points: $(\frac{\sqrt{2}}{2}, \frac{\pi}{4})$ and $(-\frac{\sqrt{2}}{2}, \frac{5\pi}{4})$. However, plotting these points shows that they are actually the same point. To make things worse, there are indeed two intersection points. In Figure 4.11, a plot of the two functions shows these two circles intersecting twice—at the point we found algebraically and also at the origin. Here's what has happened: The curve $r = \sin\theta$ includes the origin as $(0, 0)$, while the other curve, $r = \cos\theta$, includes the origin as $(0, \frac{\pi}{2})$. The lesson here is that the algebra is not sufficient. When solving systems of equations given in polar coordinates, you must look at a graph to see (literally) if there are any other intersection points that may have been hidden by alternative descriptions of the points.

The polar grid consists of circles centered at the origin and rays coming from the origin. These circles are represented by equations in the pattern $r = a$.

Notice that $a$ could be a negative number or even 0. Once again, however, we have the difficulty of nonunique representations. For example, the equations $r = 5$ and $r = -5$ represent the same circle. The equation $r = 0$ represents the "circle" consisting of just a single point. The lines from the origin also have simple equations, and again these are not unique. The equations $\theta = \frac{4\pi}{3}$, $\theta = \frac{-2\pi}{3}$, and $\theta = \frac{10\pi}{3}$ represent the same line. In fact, because point $(x, \frac{4\pi}{3})$ lies on line $\theta = \frac{4\pi}{3}$ for any (positive or negative) value of $x$, the line $\theta = \frac{\pi}{3}$ is yet another representation for this same line.

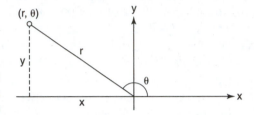

**FIGURE 4.12**
Converting Between Cartesian
and Polar Coordinates

Finding an equation for a general line—one that does not go through the origin—is a more difficult question. It is helpful to take an indirect approach. First, let us examine how polar coordinates are related to Cartesian coordinates. The diagram in Figure 4.12 shows this relationship. From basic trigonometry, we know that

$$\begin{cases} x^2 + y^2 = r^2, \\ \tan\theta = \frac{y}{x}. \end{cases}$$

We also know that

$$\begin{cases} x = r\cos\theta, \\ y = r\sin\theta. \end{cases}$$

(We chose a point in quadrant II to remind you that these relationships hold for any angle, not merely acute angles.)

Now consider the general form of an equation for a line, and substitute $r\cos\theta$ and $r\sin\theta$ for $x$ and $y$, respectively:

$$Ax + By = C$$
$$A(r\cos\theta) + B(r\sin\theta) = C$$
$$r = \frac{C}{A\cos\theta + B\sin\theta}.$$

Because $A$, $B$, and $C$ are constants (the coefficients of the equation in general form), we see that $r$ is a function of $\theta$, that is, $r = f(\theta)$. However, this is a rather awkward equation for a simple thing like a line!

We have seen that in polar coordinates, lines cannot be described by linear equations, except in the special case of lines through the origin. Linear equations in polar coordinates describe a completely different type of curve, called a *spiral*. Try graphing $r = \theta$, for instance, or $r = 2\theta + 1$. Many functions other than the linear functions also produce spirals. The function $r = \ln\theta$ is the *logarithmic spiral*, which has many interesting properties. This spiral can be seen in the growth pattern of seashells.

It is interesting to see the variety of curves produced by familiar functions when they are graphed in a polar system. You saw examples of this in Activity 11. When plotted on a Cartesian system, the function $r = \sin(3\theta)$ is a sine wave of amplitude 1 and period $\frac{2\pi}{3}$. The period means that in the standard interval $0 \le \theta \le 2\pi$, the wave will occur three times. So, it should not be surprising that the polar graph has a maximum distance of 1 from the origin and that there are three petals to the rose. The rose for $r = \sin(4\theta)$ may have surprised you. Do you see why there are eight petals instead of four? (See Figure 4.13.)

f(θ) = sin(4·θ)

**FIGURE 4.13**
A Rose with Eight Petals

It is enlightening to draw the curve $r = \sin(4\theta)$ in Sketchpad, construct a point on the plot, measure its polar coordinates, and then animate the point. (Make sure the point is moving in the direction of positive angles.) Which of the petals are produced by positive $r$-values and which are produced by negative $r$-values?

Activity 11, part c, asks for a rose with eighteen petals. This is difficult, for using a coefficient of 9 gives only nine petals and using a coefficient of 18 gives thirty-six petals. Try cutting 18 in half twice for a coefficient of $\frac{18}{4} = 4.5$. The problem with this new function is that $0 \le \theta \le 2\pi$ will not draw the entire rose. You must expand the domain (choose **Properties** | **Plot** from the Edit menu).

## THE NINE-POINT CIRCLE, REVISITED

Any $\triangle ABC$ has three altitudes. The point where a line containing an altitude intersects a line containing the opposite side of the triangle is called the *foot* of the altitude. In Figure 4.14, the feet of the altitudes are the points $D$, $E$, and $F$.

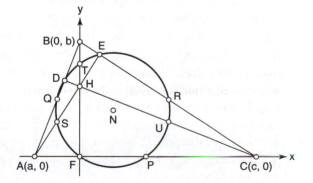

**FIGURE 4.14**
The Nine-Point Circle on a
Coordinate System

These points are the vertices of the *pedal triangle* for $\triangle ABC$. The circumcircle of this pedal triangle turns out to contain many other important points for $\triangle ABC$.

**THEOREM 4.6**   **The Nine-Point Circle Theorem**   For any triangle, the three feet of the altitudes, the three midpoints of the sides, and the three midpoints of the segments from orthocenter to vertex all lie on a common circle, known as the *nine-point circle* of the triangle.

It is possible to prove this theorem without coordinates, as we did in the proofs seen in earlier chapters. However, we can use the techniques discussed in this chapter, such as representing lines by linear equations, to give a proof that involves coordinates and algebra. This is called an *analytic proof*.

**Proof**   Let us begin with a list of the general steps needed for the proof.

1. Place the triangle on a coordinate system, preferably in a way that will help the algebra.
2. Find equations for the triangle's altitudes.
3. Find the coordinates of the feet of the altitudes and the coordinates of the orthocenter.
4. Find the center and radius of the circumcircle of the pedal triangle.
5. Write the equation for the circumcircle of the pedal triangle.
6. Verify (confirm) that the three feet lie on this circle.
7. Verify that the three midpoints of the sides lie on this circle.
8. Verify that the three midpoints of segments from orthocenter to vertex lie on this circle.

As we carry out these steps, we will not show every detail of the algebra. It is good practice to do at least some of this algebra yourself. Here we will include just enough to show you the direction of the proof.

**Step 1**   Figure 4.14 shows one way to place $\triangle ABC$ on a coordinate system. This placement has the advantage that each vertex has 0 among its coordinates. Another benefit is that the vertical axis is one of the altitudes; this will be useful in Steps 2 and 3. As the triangle is drawn here, it appears that $a < 0$, $b > 0$, and $c > 0$. However, this is not necessarily true, and we must not assume this during our work. For instance, an obtuse triangle could have both $a$ and $c$ with positive values.

**Step 2**   The altitude through point $A$ contains the point $(a, 0)$ and is perpendicular to $\overleftrightarrow{BC}$. Therefore, by the point-slope form, the altitude on $A$ is described by

$$y - 0 = \frac{c}{b}(x - a).$$

The altitude through point $B$ is the vertical axis. Therefore, the altitude on $B$ is described by

$$x = 0.$$

The altitude through point $C$ contains the point $(c, 0)$ and is perpendicular to $\overleftrightarrow{AB}$. Therefore, the altitude on $C$ is described by

$$y - 0 = \frac{a}{b}(x - c).$$

**Step 3**   The foot $D$ is the intersection of the altitude on $C$ with $\overleftrightarrow{AB}$. So, we must solve

$$\frac{a}{b}(x - c) = \frac{b}{-a}(x - a).$$

After solving and substituting into either one of the linear equations, we get

$$D = \left( \frac{a(b^2 + ac)}{a^2 + b^2}, \frac{ab(a - c)}{a^2 + b^2} \right).$$

The foot $E$ is the intersection of the altitude on $A$ with $\overleftrightarrow{BC}$. By similar work, we get

$$E = \left( \frac{c(b^2 + ac)}{b^2 + c^2}, \frac{bc(c - a)}{b^2 + c^2} \right).$$

Notice that $D$ and $E$ are in the same pattern, except that the roles of $a$ and $c$ are reversed. This is encouraging, for it suggests that we did the algebra correctly!

The foot $F$ is the intersection of the altitude on $B$ with $\overleftrightarrow{AC}$:

$$F = (0, 0).$$

The orthocenter of $\triangle ABC$ can be found by substituting $x = 0$ into one of the equations for a nonvertical altitude:

$$H = \left( 0, \frac{-ac}{b} \right).$$

**Step 4**   We need to find the center of the circumcircle for the pedal triangle $DEF$. From Chapter 2, we know that this center is the intersection of the perpendicular bisectors of the sides of $\triangle DEF$. Look first at the side $DF$. (Do you see why we want to include $F$ in our calculations?) The slope of side $DF$ is

$$\frac{b(a - c)}{b^2 + ac},$$

and the midpoint of this side is

$$\left( \frac{a(b^2 + ac)}{2(a^2 + b^2)}, \frac{ab(a - c)}{2(a^2 + b^2)} \right).$$

After some work, this gives the perpendicular bisector of $DF$ as

$$y = \frac{b^2 + ac}{-b(a - c)}x + \frac{a(b^2 + c^2)}{2b(a - c)}.$$

Now look at the side $EF$. Here we get that the slope of $EF$ is

$$\frac{b(c - a)}{b^2 + ac},$$

and the midpoint of this side is

$$\left( \frac{c(b^2 + ac)}{2(b^2 + c^2)}, \frac{bc(c - a)}{2(b^2 + c^2)} \right).$$

Thus, the perpendicular bisector of $EF$ is

$$y = \frac{b^2 + ac}{-b(c - a)} x + \frac{c(a^2 + b^2)}{2b(c - a)}.$$

The circle center we want is the intersection of $EF$ and $DF$. A good deal of careful algebra gives the following point:

$$N = \left( \frac{a + c}{4}, \frac{b^2 - ac}{4b} \right).$$

This is the *nine-point center*.

The radius of the nine-point circle is the distance from $N$ to any of the three feet. Picking the easiest choice, we find that the radius is

$$|NF| = \frac{\sqrt{(a^2 + b^2)(b^2 + c^2)}}{4b}.$$

**Step 5**    Using the information from Step 4, the desired circle can be represented by

$$\left( x - \frac{a + c}{4} \right)^2 + \left( y - \frac{b^2 - ac}{4b} \right)^2 = \frac{(a^2 + b^2)(b^2 + c^2)}{16b^2}.$$

**Step 6**    If we are confident that we have done the algebra correctly, it should not be necessary to verify that the three feet lie on this circle. After all, the equation in Step 5 was developed from these three points. But it is good practice, and it helps confirm that we did the algebra correctly.

Is $D$ on this circle? Substitute the coordinates of $D$ into the circle equation to see if it is satisfied:

$$\left( \frac{a(b^2 + ac)}{a^2 + b^2} - \frac{a + c}{4} \right)^2 + \left( \frac{ab(a - c)}{a^2 + b^2} - \frac{b^2 - ac}{4b} \right)^2.$$

If you are very patient, and are careful with minus signs, this simplifies to

$$\frac{(a^2 + b^2)(b^2 + c^2)}{16b^2},$$

the correct value for the equation. (If you are not that patient, a computer algebra system can simplify it for you.)

In an equally messy fashion, we can verify that $E$ lies on this circle. Point $F$ is more cooperative:

$$\left( 0 - \frac{a + c}{4} \right)^2 + \left( 0 - \frac{b^2 - ac}{4b} \right)^2$$

$$= \frac{(a + c)^2}{16} + \frac{(b^2 - ac)^2}{16b^2}$$

$$= \frac{(a^2 + b^2)(b^2 + c^2)}{16b^2}.$$

**Step 7**  Now we need to verify that the three midpoints of the sides of $\triangle ABC$ lie on this circle. In Figure 4.14, these are the points $P$, $Q$, and $R$. The coordinates of $P$ are $(\dfrac{a+c}{2}, 0)$. So, we get

$$\left(\frac{a+c}{2} - \frac{a+c}{4}\right)^2 + \left(0 - \frac{b^2 - ac}{4b}\right)^2$$

$$= \frac{(a+c)^2}{16} + \frac{(b^2 - ac)^2}{16b^2}$$

$$= \frac{(a^2 + b^2)(b^2 + c^2)}{16b^2}.$$

For point $Q = \left(\dfrac{a}{2}, \dfrac{b}{2}\right)$, we get

$$\left(\frac{a}{2} - \frac{a+c}{4}\right)^2 + \left(\frac{b}{2} - \frac{b^2 - ac}{4b}\right)^2$$

$$= \frac{(a-c)^2}{16} + \frac{(b^2 + ac)^2}{16b^2}$$

$$= \frac{(a^2 + b^2)(b^2 + c^2)}{16b^2}.$$

Verifying point $R = \left(\dfrac{c}{2}, \dfrac{b}{2}\right)$ is similar to the work done for $Q$.

**Step 8**  The three midpoints of segments from the orthocenter to a vertex also lie on this circle. In our figure, these are the points $S$, $T$, and $U$. The coordinates of $S$ are $\left(\dfrac{a}{2}, \dfrac{-ac}{2b}\right)$. So, we get

$$\left(\frac{a}{2} - \frac{a+c}{4}\right)^2 + \left(\frac{-ac}{2b} - \frac{b^2 - ac}{4b}\right)^2$$

$$= \frac{(a-c)^2}{16} + \frac{(-b^2 - ac)^2}{16b^2}$$

$$= \frac{(a^2 + b^2)(b^2 + c^2)}{16b^2}.$$

For point $T = \left(0, \dfrac{b^2 - ac}{2b}\right)$, we get

$$\left(0 - \frac{a+c}{4}\right)^2 + \left(\frac{b^2 - ac}{2b} - \frac{b^2 - ac}{4b}\right)^2$$

$$= \frac{(a+c)^2}{16} + \frac{(b^2 - ac)^2}{16b^2}$$

$$= \frac{(a^2 + b^2)(b^2 + c^2)}{16b^2}.$$

Verifying point $U = \left(\dfrac{c}{2}, \dfrac{-ac}{2b}\right)$ is similar to the work done for $S$.

------------------------------------------

Looking at Figure 4.14, you might see an alternative way to attack Steps 6 and 8. Instead of substituting these points into the equation of the nine-point circle, we could use the equation of an altitude and find the intersections with this circle. This works out nicely for the altitude $x = 0$:

$$\left(0 - \frac{a+c}{4}\right)^2 + \left(y - \frac{b^2 - ac}{4b}\right)^2 = \frac{(a^2 + b^2)(b^2 + c^2)}{16b^2}.$$

After clearing denominators, expanding the squared quantities, and canceling lots of terms, this yields

$$y = 0 \quad \text{or} \quad y = \frac{b^2 - ac}{2b}.$$

The first solution gives the foot $F$. The second solution gives the midpoint $T$ of segment $HB$.

Unfortunately, the other altitudes do not work this well. For instance, the altitude $CD$ is $y = \frac{a}{b}(x - c)$, which leads us to solve

$$\left(x - \frac{(a+c)}{4}\right)^2 + \left(\frac{a}{b}(x - c) - \frac{(b^2 - ac)}{4b}\right)^2 = \frac{(a^2 + b^2)(b^2 + c^2)}{16b^2}.$$

A computer algebra system will confirm the intersections at $D$ and $U$, but this equation is quite difficult to solve by hand.

Our proof is not the shortest proof of the nine-point circle, nor is it the clearest. Algebra does not always provide insight into why a theorem is true. However, algebraic methods can be very powerful. The connections between algebraic expressions and geometric objects are still being explored today.

Soon after the discovery of the nine-point circle, the German mathematician Karl Feuerbach proved something more about this circle. Because of his remarkable theorem, the nine-point circle is sometimes called *Feuerbach's circle*.

**THEOREM 4.7**　**Feuerbach's Theorem**　The nine-point circle of $\triangle ABC$ is tangent to the incircle and to each of the three excircles of $\triangle ABC$.

Feuerbach proved his theorem by analytic methods, a proof considerably longer and more complex than the proof just shown. In Chapter 6, we will present a different approach to proving Feuerbach's Theorem, using the tools of inversion.

## 4.3 EXERCISES

Give clear and complete answers to the exercises, expressing your explanations in complete sentences. Include diagrams whenever appropriate.

1. For a 60°-coordinate system, the distance between two points is given by

$$d = \sqrt{(x_1 - x_2)^2 + (y_1 - y_2)^2 - 2(x_1 - x_2)(y_1 - y_2)\cos(60°)}$$

Explain how the distance formula improves if the coordinate axes are perpendicular.

2. Write a detailed step-by-step proof of Theorem 4.4.

3. Use angles of inclination and a trigonometric identity for $\tan(\alpha_1 - \alpha_2)$ to give a different proof of Theorem 4.5. (See Figure 4.7.)

4. Using coordinates, write a detailed step-by-step proof that the set of points equidistant from two fixed points, $A$ and $B$, is the perpendicular bisector of the segment $AB$.

5. The usual distance function (the one that assumes the coordinate axes are perpendicular) is based on the Pythagorean Theorem.
   a. How can we use coordinates to represent points $A$ and $B$ in three-dimensional space?
   b. Extend the Pythagorean Theorem to three dimensions, and prove that this formula is correct.
   c. Given two points, $A$ and $B$, in three-dimensional space, what is the set of points equidistant from $A$ and $B$? Prove your answer, using coordinates.
   d. What does your answer for part c look like geometrically?

6. a. How many times can a line intersect a circle? Prove your answer algebraically.
   b. How many times can a parabola intersect a circle? Prove your answer algebraically.

7. While finding a formula for the distance from a point to a line, we had to solve the system of linear equations
$$\begin{cases} y = mx + c, \\ y = -\frac{1}{m}(x - a) + b. \end{cases}$$
Find the intersection point of these two lines.

8. Prove the midpoint formula given in Theorem 4.1.

9. Explain why the point-slope and the slope-intercept forms of the equation of a straight line cannot be used to describe a vertical line. Explain why the intercept form cannot be used to describe a vertical line or a horizontal line. Then explain how the general form of the equation for lines can be used to describe any line.

10. a. For the situation in Figure 4.15, prove that the slope between points $A_1$ and $B_1$ equals the slope between $A_2$ and $B_2$.
    b. Prove that the slope of a horizontal line is 0.

11. Prove that the shortest distance from a point to a line lies along a segment perpendicular to the line. (Try to do this without using calculus.)

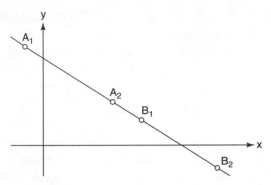

**FIGURE 4.15**
Figure for Exercise 10

12. Use the similar triangles in Figure 4.8 to give another proof for the formula for the distance from a point to a line.

13. Consider $\triangle ABC$ as shown in Figure 4.10. If $a > 0$, where is the orthocenter $H$? If $a = 0$, where is $H$? Justify your answers using the coordinates of $H$.

14. Here are general equations for two lines:
$$\begin{cases} y = m_1 x + b_1, \\ y = m_2 x + b_2. \end{cases}$$
    a. Find the $x$-coordinate for the intersection point of these two lines.
    b. Find the $y$-coordinate for the intersection point of these two lines.
    c. If $Q$ is the point of intersection of these two lines, for what situation will $Q$ be undefined? Explain.

15. Using coordinates, prove that the diagonals of a rhombus are perpendicular bisectors of each other. (*Note:* There are two things to prove.)

16. Use coordinates to prove that the midpoint of the hypotenuse of a right triangle is equidistant from all of the vertices.

17. Suppose we draw the line segment connecting the midpoints of the nonparallel sides of a trapezoid. Use coordinates to prove that this segment is parallel to the parallel sides and that its length is the average of their lengths. In addition, prove that the length of this segment times the height of the trapezoid gives its area.

18. a. Using coordinates, prove that the medians of a triangle are concurrent.
    b. If the vertices of a triangle are the points $(x_1, y_1)$, $(x_2, y_2)$, and $(x_3, y_3)$, prove that the centroid of the triangle is the point
$$\left(\frac{x_1 + x_2 + x_3}{3}, \frac{y_1 + y_2 + y_3}{3}\right).$$

19. Use coordinates to prove that the nine-point center is the midpoint of the segment from the orthocenter to the circumcenter. This shows that the nine-point center lies on the Euler line.

20. Use coordinates to prove that the radius of the nine-point circle is half the length of the radius of the circumcircle.

21. Create a diagram in Sketchpad illustrating Feuerbach's Theorem.

22. Convert each of the following Cartesian equations to a polar equation.
    a. $3x + y = 0$
    b. $x - 6 = y^2$
    c. $16x^2 + 9y^2 = 144$

23. Convert each of the following polar equations to a Cartesian equation.
    a. $\theta = \frac{\pi}{3}$
    b. $r = 5$
    c. $r \sec \theta = 3$
    d. $r^2 = \cos(2\theta)$

24. Find all points where the curves $r = 3 \sin \theta$ and $r = 3 \cos(2\theta)$ intersect.

25. The graph of the equation $r = 2 \cos \theta + 1$ is an example of a *limaçon*. Create a graph of this equation. Now consider the circle $r = a$. Find values of $a$ so that the two curves have the following number of intersection points. In each case, find the coordinates of the intersection points algebraically.
    a. 4 intersection points
    b. 3 intersection points
    c. 2 intersection points
    d. 1 intersection point
    e. 0 intersection points

26. In rectangular coordinates, if $y = f(x) = f(-x)$, the graph will be symmetric with respect to the $y$-axis.

a. What are the graphical consequences of $f(-x) = -f(x)$ (in rectangular coordinates)?
b. Consider the situation in polar coordinates. If $f(\theta) = f(-\theta)$, what can you say about the graph of the equation $r = f(\theta)$? Justify your answer and give examples.
c. If $f(-\theta) = -f(\theta)$, what can you say about the graph of the equation $r = f(\theta)$? Justify your answer and give examples.

27. The equation $r = f(\theta) = e^{a\theta}$ describes one type of spiral.
    a. Does this spiral include the origin? Explain why or why not.
    b. With a great deal of trigonometry, it can be proven that the angle $\psi$ between a tangent line at a point on $r = f(\theta)$ and the segment from the origin to that point satisfies
$$\tan \psi = \frac{f(\theta)}{f'(\theta)},$$
    where $f'(\theta)$ is the derivative of the function $f$. For $f(\theta) = e^{a\theta}$, show that $\psi$ is constant.
    c. Explain the role of the constant $a$ in part b.
    d. Explain what happens in part b if $a = 0$.

28. *Another look at rose curves:* Experiment with various values, positive and negative, for $a$, $n$, and $b$ in the equation $r = a \sin^n(5\theta) + b$. Then describe what effect each of $a$, $n$, and $b$ has on the graph. Also, what would be different if cosine were used instead of sine?

Exercises 29–30 are especially for future teachers.

29. In the *Principles and Standards for School Mathematics,* the National Council of Teachers of Mathematics (NCTM) recommends that "Instructional programs from prekindergarten through grade 12 should enable all students to . . . specify location and describe spatial relationships using coordinate geometry and other representational systems" (NCTM, 2000, 41). What does this mean for your future students?
    a. Find a copy of the *Principles and Standards,* and study the discussion of the second Geometry Standard (NCTM, 2000, 41–43).

What are the specific NCTM recommendations with regard to developing a sense of spatial relationships and coordinate geometry as children progress in their study of geometry from prekindergarten through grade 12?

b. Find some mathematics textbooks for one of the grade levels for which you are seeking teacher certification. How is the second Geometry Standard implemented in these textbooks? Cite specific examples.

c. Write a report in which you present and critique what you learn in studying the second Geometry Standard in light of your experiences in this course. Your report should include your answers to parts a and b.

**30.** Design several classroom activities involving coordinate geometry or analytic geometry that would be appropriate for students in your future classroom. Write a short report, explaining how the activities you design reflect both the NCTM recommendations and what you are learning about analytic geometry in this class (Chapters 1–4).

Reflect on what you have learned in this chapter.

**31.** Review the main ideas of this chapter. Describe, in your own words, the concepts you have studied and what you have learned about them. What are the important ideas? How do they fit together? Which concepts were easy for you? Which were hard?

**32.** Reflect on the learning environment for this course. Describe aspects of the learning environment that helped you understand the main ideas in this chapter. Which activities did you like? Dislike? Why?

## 4.4 CHAPTER OVERVIEW

*Analytic geometry* provides a bridge between geometry and algebra. The use of a coordinate system provides a way of representing geometric objects—points, lines, circles, curves—using algebraic tools. It also makes it possible for us to represent algebraic functions visually as graphs. Coordinate systems make it possible to use algebraic tools to prove many theorems about geometric objects.

In this chapter, we developed a formula for the distance between points $P$ and $Q$. If $P = (p_1, p_2)$ and $Q = (q_1, q_2)$ are points in the plane, their distance is given by

$$d(P, Q) = \sqrt{(p_1 - q_1)^2 + (p_2 - q_2)^2}.$$

This formula, which is based on the Pythagorean Theorem, actually works for points $P$ and $Q$ in any number of dimensions. If $P$ and $Q$ are points on a number line (a one-dimensional space), then each point has only one coordinate: $P = (p_1)$ and $Q = (q_1)$. So, the distance formula collapses to

$$d(P, Q) = \sqrt{(p_1 - q_1)^2}.$$

If $P$ and $Q$ are points in three-dimensional space—that is, $P = (p_1, p_2, p_3)$ and $Q = (q_1, q_2, q_3)$—we can expand the distance formula to

$$d(P, Q) = \sqrt{(p_1 - q_1)^2 + (p_2 - q_2)^2 + (p_3 - q_3)^2}.$$

In general, if $P$ and $Q$ are points in $n$-dimensional space with $P = (p_1, p_2, \ldots, p_n)$ and $Q = (q_1, q_2, \ldots, q_n)$, we calculate the distance from $P$ to $Q$ using the formula

$$d(P, Q) = \sqrt{(p_1 - q_1)^2 + (p_2 - q_2)^2 + \cdots + (p_n - q_n)^2}$$

$$= \sqrt{\sum_{i=1}^{n}(p_i - q_i)^2}.$$

In this chapter, we explored the use of several coordinate systems. We used the usual Cartesian coordinate system, in which the coordinate axes are perpendicular to each other. We also looked at a skew coordinate system, in which the coordinate axes meet at an angle of $\theta$. In this skew coordinate system, we used the *Law of Cosines* to develop a distance formula:

$$d(P, Q) = \sqrt{(p_1 - q_1)^2 + (p_2 - q_2)^2 - 2(p_1 - q_1)(p_2 - q_2)\cos\theta}.$$

For the examples given in this chapter, we saw that we could use two applications of the Pythagorean Theorem in place of the Law of Cosines to get the same result. Extending the distance formula to higher dimensions when we are using a skew coordinate system becomes much more complicated, because we must consider the angles between each pair of coordinate axes, which do not intersect at right angles.

Lines are two-dimensional objects, requiring two pieces of information—two parameters—to specify each particular line. The required pieces of information are *location* (which can be specified by giving a point on the line) and *direction* (or slope) of the line. Parallel lines have the same slopes, while the slopes of perpendicular lines are negative reciprocals of each other. We have explored several different forms for representing linear equations (see page 93). The choice of a particular form of the linear equation is usually dictated by the situation in the problem we are trying to solve.

The distance from a point $P$ to another point $Q$ can easily be calculated using the Pythagorean Theorem or the Law of Cosines. The distance from a point $P$ to a line $\ell$, however, requires us to develop a strategy for finding a point $Q$ on the line $\ell$ that is closest to $P$. One strategy for doing this is to think about the family of circles centered at $P$. Some of these circles will intersect the line $\ell$, and some will not. One of these circles, the first one to touch the line $\ell$, will be tangent to $\ell$ at the desired point $Q$. In other words, the line segment $PQ$ will be a radius of the desired circle, so $PQ$ will be perpendicular to line $\ell$.

Polar coordinates are another example of a coordinate system. The polar coordinate system is probably more familiar to you than the skew coordinate system presented at the beginning of the chapter. In a rectangular coordinate system, such as the familiar Cartesian coordinate system, and even in a skew coordinate system, each point has a unique coordinate representation. This makes it easy to solve problems involving points of intersection by simply solving a system of equations. In a polar coordinate system, however, each point can be represented in many different ways. In fact, each point has infinitely many different representations! Therefore, it is necessary to continually check the reasonableness of an algebraic solution by looking at the actual geometry of the situation in a graph. Using algebra alone will often lead to missed intersection points.

Polar coordinates are particularly useful in situations where we are interested in representing curves. Circles, spirals, rose curves, limaçons, and many other curves have polar equations that are much simpler than their corresponding equations in a rectangular coordinate system. The basic components of a rectangular system are lines, which is why linear equations are simpler in that system. The basic components of the polar coordinate system are circles and rays through the origin. So, curves that loop around the origin—in fact, any kind of motion involving circular movement—can usually be expressed more simply using polar equations.

This chapter closed with a fairly intricate proof of the Nine-Point Circle Theorem. This proof uses many of the tools from earlier in the chapter: coordinates of points, representing lines by linear equations, slopes of perpendicular lines, midpoints of segments, finding intersection points by solving systems of equations, and the equation of a circle. Although the algebra involved is not difficult, it takes patience and care to work it out correctly.

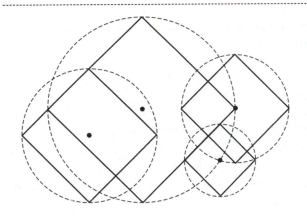

## CHAPTER FIVE

# Taxicab Geometry

The ordinary Euclidean way of measuring distance makes use of the Pythagorean Theorem. If $P = (x_P, y_P)$ and $Q = (x_Q, y_Q)$, the distance from $P$ to $Q$ is given by the metric formula

$$d(P, Q) = \sqrt{(x_P - x_Q)^2 + (y_P - y_Q)^2}.$$

We can think of this as the *straight-line distance,* or the distance from point $P$ to point $Q$ as a crow would fly from one point to the other.

    In this chapter, we will examine properties (axioms) for rules or formulas that we can use to measure distance. Such a distance formula is called a *metric.* We will investigate the *taxicab metric,* an alternative to the usual Euclidean metric. To see the effect that our assumptions about how distance is measured have on the shape of space, we will look at some familiar curves—circles, ellipses, and parabolas—using both these metrics. We can define these curves in terms of distances of points from fixed points or lines. Because the taxicab and Euclidean metrics measure these distances differently, these curves will take on different shapes, depending on our choice of metric.

Do the following activities, writing your explanations clearly in complete sentences. Include diagrams whenever appropriate. Save your work for each activity, as later work sometimes builds on earlier work. You will find it helpful to read ahead into the chapter as you work on these activities.

For most of the activities in this chapter, you will be working on a rectangular coordinate system. Choosing to work with the **Square Grid** of Sketchpad rather than with the **Rectangular Grid** may make it easier to see what is happening.

1. Open a worksheet in Sketchpad, and define a coordinate system with a square grid. In a convenient place near the edge of this worksheet, construct a line segment, $AB$, and measure its length. Choose **Edit | Properties | Label** to rename the result of this measurement as $r$.

   a. Construct a point, $C$, anywhere on the coordinate plane, and measure its coordinates.

   b. Select point $C$ and measurement $r$, and construct a circle centered at $C$ with radius $r$. You can use a command on the Measure menu to find the equation of this circle.

   c. Vary the point $C$ and the length of $AB$. What do you observe about the equation of this circle?

   d. What is the equation of a circle centered at $(h, k)$ with radius $r$?

2. Open a worksheet in Sketchpad. In a convenient place near the edge of this worksheet, construct a line segment, $AB$. Construct point $P$ on $AB$, then construct two subsegments, $AP$ and $BP$, and measure the length of each. Choose **Edit | Properties | Label** to rename the results of these measurements as $r_1$ and $r_2$. You might also find it helpful to use different colors for each segment.

   a. Construct two points, and label them $F_1$ and $F_2$. Construct a circle centered at $F_1$ with radius $r_1$, and a circle centered at $F_2$ with radius $r_2$. Color these circles to match the colors of $AP$ and $BP$.

   b. Adjust the length of $AB$ and/or the distance between points $F_1$ and $F_2$ (if necessary) so that these circles intersect. Construct these intersections.

   c. Trace the intersections from part b as you vary (or animate) point $P$. What do you observe?

   SKETCHPAD TIP   You can turn **Trace** on and off by selecting the point(s) and choosing **Display | Trace.** The checkmark in front of **Trace** tells you that it is turned on for the points you have selected.

   d. Turn off **Trace,** erase the traces, and hide the two circles. Select one of the intersections and point $P$, and choose **Construct | Locus.** Select the other intersection point and point $P$, and construct the other part of this locus. Vary points $F_1$ and $F_2$. What do you observe about the shape of this locus? (*Note:* When $F_1$ and $F_2$ are very close together, there may be some distortion in the curve represented by this locus.)

3. Open a worksheet in Sketchpad. In a convenient place near the edge of this worksheet, construct a line, $j$. Construct a segment, $AB$, on this line, and measure the length of $AB$. Rename the result of this measurement as $d$.

a. Construct a line, $k$, and a point, $F$, not on $k$. We want to construct the locus of points that are equidistant from $k$ and $F$.

   The set of points at a fixed distance from point $F$ will form a circle centered at $F$. The set of points at a fixed distance from line $k$ will form a pair of parallel lines on either side of $k$. Using these ideas, construct the set of points that are at a distance $d$ from $k$ and at a distance $d$ from $F$. Vary the length of $AB$, and observe where these two sets intersect.

b. Select one of these intersection points and point $B$, and construct one-half of the desired locus. Then select the other intersection point and point $B$ to construct the other half of the locus.

   i. Vary point $B$. What do you observe?
   ii. Vary point $F$. What do you observe?
   iii. Vary line $k$. What do you observe?
   iv. What happens when you move $F$ to the other side of line $k$? Can you see a way to make your construction robust enough so that the locus doesn't disappear when you move $F$ to the other side of $k$?

4. Open a worksheet in Sketchpad. Construct three points, $P$, $Q$, and $R$. Measure the distances $PQ$, $PR$, and $QR$.

a. Vary point $P$. What happens to distance $PQ$ as $P$ gets closer to (or farther from) $Q$? How small can you make distance $PQ$? What can you say about points $P$ and $Q$ when the distance between them is as small as you can make it?

b. Does it make a difference if you measure distance $PQ$ or distance $QP$?

c. Calculate $PQ + QR$, and compare this with $PR$. What do you observe? Vary points $P$, $Q$, and $R$. Does your observation continue to hold?

d. Calculate the ratio $\dfrac{PR}{PQ + QR}$. Vary points $P$, $Q$, and $R$ to make this ratio as large as possible. What do you observe? Can you explain what is happening?

5. The Euclidean distance between the points $P$ and $Q$ is calculated by the distance formula:

$$\sqrt{(x_P - x_Q)^2 + (y_P - y_Q)^2},$$

which is based on the Pythagorean Theorem. The **Distance** and **Length** tools on the Measure menu calculate the Euclidean distance between two points.

   Let's define a new rule for measuring distance—that is, a new *metric*. Open a worksheet in Sketchpad, and define a square coordinate system.

   In the *taxicab metric*, distances are measured as though you have to travel like a taxicab, along the streets—either vertically or horizontally, never diagonally. On a square coordinate system, you can measure the taxicab distance easily by counting the blocks from one intersection of the grid to the next.

a. Plot points $P(3, 4)$, $A(2, 2)$, $B(3, 7)$, $C(2, 5)$, and $D(5, 5)$. By counting the number of blocks from $P$ to $A$, we find that the *taxi-distance PA* is 3 units. Find the taxi-distances $PB$, $PC$, and $PD$. Two of these points are the same taxi-distance from $P$ as $A$ is. Which two?

b. The set of all points that are at the same taxi-distance from $P$ form a *taxi-circle* centered at $P$. In part a, three of the points lie on a taxi-circle of radius 3 centered at $P$. Find several additional points on this taxi-circle. Describe the set of all points that are at a taxi-distance of 3 units from a fixed point $P$. How is the shape of a taxi-circle different from (or similar to) the shape of an ordinary Euclidean circle?

c. If you are given a point $Q(x_Q, y_Q)$ and a radius $r$, how could you quickly sketch a taxi-circle of radius $r$ centered at $Q$?

d. Experiment with the **Taxi-Circle** tool, which is provided on the CD that accompanies this text. To use this tool, you need to select points that represent the center of the circle and the endpoints of a segment that gives the radius.

SKETCHPAD TIP    To use the **Taxi-Circle** tool, open **TaxiCircleTool.gsp.** The **Taxi-Circle** tool is on the bottom button of the Toolbox. As long as **TaxiCircleTool.gsp** is open, you will be able access the **Taxi-Circle** tool from other Sketchpad documents.

6. Instead of counting blocks by hand, we want Sketchpad to calculate the taxi-distances for us. Let's create a new tool that will measure distance according to this new rule.

a. As before, construct two points, $P$ and $Q$, and find their coordinates. Calculate

$$d_T(P, Q) = |x_P - x_Q| + |y_P - y_Q|.$$

(*Hint:* To calculate the absolute value, choose **abs** from **Measure | Calculate | Functions.**)

Notice that this rule allows us to find the taxi-distance between points that do not lie at the integer coordinates of the grid—that is, the $x$- and $y$-coordinates of a point can be any real numbers.

b. Select the points $P$ and $Q$ and the result of the calculation $|x_P - x_Q| + |y_P - y_Q|$. Then click on the bottom button of the Toolbox to create a new tool. Name this tool **Taxi-distance.** Now, whenever you want to calculate the taxi-distance between two points, select this tool, then click on the two points.

SKETCHPAD TIP    A **Custom** tool is associated with the worksheet in which it is created. To save the **Taxi-distance** tool, be sure to save this worksheet. If you close the worksheet, you will also be closing the tool. As you create your own set of customized tools, you can save several of them to one worksheet. When this worksheet is open on your desktop, you will be able to access these tools from other worksheets.

c. Does taxicab distance have the same properties as Euclidean distance? Construct another point $R$ on your worksheet, and calculate the taxi-distances $d_T(P, Q)$, $d_T(P, R)$, and $d_T(Q, R)$. Vary the points, and answer the questions of Activity 4 for these taxi-distances. For which of these questions does taxicab distance give results similar to Euclidean distance?

7. Taxi-circles are not round in the ordinary Euclidean sense. In Activity 2, we constructed ellipses by finding the locus of intersection points for particular pairs of circles. We would now like to investigate the shape of taxi-ellipses, using an adaptation of this method. The shape of taxi-circles will affect the shape of taxi-ellipses.

Open a worksheet in Sketchpad, and define a coordinate system with a square grid. In a convenient place near the edge of this worksheet, construct a line parallel to the $y$-axis. Construct a line segment, $AB$, on this line. Construct point $P$ on $AB$, then construct two subsegments, $AP$ and $BP$. As before, you might find it helpful to use contrasting colors for each subsegment.

a. Construct two points in the coordinate plane, and label these $F_1$ and $F_2$. Construct a taxi-circle centered at $F_1$ with radius $AP$ and a taxi-circle centered at $F_2$ with radius $BP$. Color these circles to match the colors of segments $AP$ and $BP$.

b. Adjust the length of $AB$ and/or the distance between points $F_1$ and $F_2$ (if necessary) so that

$$d_T(A, B) > d_T(F_1, F_2).$$

Construct the points where the taxi-circles intersect. (*Hint:* Each taxi-circle consists of four line segments. To construct the intersection points for a pair of taxi-circles, it is necessary to construct intersections of each pair of intersecting segments.)

c. Select one of the intersection points and the point $P$, and choose **Construct | Locus.** Select another intersection point and point $P$, and construct another part of this locus. Continue in this manner until you have constructed the several segments of the locus for a taxi-ellipse. (*Note:* Depending on the locations of points $F_1$ and $F_2$, this part of the locus will consist of either two or four segments.)

d. Unfortunately, this construction does not give the entire taxi-ellipse. Vary point $P$, and observe the taxi-circles changing size as their radii get longer and shorter. When you choose **Construct | Intersection,** Sketchpad will construct *points* of intersection, but not *segments* of intersection. Can you see what needs to be done to complete the taxi-ellipse?

e. Vary points $F_1$ and $F_2$. What do you observe about the shape of this locus? What if $F_1$ and $F_2$ lie on diagonally opposite corners of a rectangle? What if $F_1$ and $F_2$ lie along a vertical line? What if $F_1$ and $F_2$ lie along a horizontal line?

8. Recall that in Chapter 4 we investigated what it means to talk about the distance from point $P$ to line $\ell$. (See Activity 5, Chapter 4, and the discussion on page 98.) One way to visualize the idea of finding the distance from $P$ to $\ell$ is to think of a circle centered at $P$ with radius $r$. If $r$ is small,

the circle will not touch $\ell$. If $r$ is large enough, the circle will intersect $\ell$ twice. In the Euclidean plane, there will be just one circle that touches $\ell$ in exactly one point. The radius of this circle gives the distance from $P$ to $\ell$. Since taxi-circles have flat sides, we need to reconsider this method of finding the distance from a point to a line.

Open a worksheet in Sketchpad, and define a coordinate system with a square grid. In a convenient place near the edge of this worksheet, construct a line parallel to the $y$-axis. Construct a line segment, $AB$, on this line.

a. Construct a line $\ell$ and a point $P$ not on $\ell$. Construct a taxi-circle centered at $P$ with radius $AB$. Vary the length of $AB$, and observe whether the taxi-circle touches $\ell$. Try to find the radius of the smallest taxi-circle centered at $P$ that touches $\ell$.

b. Vary the line $\ell$ so that its slope is greater than 1, equal to 1, or less than 1. Try some negative values of the slope as well. Formulate a rule for finding the distance from a point $P$ to a line $\ell$ using the taxicab metric.

9. Now let's investigate taxi-parabolas. Open a worksheet in Sketchpad, and define a coordinate system with a square grid. In a convenient place near the edge of this worksheet, construct a line parallel to the $y$-axis. Construct a line segment, $AB$, on this line.

a. Construct a line $k$ and a point $F$ not on $k$. We want to construct the locus of points that are equidistant from $k$ and $F$.

b. The set of points at a fixed taxi-distance from the point $F$ will form a taxi-circle centered at $F$. The set of points at a fixed distance from the line $k$ will form a pair of parallel lines on either side of $k$. Because taxi-circles have flat sides, you will have to think carefully about how to construct the set of points that are at a distance $d$ from line $k$. This construction will depend on whether the absolute value of the slope of line $k$ is less than, equal to, or greater than 1.

Using these ideas, construct the locus of points that are at a distance $d$ from $k$ and at a distance $d$ from $F$. The taxi-parabola will be composed of several segments, which you will have to construct separately.

To get a complete picture of taxi-parabolas, you will have to consider various cases based on the slope of $k$. What happens when $|m| < 1$? $|m| = 1$? $|m| > 1$? What if $k$ is vertical or horizontal?

  i. Vary point $B$. What do you observe?

 ii. Vary point $F$. What do you observe?

iii. Vary line $k$. What do you observe?

## 5.2 DISCUSSION

In Chapter 4, we discussed the idea of measuring distance. If the points lie on a line (e.g., a number line), the location of each point can be specified by a single coordinate; for example, $P$ and $Q$ can be specified by coordinates $x_P$ and $x_Q$, respectively. We can calculate the distance between these points by taking the difference between their coordinates:

$$d(P, Q) = |x_P - x_Q| = \sqrt{(x_P - x_Q)^2}.$$

When the points lie in a plane, we need two coordinates for each point. In Chapter 4, we investigated three different coordinate systems: a skew coordinate system, in which the axes were set at a 60° angle to each other; a rectangular coordinate system, in which the axes were at right angles; and a polar coordinate system, in which the coordinates of a point $P$ tell the distance of $P$ from the origin and the angle that a ray through $P$ makes with the polar axis.

When points $P$ and $Q$ are specified by coordinates in either a skew coordinate system or a rectangular coordinate system, we can calculate the distance $d(P, Q)$ between them by the formula

$$\sqrt{(x_P - x_Q)^2 + (y_P - y_Q)^2 - 2(x_P - x_Q)(y_P - y_Q)\cos(\theta)},$$

where $\theta$ is the angle between the axes (measured counterclockwise from the $x$-axis to the $y$-axis). In the ordinary rectangular coordinate system, $\theta$ is a right angle, so that $\cos(\theta) = 0$; the distance formula thus reduces to

$$\sqrt{(x_P - x_Q)^2 + (y_P - y_Q)^2}.$$

This is the usual Euclidean distance formula and is based on the Pythagorean Theorem.

## AN AXIOM SYSTEM FOR METRIC GEOMETRY

A formula or a rule for measuring distance is called a *metric*. The formula for measuring distance in a skew coordinate system and the ordinary Euclidean distance formula are examples of metrics.

We have certain expectations about distance. For instance, we expect the distance between two different points to be positive. However, there are certain situations where we might allow a negative number to be given for distance, with the understanding that the negative sign means that one is backing up or going in the other direction. For example, in a coordinate system, the sign—positive or negative—of the coordinate tells us on which side of the origin a particular point may lie; that is, the signed number gives information about the *direction* as well as the *distance*. We are not considering such directed distances in this chapter.

The idea of distance from point $P$ to point $Q$ is closely related to the idea of the length of line segment $PQ$. Sketchpad uses the notation $PQ$ for distance between points, and $m\overline{PQ}$ for the length (i.e., the measure) of the line segment. In this text, we have been using $PQ$ to designate the line segment from $P$ to $Q$. When we want to emphasize that we are talking about the distance between points, we will use $d(P, Q)$. In Activity 4, we used the simpler notation $PQ$ for $d(P, Q)$, following the notation in Sketchpad.

As you worked on Activity 4, you probably observed that $d(P, Q)$ is usually positive, and $d(P, Q) = 0$ only when $P$ and $Q$ are in the same location. We ordinarily expect the distance from $P$ to $Q$ to be the same as the distance from $Q$ to $P$.

As you worked on Activity 4, did you observe that $d(P, Q) + d(Q, R) \geq d(P, R)$? This is called the *triangle inequality*. If $PQ$, $QR$, and $PR$ are considered as line segments, then the points $P$, $Q$, and $R$ might be collinear, or they might form the vertices of a triangle. The sum of the lengths of two sides of a triangle cannot be smaller than the length of the third side. In fact, if $d(P, Q) + d(Q, R) = d(P, R)$, the three points $P$, $Q$, and $R$ are collinear, with $Q$ between $P$ and $R$. As a consequence of the triangle inequality, $\dfrac{PR}{PQ + QR} \leq 1$. If this ratio equals 1, the points $P$, $Q$, and $R$ are collinear, and $Q$ lies on line segment $PR$.

These ideas about distance are so important, so fundamental to the geometry of spaces where distances can be measured, that we will formulate a set of axioms for a metric space.

Let $P$, $Q$, and $R$ be points, and let $d(P, Q)$ denote the distance from $P$ to $Q$.

**Metric Axiom 1**  $d(P, Q) \geq 0$, and $d(P, Q) = 0$ if and only if $P = Q$.

**Metric Axiom 2**  $d(P, Q) = d(Q, P)$.

**Metric Axiom 3**  $d(P, Q) + d(Q, R) \geq d(P, R)$.

We will examine several different rules for measuring distance. In each case, we must be sure that the rule we propose meets these three criteria before we can call it a metric.

## The Euclidean Distance Formula

**THEOREM 5.1**  The ordinary Euclidean distance formula,

$$d(P, Q) = \sqrt{(x_P - x_Q)^2 + (y_P - y_Q)^2},$$

satisfies all three of the metric axioms. Hence, the Euclidean distance formula is a metric in $\Re^2$.

**Proof**

1. The expression $(x_P - x_Q)^2 + (y_P - y_Q)^2$ is the sum of numbers that have been squared. By convention, when we take the square root of a number, we consider the positive square root. So, the expression $\sqrt{(x_P - x_Q)^2 + (y_P - y_Q)^2}$ will always give a result greater than or equal to zero. The only way that $\sqrt{(x_P - x_Q)^2 + (y_P - y_Q)^2}$ can equal zero is for both $(x_P - x_Q)^2 = 0$ and $(y_P - y_Q)^2 = 0$, in which case both $x_P = x_Q$ and $y_P = y_Q$; in other words, $P = Q$. So, the first metric axiom is satisfied.

2. Observe that $(x_P - x_Q)^2 = (x_Q - x_P)^2$ and $(y_P - y_Q)^2 = (y_Q - y_P)^2$. So, $d(P, Q) = d(Q, P)$, satisfying the second metric axiom.

3. Finally,

$$\begin{aligned}
d(P, Q) + d(Q, R) &= \sqrt{(x_P - x_Q)^2 + (y_P - y_Q)^2} \\
&\quad + \sqrt{(x_Q - x_R)^2 + (y_Q - y_R)^2} \\
&\geq \sqrt{(x_P - x_R)^2 + (y_P - y_R)^2} \\
&= d(P, R).
\end{aligned}$$

So, the third axiom is also satisfied.

Because all three metric axioms are satisfied for points in the plane, the Euclidean distance formula is a metric in $\mathfrak{R}^2$.

The **Distance** and **Length** tools on the Measure menu calculate the Euclidean distance between two points. To verify this, construct two points, $P$ and $Q$, and find their coordinates. Calculate

$$\sqrt{(x_P - x_Q)^2 + (y_P - y_Q)^2},$$

and compare this with the distance $PQ$ as calculated by Sketchpad. Vary points $P$ and $Q$ to see if your observation continues to hold.

## The Taxicab Distance Formula

**DEFINITION 5.1**  The *taxicab distance* from $P(x_P, y_P)$ to $Q(x_Q, y_Q)$ is given by the formula

$$d_T(P, Q) = |x_P - x_Q| + |y_P - y_Q|.$$

In Activity 6, you created a tool to measure the taxi-distance between two points. Does taxicab distance satisfy all three of the metric axioms?

1.  The expression $|x_P - x_Q| + |y_P - y_Q|$ is the sum of absolute values; thus, it is always greater than or equal to zero. The only way that $|x_P - x_Q| + |y_P - y_Q|$ can equal zero is for both $|x_P - x_Q| = 0$ and $|y_P - y_Q| = 0$, in which case $P = Q$. So, $d_T(P, Q) \geq 0$, and $d_T(P, Q) = 0$ iff $P = Q$. The first metric axiom is satisfied.

2.  Observe that $|x_P - x_Q| = |x_Q - x_P|$ and $|y_P - y_Q| = |y_Q - y_P|$. So, $d(P, Q) = d(Q, P)$, satisfying the second metric axiom.

3.  Finally,
$$\begin{aligned}
d_T(P, Q) + d_T(Q, R) &= |x_P - x_Q| + |y_P - y_Q| + |x_Q - x_R| + |y_Q - y_R| \\
&= (|x_P - x_Q| + |x_Q - x_R|) + (|y_P - y_Q| + |y_Q - y_R|) \\
&\geq (|x_P - x_Q + x_Q - x_R|) + (|y_P - y_Q + y_Q - y_R|) \\
&= (|x_P - x_R|) + (|y_P - y_R|) \\
&= d_T(P, R).
\end{aligned}$$

So, $d_T(P, Q) + d_T(Q, R) \geq d_T(P, R)$, and the third axiom is also satisfied.

Since taxicab distance satisfies all three of the metric axioms, we can call this formula a metric. We have proven the following theorem, which we now state formally.

**THEOREM 5.2**  The taxicab distance formula is a metric in $\mathfrak{R}^2$.

## CIRCLES

**DEFINITION 5.2**  A *circle* is defined as the set of all points at a given distance, $r$, from a fixed center, $C$. We can express this symbolically as

$$\text{circle} = \{P : d(P, C) = r, \text{ where } r > 0 \text{ and } C \text{ is fixed}\}.$$

The fixed point $C$ is the *center* of the circle, and the length $r$ is its *radius*.

This definition describes a circle in terms of distances from a fixed point, but it does not tell us which metric to use. If we choose the Euclidean metric, we get ordinary round circles. (See pages 97–98.) As you worked on Activity 1, you probably observed that a Euclidean circle centered at $(h, k)$ with radius $r$ is described by the equation

$$(x - h)^2 + (y - k)^2 = r^2.$$

## Taxi-Circles

Let's see what happens when we choose the taxicab metric. In Activity 5, you worked with an example of a taxi-circle centered at $P(3, 4)$ with a radius of 3 units. You probably observed (to your surprise!) that this circle has flat sides—line segments with slopes of $\pm 1$. Let's analyze this carefully.

We can simplify the algebra by choosing $C$ at the origin. The taxi-circle centered at $C = (0, 0)$ with radius $r > 0$ is the set

$$\{P : d_T(P, C) = r\} = \{(x_P, y_P) : |x_P - 0| + |y_P - 0| = r\}$$
$$= \{(x_P, y_P) : |x_P| + |y_P| = r\}.$$

What does this look like graphically? The expression $|x_P| + |y_P| = r$ is a linear equation (four linear equations, actually). We can rewrite these in slope-intercept form as $|y_P| = r - |x_P|$. This gives us

$$y_P = \pm(r - |x_P|) = \pm(r \mp x_P).$$

Let's consider one of these lines. The line $y_P = r - x_P$ has a $y$-intercept of $+r$ and a slope of $-1$. The set of points on this line that are at a taxi-distance from the origin of $r$ will be the points on the line segment from $(0, r)$ to $(r, 0)$. We can carry out a similar analysis for each of the other four equations, giving three additional line segments: from $(r, 0)$ to $(0, -r)$, from $(0, -r)$ to $(-r, 0)$, and from $(-r, 0)$ to $(0, r)$. In other words, these equations give us four lines; the requirement that $d_T(P, C) = r$ limits both $x_P$ and $y_P$, so that $-r \leq x_P \leq +r$ and $-r \leq y_P \leq +r$. So, a taxi-circle is formed by four line segments, forming a diamond shape standing on one vertex.

If the center of the circle is at $C = (a, b)$, the calculations are similar, just a little more complicated. You will have an opportunity to work out these calculations in the exercises.

## ELLIPSES

**DEFINITION 5.3**  An *ellipse* is defined as the set of points $P$, the sum of whose distances from two fixed points, $F_1$ and $F_2$, is constant. We can write this symbolically as

$$\text{ellipse} = \{P : d(P, F_1) + d(P, F_2) = d,$$
$$\text{where } d > 0 \text{ and } F_1, F_2 \text{ are fixed points}\}.$$

The fixed points are called the *foci* (singular, *focus*) of the ellipse.

This definition describes an ellipse in terms of distances from two fixed points, but it does not tell us which metric to use. If we choose the Euclidean metric, we get ordinary rounded ellipses. As you experimented with your diagram in Activity 2, you probably observed that as you moved the foci closer together, the ellipse became rounder, more like a circle; as you moved the foci farther apart, the ellipse became more elongated. At one extreme case, if $d(F_1, F_2) = 0$ so that $F_1 = F_2$, the ellipse is a circle centered at $F_1$ with radius $\frac{d}{2}$. At the other extreme case, if $d(F_1, F_2) = d$, the ellipse shrinks (or collapses) to the line segment $F_1F_2$.

What happens if we choose the taxicab metric? This situation is what you investigated in Activity 7. Instead of being rounded, taxi-ellipses are either octagonal or hexagonal. One or two pairs of sides of the ellipse will be horizontal and/or vertical, and the remaining four sides will follow the sides of taxi-circles, which are line segments with slopes of ±1. If the foci lie on diagonally opposite corners of a rectangle, the taxi-ellipse will be octagonal. However, if the foci lie on the same vertical or horizontal line, one pair of horizontal or vertical segments disappears, and the ellipse is hexagonal. (See Figure 5.1. An interactive version of this figure is provided on the CD that accompanies this text.)

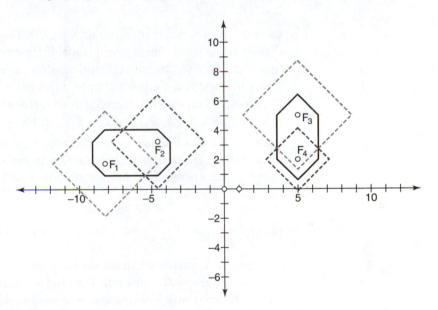

**FIGURE 5.1**
Taxi-Ellipses

## MEASURING DISTANCE FROM A POINT TO A LINE

To measure the distance from a point $P$ to a line $\ell$, we need a strategy for finding the point on $\ell$ that is closest to $P$. As we discussed in Chapter 4, one possible strategy is to think about the family of circles centered at $P$. When $r$ is very small, the circle around $P$ will not intersect $\ell$. As $r$ increases, eventually the circle and the line intersect.

We are used to the idea that in the ordinary Euclidean plane, a circle can intersect a line in two points, one point, or no points. In Euclidean space, when we drop a perpendicular from a point $P$ to a line $\ell$ (or, equivalently, when we find the foot of the perpendicular from $P$ to $\ell$), we are finding the point $X$ on $\ell$ that is as close as possible to $P$. A circle centered at $P$ with radius $PX$ will be tangent to $\ell$.

The situation in the taxicab plane is a bit different, as you saw in Activity 8. A taxi-circle might still intersect a line in two points, one point, or no points. However, because the sides of a taxi-circle are segments with slope ±1, it is also possible for a taxi-circle to intersect a line in a segment along one of its sides. So, after locating points of intersection, we also have to look for segments of intersection.

In the taxicab plane, if we are looking for a point $X$ on line $\ell$ that is as close as possible to point $P$ (which is assumed not to be on $\ell$), there are essentially three cases to consider: (1) situations where the slope of $\ell$ is between $-1$ and $+1$, (2) situations where the slope of $\ell$ is equal to $\pm 1$, and (3) situations where the absolute value of the slope of $\ell$ is greater than 1. Let's consider each case separately. Suppose we are given a line $\ell$ with slope $m$ and a point $P$ in the taxicab plane.

**Case 1**  If $-1 < m < +1$, then a taxi-circle centered at $P$ will first touch $\ell$ at a point directly above or below $P$. To find the point $X$ on $\ell$ that is closest to $P$, construct a vertical line through $P$. Point $X$ will be the point where this vertical line intersects $\ell$.

**Case 2**  If $m = \pm 1$, then the point $X$ on $\ell$ that is closest to $P$ is not unique; there will be a whole line segment containing points $X$ for which $d_T(P, X)$ is a minimum. To find this line segment, construct a vertical (or horizontal) line through $P$, and find the point $Y$ where this vertical (horizontal) line intersects $\ell$. Construct the taxi-circle centered at $P$ with radius $d_T(P, Y)$. Because the sides of this circle have slopes $\pm 1$, one of its sides will lie along line $\ell$.

**Case 3**  If $|m| > 1$, then a taxi-circle centered at $P$ will first touch $\ell$ at a point directly to the left or right of $P$. To find the point $X$ on $\ell$ that is closest to $P$, construct a horizontal line through $P$. Point $X$ will be the point where this horizontal line intersects $\ell$.

## PARABOLAS

You may be familiar with parabolas from your study of high school or college algebra. The graph of a quadratic equation, $y = ax^2 + bx + c$ (where $a$, $b$, and $c$ are constants), will have the shape of a parabola. If $b = 0$, the parabola will be symmetric with respect to the $y$-axis and will intersect the $y$-axis at the point $(0, c)$. If $a$ is positive, the parabola will open up (so that it could hold water); if $a$ is negative, the parabola will open down (or be shaped like a hill). By varying the size of $a$, you can make the parabola narrower and steeper, or wider with a more gradual slope.

In this course, we are considering the parabola as a geometric object, rather than as the graph of an algebraic equation. In other words, we are considering the parabola as a locus of points with a specific relationship to a given line and a given point.

**DEFINITION 5.4**  Given a fixed line, $k$, and a fixed point, $F$, a *parabola* is defined as the set of points $P$ that are equidistant from $k$ and $F$. We can write this symbolically as

$$\text{parabola} = \{P : d(P, F) = d(P, k)\}$$

Line *k* is called the *directrix* of the parabola, and point *F* is called the *focus* of the parabola.

This definition describes a parabola in terms of distances from two fixed objects, and again it does not tell us which metric to use. If we choose the Euclidean metric, we get an ordinary rounded parabola. As you experimented with your diagram in Activity 3, you probably observed that as you moved the focus closer to the directrix, the parabola became narrower, and as you moved the focus farther from the directrix, the parabola became wider.

In geometry we think of a parabola as a locus of points, while in algebra we think of a parabola as a graph of a quadratic function. Geometry and algebra are two different lenses through which we can view the same object, in this case, parabolas. You will have an opportunity to investigate this more deeply in the exercises.

In Activity 9, you investigated what happens if we choose the taxicab metric. Let point *F* be the focus and line *k* be the directrix. We are interested in the locus of points that are the same distance from *F* as from *k*. Let *m* denote the slope of line *k*. Figure 5.2 illustrates the case for $|m| < 1$. (An interactive version of this figure is provided on the CD that accompanies this text.) In this case, the taxi-parabola consists of two segments and two rays. The corners (or turning points) of the parabola occur where the point of the locus is formed at one of the vertices (or corners) of the circle centered at *F*. If $|m| > 1$, the parabola will open horizontally to the right or to the left, depending on the location of *F* with respect to line *k*.

If $|m| = 1$, the parabola will consist of one segment and two rays. The segment will be parallel to *k*, because it will be formed by the side of a taxi-circle centered at *F* whose radius is one-half the taxi-distance from *F* to *k*.

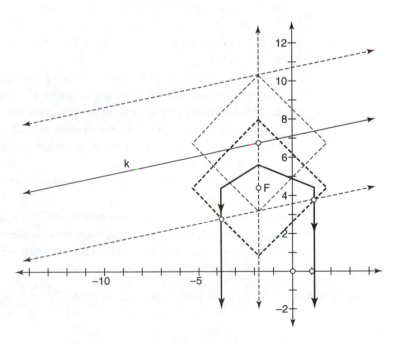

**FIGURE 5.2**
A Taxi-Parabola

# HYPERBOLAS

Hyperbolas are defined similarly to ellipses, with one very small difference—a change in sign. This change in sign, from positive to negative, makes a great deal of difference in the shape of the figure.

**DEFINITION 5.5**  A *hyperbola* is defined as the set of points $P$, the difference of whose distances from two fixed points, $F_1$ and $F_2$, is constant. We write this symbolically as

$$\text{hyperbola} = \{P : |d(P, F_1) - d(P, F_2)| = d,$$

$$\text{where } d > 0 \text{ and } F_1, F_2 \text{ are fixed points}\}.$$

Notice that because we are subtracting, we take the absolute value, which gives rise to two branches of the hyperbola. As with ellipses, the fixed points are called the *foci* (singular, *focus*) of the hyperbola.

You will have an opportunity to investigate both Euclidean hyperbolas and taxi-hyperbolas in the exercises.

# AXIOM SYSTEMS

A distinctively mathematical way of thinking is to express our assumptions in a formal way and then use rules of logical reasoning to develop proofs (or disproofs) of conjectures that may have been based on observations of particular phenomena. This is a process of deductive reasoning, and it is what you have been invited to do throughout this course.

An *axiom system* is a formal statement of our most basic expectations about a particular concept. As you worked on Activity 4, you may have wondered why we raised such a basic issue as whether the distance $PQ$ is the same as the distance $QP$. Yet, in an urban setting with one-way streets, the taxi-distances $d_T(P, Q)$ and $d_T(Q, P)$ may be very different. So Metric Axiom 2 is more restrictive than it may appear at first.

So far in this course, you have seen Euclid's postulates, which are basic assumptions—axioms—about the nature of plane space. The metric axioms given in this chapter are another example of an axiom system. These metric axioms formalize our ideas about distance, stating the obvious so that we can study our assumptions and investigate the effect these assumptions have on the shapes of elementary curves in the plane.

There are other concepts that we might formalize axiomatically. For example, what do we mean when we say that point $B$ is *between* points $A$ and $C$? In the real world, is there a gas station between your home and the grocery store? What exactly would this mean? In other words, what are your most basic assumptions—your axioms—about the concept of betweenness? You will have a chance to explore this question in the exercises.

## 5.3 EXERCISES

Give clear and complete answers to the exercises, expressing your explanations in complete sentences. Include diagrams whenever appropriate.

1. The equation $x^2 - 4x + y^2 + 6y = 12$ describes a Euclidean circle. Find the center and the radius of this circle.

2. Which of the following equations describe a Euclidean circle? If the equation does describe a circle, find its center and radius. If the equation does not describe a circle, explain what it does describe.
   a. $x^2 - 6x + y^2 + 2y - 6 = 0$
   b. $x^2 - 2x + y^2 + 4y + 12 = 0$
   c. $x^2 - 3x + y^2 + 5y = 0$
   d. $x^2 - 6x + y^2 + 2y + 10 = 0$

3. Prove that the formula for distance in a skew coordinate system meets the conditions specified by the metric axioms.

4. Let $g(P, Q) = \max(x_P, x_Q) + \max(y_P, y_Q)$. (*Note:* $\max(a, b)$ denotes the larger of the two numbers $a$ and $b$.) Prove or disprove that $g$ is a metric.

5. Let $h(P, Q) = \min(x_P, x_Q) + \min(y_P, y_Q)$. (*Note:* $\min(a, b)$ denotes the smaller of the two numbers $a$ and $b$.) Prove or disprove that $h$ is a metric.

6. Suppose you have a set of points $S = \{a, b, c, d, e\}$. Let $g(x, y)$ be defined on $S$ so that
$$g(x, y) = \begin{cases} 1 & x \neq y \\ 0 & x = y. \end{cases}$$
Prove or disprove that $g$ is a metric on $S$.

7. Starting with the definition of a taxi-circle, find the equations of the four line segments that describe the taxi-circle centered at $C = (a, b)$ with radius $r$.

8. Explain in detail why a taxi-circle consists of four line segments, and not four lines.

9. In the Euclidean plane, if point $Q$ lies between points $P$ and $R$, then $Q$ lies on the line segment $PR$. What can you say about the set of points that lie between points $A$ and $B$ in the taxicab plane?

10. Think about the concept of *betweenness*.
    a. What does it mean for point $B$ to lie between points $A$ and $C$? Formalize your assumptions as a set of axioms for *betweenness of points*.
    b. What does it mean for a ray, $\overrightarrow{AD}$, to lie between $\overrightarrow{AB}$ and $\overrightarrow{AC}$? Formalize your assumptions as a set of axioms for *betweenness of rays*.

11. Develop a method to investigate Euclidean hyperbolas, similar to the method you used in Activity 2 to investigate Euclidean ellipses. In this case, you will need to construct a pair of segments whose lengths maintain a constant difference while one of their endpoints is varied.

    One way to do this is to construct three points in a line $j$ and label the points $A$, $B$, and $X$. Construct two circles centered at $X$, one with radius $XA$ and the other with radius $XB$. As you vary $X$ (but neither $A$ nor $B$), $|d(X, A) - d(X, B)|$ should remain constant.

12. Repeat Exercise 11 using the taxicab metric.

13. Open a worksheet in Sketchpad, and define a coordinate system with a square grid. In a convenient place near the edge of this worksheet, construct a line parallel to the $y$-axis. Construct a line segment, $AB$, on this line.
    a. Construct point $F$ on the $y$-axis. $F$ will be the focus of a parabola. Measure the coordinates of $F$.
    b. Select the $x$-axis, then choose **Transform | Mark Mirror** (or double-click on the $x$-axis). Then select the point $F$ and choose **Transform | Reflect** to construct a new point $F'$ that is the reflection of point $F$ across the $x$-axis.
    c. Construct a line through $F'$ perpendicular to the $y$-axis. This will be the directrix of the parabola; label it $k$. Select $k$, and choose **Measure | Equation.**
    d. Use the method of Activity 3 to construct the parabola with focus $F$ and directrix $k$. This will give a parabola through the origin

opening up or down (depending on the location of the point $F$).

e. Choose **New Function** from the Graph menu. Define the function $f(x) = \frac{1}{4}x^2$. Then choose **Graph | Plot Function.** This, too, will give a parabola through the origin.

f. Vary point $F$, and try to make the two parabolas—the one created as a locus of points, and the one created as the graph of a function—coincide. What are the coordinates of $F$ when this happens?

g. Experiment with different functions $g(x) = ax^2$, each time observing the coordinates of $F$ for the particular value of $a$ when the locus and the graph coincide. Some good values of $a$ to try might be $\frac{1}{3}$, $\frac{1}{5}$, and $\frac{1}{8}$. What is the relationship between the value of $a$ and the coordinates of $F$?

14. The parabolas you experimented with in Exercise 13 open up or down (depending on the sign of $a$). Parabolas can also open to the right or left, in which case the form of the equation is $x = f(y)$. Repeat Exercise 13, investigating this type of parabola.

Exercises 15 and 16 are especially for future teachers.

15. In the *Principles and Standards for School Mathematics,* the National Council of Teachers of Mathematics (NCTM) recommends that "Instructional programs from prekindergarten through grade 12 should enable all students to . . . specify locations and describe spatial relationships using coordinate geometry and other representational systems" (NCTM, 2000, 41). Furthermore, these instructional programs "should enable all students to . . . understand measurable attributes of objects and the units, systems, and processes of measurement" (NCTM, 2000, 44). What does this mean for you and your future students?

a. Find a copy of the *Principles and Standards for School Mathematics,* and read through the overview of the Geometry Standard and the Measurement Standard (pp. 41–47). Then study the specific recommendations for one grade band (i.e., pre-K–2, 3–5, 6–8, or 9–12). What are the specific NCTM recommendations regarding coordinate geometry and ideas about measurement for that grade band?

b. Find copies of school mathematics textbooks for these same grade levels. How are the NCTM standards implemented in those textbooks? Cite specific examples.

c. Write a report in which you present and critique what you learn. Your report should include your answers to the questions in parts a and b.

16. Design several classroom activities involving coordinate geometry and/or measurement that would be appropriate for students in your future classroom. Write a short report explaining how the activities you design reflect both what you have learned in studying this chapter and the recommendations of the NCTM.

Reflect on what you have learned in this chapter.

17. Review the main ideas of this chapter. Describe, in your own words, the concepts you have studied and what you have learned about them. What are the important ideas? How do they fit together? Which concepts were easy for you? Which were hard?

18. Reflect on the learning environment for this course. Describe aspects of the learning environment that helped you understand the main ideas in this chapter. Which activities did you like? Dislike? Why?

## 5.4 CHAPTER OVERVIEW

In this chapter, we investigated some ideas of metric geometry. A metric space is a set of points with a rule or formula for measuring distance that satisfies three metric axioms.

Let $P$, $Q$, and $R$ be points, and $d(P, Q)$ denote the distance from $P$ to $Q$.

**Metric Axiom 1**    $d(P, Q) \geq 0$, and $d(P, Q) = 0$ if and only if $P = Q$.

**Metric Axiom 2**    $d(P, Q) = d(Q, P)$.

**Metric Axiom 3**    $d(P, Q) + d(Q, R) \geq d(P, R)$.

We approached the concept of a metric space by investigating some familiar curves, using two different rules for measuring distances: the ordinary Euclidean metric and the taxicab metric. We investigated circles, ellipses, parabolas, and hyperbolas, using their metric definitions, and found that these ordinary curves have shapes in taxicab space that are surprisingly different from their shapes in ordinary Euclidean space.

By using their metric definitions, we have treated these curves as geometric objects; that is, as sets of points with specified distances from certain fixed points or fixed lines. Using the analytic tools from Chapter 4, we can develop formulas for these geometric objects. As you continue your study of geometry, you will be able to use either geometric tools or algebraic tools to study the properties of these and other curves. In this text, we have restricted our study to objects in the plane, that is, to two-dimensional objects. With very little additional machinery, it would be possible to extend your investigations to three dimensions (and to even higher dimensions!).

In Chapters 1 and 3, we presented Euclid's postulates, which are axioms that express our basic assumptions about the nature of space. In this chapter, we have used axioms to express our most basic expectations about what we need to do when measuring distances from one point to another. Perhaps you are beginning to get some insight into how mathematicians think about problems. We try to express our most basic assumptions—those things that we want to take for granted—as axioms. Axioms give mathematicians a way of formalizing assumptions about a particular situation. Working from the axioms, we build theorems and eventually a whole theory. In this chapter, we just touched on the very beginnings of a theory of metric spaces. By investigating some unexpected results about some very familiar curves, we hope that we have opened your mind to some new vistas in geometric reasoning!

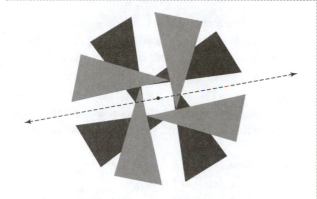

## CHAPTER SIX

# Transformational Geometry

From your work in previous courses, you are familiar with functions whose inputs and outputs are real numbers. In this chapter, you will work with functions that use points of the plane as inputs and outputs. If a set of points forms a geometric figure in the plane, the functions we will study in this chapter will transform that figure in various ways. A figure may simply be moved from one place to another, or it may be deformed in some way—perhaps stretched or twisted.

Of particular interest are transformations of the plane that do not alter the distance between points. These special functions, called *isometries,* are the focus of this chapter. We introduce the four basic types of isometry and prove that every isometry is one of these basic types. We show how to use isometries to study congruence of geometric figures. As functions, isometries can be combined by composition, which leads to notions of closure, identity, inverse, and eventually to the important concept of *group*.

This chapter concludes with a deeper look at inversion, an interesting transformation that is not an isometry. Some basic properties of inversion are presented, and these are used to give another proof of the Feuerbach Theorem.

Do the following activities, writing your explanations clearly in complete sentences. Include diagrams whenever appropriate. Get into the habit of saving your work for each activity, as later work sometimes builds on earlier work. You will find it helpful to read ahead into the chapter as you work on these activities.

1. Consider the function $f_1(x, y) = (x, \frac{1}{y})$. This function takes points in the coordinate plane as inputs and returns points in the plane as outputs. Let us examine the effect this function has on its input. Carry out the following steps:

   i. Create an $xy$-coordinate system.
   ii. Construct a circle of radius 2 centered at the origin.
   iii. Construct a point, $P$, on this circle. Measure the abscissa and ordinate of $P$.
   iv. Calculate $\frac{1}{y}$ using the measured $y$-coordinate.
   v. Select the coordinates $(x, \frac{1}{y})$, in the correct order.
   vi. Choose **Plot as (x,y)** from the Graph menu to get a new point $P'$.
   vii. Drag the original point $P$ around the circle, and watch what happens to the new point $P'$. You may find it helpful to use an option on the Display menu to trace $P'$ as you drag $P$.

   Now answer the following questions about $f_1$.

   a. Does every point on the circle have an image point, that is, an output? Use the expression for the function to explain why or why not.
   b. Which points from this circle are unchanged by the function, that is, which points are in the same location on the output as on the input?
   c. What does the function do to a point if its $y$-value is large, for example, near 2? What happens to a point if its $y$-value is small, for example, near 0? Use the expression of the function to explain your answers.
   d. What is different for points below the horizontal axis?
   e. Try to answer parts a–d when the domain is the entire plane. Use the expression of the function to justify your answers. Discuss the domain and range of the function, as well as its geometric effects. (You may wish to select the point $P$, then choose **Split Point from Circle** from the Edit menu. This will allow you to observe $P'$ as you move $P$ about the entire Sketchpad window.)

2. Repeat Activity 1 for the following functions.
   a. $f_2(x, y) = (0, y)$
   b. $f_3(x, y) = (y, x)$
   c. $f_4(x, y) = (x^2, -y)$
   d. $f_5(x, y) = (x - 2, y + 1)$
   e. $f_6(x, y) = \left( \dfrac{4x}{x^2 + y^2}, \dfrac{4y}{x^2 + y^2} \right)$

3. Which of the functions in Activities 1 and 2 are one-to-one? Which are onto? Explain why.

4. When a set of points is put through a function, such as those in Activities 1 and 2, a variety of things can happen. Figures can move, stretch, and twist in many ways. One kind of function that behaves better than others is the *isometry,* a transformation that preserves distances. If $f$ is a distance-preserving function, then $|AB| = |f(A)f(B)|$; that is, the distance between points $A$ and $B$ is the same as the distance between points $f(A)$ and $f(B)$. Which of the functions from Activities 1 and 2 are isometries? Justify your decisions.

5. In this activity, you will see some basic isometries.
   a. Using Sketchpad, draw a scalene triangle. Draw a line not intersecting the triangle. Select this line, then choose **Mark Mirror** from the Transform menu. Next, select your triangle, and choose **Reflect** from the Transform menu. Drag your triangle, and observe what happens.
   b. In a new sketch, draw another scalene triangle. Draw two other points (or draw a line segment). Choose **Mark Vector** from the Transform menu to mark a vector between these two points. Then select your triangle, and choose **Translate** from the Transform menu. Drag your original triangle, and observe what happens.
   c. In a third sketch, draw a scalene triangle. Draw a new point, and choose **Mark Center** from the Transform menu to mark this point as the center of a rotation. Then select your triangle, and choose **Rotate** from the Transform menu. Choose any (nonzero) angle you wish. As before, drag your original triangle, and observe what happens.
   d. How is the result in part a different from the results in parts b and c?

6. A *fixed point* for a transformation is a point that is not changed by the transformation. In other words, the output point from the function is in exactly the same position as the input point. What are the fixed points of a reflection? Of a translation? Of a rotation?

7. Every isometry is either direct or opposite. (This is the difference in Activity 5.) Every isometry either has fixed points or it doesn't. This leads to four possibilities. List the four possibilities, and decide what kind of isometry each represents. (You need to know that there is a fourth kind of isometry, the *glide reflection,* which is a reflection followed by a translation parallel to the mirror line.)

8. Because isometries are functions, it is possible to form compositions of isometries in which one transformation is followed by another.
   a. Draw a scalene triangle. Reflect it across one line, $\ell$, then reflect this image across a second line, $m$, that is parallel to $\ell$. (*Note:* It will be helpful to make the three triangles different colors.) Drag the original triangle, and observe what happens. What kind of isometry is produced by this composition? Explain how you decide this.
   b. Draw a scalene triangle. Reflect it across one line, $\ell$, then reflect this image across a second line, $m$, that intersects $\ell$. Drag the original triangle, and observe what happens. What kind of isometry is produced by this composition? Explain how you decide this.

c. Draw a scalene triangle. Rotate it $180°$ around one point, $P$, then rotate this image $180°$ around a second point, $Q$. Drag the original triangle, and observe what happens. What kind of isometry is produced by this composition? Explain how you decide this.

d. Draw a scalene triangle. Rotate it $90°$ around one point, $P$, then rotate this image $90°$ around a second point, $Q$. Drag the original triangle, and observe what happens. What kind of isometry is produced by this composition? Explain how you decide this.

9. Can you find an isometry that is the inverse of a translation, that is, an isometry that has the reverse effect of a translation? If so, what kind of isometry is it? If not, why not? Repeat this question for reflections, for rotations, and for glide reflections.

10. In this activity, we will examine a function that is not an isometry. Follow these steps:

   i. Create an $xy$-coordinate system.
   ii. Construct a unit circle $C$ (of radius 1) centered at the origin.
   iii. Construct a point $P$ not on this circle. Measure the abscissa $x$ and ordinate $y$ of $P$.
   iv. Calculate and plot the new point $P' = \left( \dfrac{x}{x^2 + y^2}, \dfrac{y}{x^2 + y^2} \right)$.

   Now answer the following questions.

   a. Drag the original point $P$, and watch what happens to the new point $P'$. Can $P'$ ever equal $P$? If so, where? Can $P'$ go anywhere in the plane? Why is the unit circle $C$ important for this function?

   b. Construct a line, $\ell$. Select $P$ and $\ell$, then choose **Merge Point to Line** from the Edit menu. Drag $P$ along $\ell$, and observe $P'$. To get a better picture of what this function is doing to line $\ell$, select $P'$ and $P$—in that order—then choose **Construct | Locus** to see the images of points from the line. Describe this locus.

   c. Drag line $\ell$, and observe what happens to its locus. What can you say about the locus if $\ell$ intersects the unit circle $C$? What can you say if $\ell$ passes through the origin?

   d. Construct another circle $D$ somewhere in this plane. You need to move $P$ to this new circle. To do so, first separate $P$ from the line by selecting $P$ and the line, then choosing **Split Point from Line** from the Edit menu. Then select $P$ and the new circle $D$, and choose **Merge Point to Circle** from the Edit menu. (At this juncture, you may want to hide line $\ell$ to avoid confusion in the sketch.) Once again, select $P'$ and $P$, and construct the locus of the circle $D$. Move $D$ about the plane, and observe what happens to its locus. Can the locus ever intersect circle $D$? Can the locus ever equal $D$? What happens to the locus if $D$ contains the origin?

You are accustomed to functions that use a real number as input and produce a unique real number as output. However, the definition of function allows any kind of mathematical object to be used as input or as output. These objects could be numbers, points, lines, sets, or even other functions. In this chapter, we are interested in functions that use points of the plane as both inputs and outputs.

Let us review the idea of *function*. A function is a mapping from a set $A$ (which can be called the *source* of the function) to a set $B$ (sometimes called the *target* of the function). To be called a function, this mapping must take each element of $A$ to a unique element of $B$. In other words,

1.  Every element of $A$ is allowable as an input for the function.

2.  Each input generates a unique output.

The set of allowable inputs is call the *domain*. Most of the time, the domain of a function is the same as its source, but it can happen, as in Activities 1 and 2, that some elements of the source cannot be used. In these activities, we asked you to think of the entire $xy$-plane as possible inputs and possible outputs for these functions; that is, both the source and the target for each of these functions is the set $\Re^2$. In Activity 1, none of the points on the $x$-axis can serve as an input for $f_1$. Thus, although the set $\Re^2$ is the source, the domain is $\Re^2 - \{x\text{-axis}\}$. Similarly, the output of $f_1$ cannot be on the $x$-axis; in this case, the set of possible outputs is a subset of the target. The target of $f_1$ is the set $\Re^2$, while the *range* of $f_1$ is $\Re^2 - \{x\text{-axis}\}$.

## TRANSFORMATIONS

In Activities 1 and 2, you saw several functions that take a point from the plane as input and return another point in the plane. What does $f_1$ do visually? Because the $x$-coordinate is unchanged by $f_1$, the output does not move horizontally from the input; it only moves vertically. However, $\frac{1}{y}$ has drastic effects on the $y$-coordinate. Small values are changed into large ones; large values are changed into small ones. Visually, this means that points near the horizontal axis are moved away from it, while points away from the axis are moved close to it. Points above the horizontal axis stay above it, and points below stay below. Only those points with $y = \pm 1$ are unaffected by the function $f_1$.

Because there are points of the plane that are not in the range of $f_1$, this function is not onto. Informally, a function is *onto* if every point in the target occurs as part of the range. Here is a more formal statement of this definition.

**DEFINITION 6.1**    A function $f : D \to T$ is *onto* if for every $Y \in T$ there is an $X$ in the domain $D$ so that $f(X) = Y$. In symbols, we write $f$ is onto if $(\forall Y \in T)\,(\exists X \in D)$ such that $f(X) = Y$.

Being onto means that the range is the same as the target. Because there are points in the plane that cannot be obtained as outputs of $f_1$, the function is not onto. The function $f_3$, however, is onto, as is one other function in Activity 2.

Another property of functions is that of being *one-to-one,* which says that different inputs must produce different outputs. Consider the function $f_1$. The input point $(x_1, y_1)$ gets mapped to $(x_1, \frac{1}{y_1})$, and the point $(x_2, y_2)$ gets mapped to $(x_2, \frac{1}{y_2})$. If the outputs are the same so that $(x_1, \frac{1}{y_1}) = (x_2, \frac{1}{y_2})$, then $x_1 = x_2$ and $\frac{1}{y_1} = \frac{1}{y_2}$. So clearly, $(x_1, y_1) = (x_2, y_2)$; in other words, the inputs must have been the same. Here is a more formal statement of the one-to-one property.

**DEFINITION 6.2**   A function $f : D \to T$ is *one-to-one* if $X_1 \neq X_2$ implies that $f(X_1) \neq f(X_2)$. Another way to say this, using the contrapositive, is that a function is one-to-one if $f(X_1) = f(X_2)$ implies $X_1 = X_2$.

Four of the functions in Activities 1 and 2 are one-to-one. A function on the plane that is both one-to-one and onto is called a *transformation* of the plane. Only two of the functions in Activities 1 and 2 are transformations.

## ISOMETRIES

The two transformations among the functions of Activities 1 and 2 have an additional property—they preserve distances. This means that the distance between the points $A$ and $B$ is the same as the distance between their images $f(A)$ and $f(B)$.

**DEFINITION 6.3**   A function $f : \Re^2 \to \Re^2$ is a *distance-preserving* function if for any points $A$ and $B$, the distance between $A$ and $B$ is the same as the distance between their images $f(A)$ and $f(B)$. In other words, $f$ preserves distances if $|AB| = |f(A)f(B)|$.

For a given function, this property is not hard to check. Suppose $(a, b)$ and $(c, d)$ are two points. The distance between them is $\sqrt{(a - c)^2 + (b - d)^2}$. Now consider the function $f_2$ as an example:

$$f_2(a, b) = (0, b) \qquad \text{and} \qquad f_2(c, d) = (0, d).$$

The distance between the images $f_2(a, b)$ and $f_2(c, d)$ is $\sqrt{(0 - 0)^2 + (b - d)^2}$, which is not the same as the distance between the original points. Thus, $f_2$ does not preserve distances.

A function with all three of these properties—onto, one-to-one, and distance preserving—is called an *isometry.* Figures that are put through an isometry emerge looking the same, without stretching, twisting, or any other distortion. However, they may come out in a different position. The word *iso-metry* literally means "same measure." Other names for isometries are *rigid motions* and *Euclidean motions.*

Activity 5 introduced three basic kinds of isometry. The first is the reflection. For a *reflection,* one line $\ell$ acts as the mirror. For any input point $A$, the output point $f(A)$ is directly across the line and is the same distance from $\ell$ as $A$ is. In more precise terms, $\ell$ is the perpendicular bisector of the segment between $A$ and $f(A)$. An object, such as a triangle, drawn on one side of line $\ell$ will be reflected to its mirror image on the other side of $\ell$. What happens if an object touches or overlaps the mirror line?

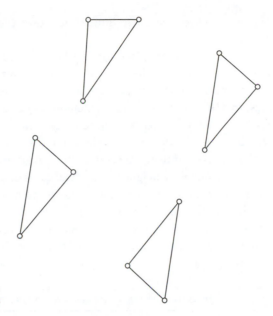

**FIGURE 6.1**
The Basic Isometries

The isometry in Activity 5b is a *translation*. This isometry is simpler than a reflection. An input point is moved (dragged) to a new location by translation. This motion is always in a constant direction and for a constant distance. Thus, a vector is an ideal way to represent the motion of a translation.

Activity 5c introduces the *rotation*, a third kind of isometry. You should experiment with cases where the center of rotation is inside your triangle or outside your triangle. In either case, an input point does not change its distance from the center, but its angle with respect to the center is changed. Using the language of polar coordinates, if the center of rotation is at the origin, the $r$-coordinate of a particular point is not changed by the rotation, but a constant is added to (or subtracted from) the $\theta$-coordinate.

In Figure 6.1, one of the triangles is the original (the input) and the other three are the results of a translation, a reflection, and a rotation. Which is which?

Finally, Activity 5d asks how the reflection is different from the translation and the rotation. It may help if you label the vertices of your triangle $A$, $B$, $C$ in a clockwise fashion. Then decide what the isometry does to each vertex, and label the images accordingly. Sometimes the orientation of the image points will also be clockwise, and sometimes it will reverse. Because translations and rotations keep the same orientation, they are called *direct* isometries. However, a reflection reverses the orientation; for this reason, reflections are called *opposite* isometries.

Here is a physical interpretation that may help you understand this distinction: Imagine that your triangle is cut from cardboard and is lying on a table. Label the vertices if you wish. You can translate the triangle without picking it up; simply drag it along the surface. You also can rotate the triangle without picking it up (although this rotation is a bit tricky if the center of rotation is not inside the triangle). However, to reflect the triangle, you must pick it up and turn it over.

This extra motion in a third dimension causes the opposite orientation of the vertices.

Now you should be able to sort out Figure 6.1. Look at the orientations of the triangles. (Need a hint? The original triangle is on the upper right.)

Another way to classify isometries is to consider their *fixed points*. In mathematics, *fixed* means constant or unchanging. Therefore, we are looking for points $X$ for which $f(X) = X$. A translation does not have fixed points; every point moves the same distance and direction. (For the moment, we will ignore the trivial case of a translation using the zero vector.) A rotation has only one fixed point: its center. (Again, we will ignore the trivial cases of a rotation through angles of 0, $2\pi$, and multiples of $2\pi$.) A reflection, however, has an entire line of fixed points. Any point on the mirror line is unchanged by the reflection.

If we combine the last two concepts—direct versus opposite, fixed points or no fixed points—we see that there are four possibilities. These possibilities are shown in Table 6.1.

**TABLE 6.1**

|          | Has fixed points | Has no fixed points |
|----------|------------------|---------------------|
| Direct   | rotation         | translation         |
| Opposite | reflection       | ?????               |

As you can see, there is a fourth possibility that has not appeared yet. This is the glide reflection, which we will discuss after we examine composition of isometries.

Working from the definition, there are some useful things we can prove about isometries. First, we state the precise definition.

**DEFINITION 6.4**  A function $f : \Re^2 \to \Re^2$ is an *isometry* of the Euclidean plane if $f$ is one-to-one, $f$ is onto, and $f$ preserves distances.

One important fact about isometries is that we need only a little information to figure out exactly what an isometry does to the whole plane. In fact, three points tell us enough.

**THEOREM 6.1**  In the Euclidean plane, the images of three noncollinear points completely determine an isometry. In other words, if we know the outputs for three noncollinear points $A$, $B$, $C$, we can figure out what the isometry does to any point $X$.

**Outline of a Proof**  The key to this proof is the distance-preserving property. A picture may make it clear (see Figure 6.2). Point $X$ is a certain distance from point $A$, so $f(X)$ must be that same distance from $f(A)$. This compels $f(X)$ to lie on a circle centered at $f(A)$ with radius $|AX|$. Furthermore, $X$ is a certain distance from point $B$, so $f(X)$ must lie on a circle centered at $f(B)$ with radius $|BX|$. These two circles intersect at two points. However, we also know that $f(X)$ must lie on a third circle, centered at $f(C)$ with radius $|CX|$. The three circles will have exactly one common intersection point.

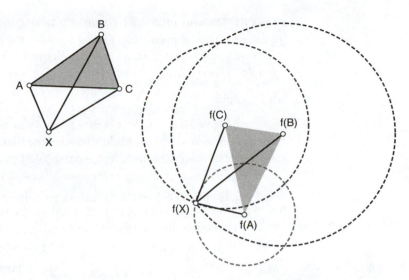

**FIGURE 6.2**
Isometry Uniquely Determined

There are some details of this proof that we have glossed over. How can we be sure that the first two circles intersect at all? Is it possible that the first two circles intersect only once, and how would that affect the proof? Can we be sure that the third circle shares an intersection point with the first two and that it does not share both intersection points? These are not difficult questions, but they deserve some thought.

It is natural to visualize three noncollinear points as the vertices of a triangle. This is why you were asked to use triangles in Activity 5. It is important to use scalene triangles to avoid any confusion that might be caused by accidental symmetries of the triangle.

**THEOREM 6.2**  An isometry preserves collinearity. In other words, three points that are collinear will still be collinear after going through an isometry.

**Proof**  First, recall the triangle inequality:

For any three points $A$, $B$, $C$, we have $|AB| + |BC| \geq |AC|$.

If the three points are collinear, there will be one way to list the points along their line that makes this an equality. Suppose that $A$, $B$, $C$ is that correct ordering. Because the isometry $f$ preserves distances, we have

$$|f(A)f(B)| + |f(B)f(C)| = |f(A)f(C)|.$$

Therefore, the three points $f(A)$, $f(B)$, and $f(C)$ must also be collinear.

------------------------------------------

We can also use the triangle inequality to prove that the order of the points along the line must stay the same.

**THEOREM 6.3**  An isometry preserves betweenness. In other words, if point $B$ is between points $A$ and $C$ along a line, then point $f(B)$ will be between points $f(A)$ and $f(C)$ along their line.

**Proof of Theorem 6.3**   By Theorem 6.2, we know that $f(B)$ is collinear with $f(A)$ and $f(C)$. Suppose that $f(A)$ is between the other two points. This makes

$$|f(A)f(B)| + |f(B)f(C)| > |f(A)f(C)|.$$

However, because $B$ is between $A$ and $C$, we know that $|AB| + |BC| = |AC|$, which implies that

$$|f(A)f(B)| + |f(B)f(C)| = |f(A)f(C)|.$$

Hence, $f(A)$ cannot be the middle point. A similar argument shows that $f(C)$ cannot be the middle point either, leaving $f(B)$ in the middle.

------------------------------------------------

Together, Theorems 6.2 and 6.3 give us a very important fact.

**COROLLARY 6.4**   Under an isometry, the image of

$$\left.\begin{array}{r} \text{a line segment} \\ \text{a triangle} \\ \text{an angle} \\ \text{a circle} \end{array}\right\} \text{ is a congruent } \left\{\begin{array}{l} \text{line segment,} \\ \text{triangle,} \\ \text{angle,} \\ \text{circle.} \end{array}\right.$$

------------------------------------------------

You should think carefully about why each part of this corollary is true.

Corollary 6.4 says that isometries are another way to study congruence. The image of a geometric figure under an isometry will be congruent to the original figure.

## COMPOSITION OF ISOMETRIES

Because isometries are functions—even though they use types of inputs and outputs that are different from what you are accustomed to—we can combine isometries by composition. Here is a quick review of how composition works: Suppose we have two functions, $f$ and $g$, for which the range of $f$ is a subset of the domain of $g$. This means that an output value $f(x)$ makes sense as an input for $g$. The composition $(g \circ f)(x)$ means that we are to compute $g(f(x))$; that is, we start with the input $x$, perform the function $f$, then use the output value $f(x)$ as an input to perform the function $g$. We can represent this process in a diagram:

$$x \longrightarrow f(x) \longrightarrow g(f(x)).$$

It is necessary, of course, that the output of $f$ is a legitimate input for $g$, so that $f(x)$ is in the domain of $g$. Otherwise, the composition is undefined. For isometries of the plane, this is not a problem, because any output is a point in the plane and can serve as an input for the next function.

In Activity 8, you investigated some of the things that can happen with a composition of two isometries. Consider, for example, a reflection across a line followed by a second reflection across a line parallel to the first line. Each reflection reverses the orientation of a triangle; this double reversal results in a direct isometry. Further, this composition of reflections has no fixed points. (Can you explain why?) Therefore, the composition of two reflections in parallel lines is a translation.

The preceding paragraph is not really a proof; rather, it is a plausibility argument. Here is a more formal proof that the composition of two reflections in parallel lines is a translation.

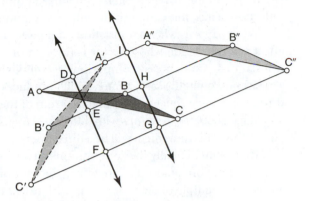

**FIGURE 6.3**
A Composition of Two
Reflections

**Proof**  By Theorem 6.1, we know that it will be enough to show what happens to three noncollinear points. In Figure 6.3, the first reflection uses the line on the left as its mirror, and the second reflection uses the line on the right. Point $A$ is reflected to point $A'$ and then to $A''$; points $B$ and $C$ are reflected similarly. Notice that the first line of reflection is the perpendicular bisector of segment $AA'$ and that the second line of reflection is the perpendicular bisector of the segment $A'A''$. Thus, point $A$ moves in a direction perpendicular to the mirrors, and the total distance it moves, $|AA''|$, equals

$$|AD| + |DA'| + |A'I| + |IA''| = 2(|DA'| + |A'I|) = 2|DI|,$$

which is twice the distance between the two mirrors. For point $B$ between the mirrors, we have the same phenomena of perpendicular bisectors. In this case, the motion is again perpendicular to the mirrors. The total distance can be calculated as follows.

$$
\begin{aligned}
|BB''| &= |BH| + |HB''| \\
&= |BH| + |B'H| \\
&= |BH| + |B'E| + |EB| + |BH| \\
&= |BH| + |EB| + |EB| + |BH| \\
&= 2(|EB| + |BH|) = 2|EH|.
\end{aligned}
$$

As happened with point $A$, the composition of these reflections moves point $B$ in the direction perpendicular to the mirror lines, and the distance moved is twice the distance between the mirrors. The argument for point $C$ is similar to this. Notice that we have omitted the two special cases when the images fall on a line; these are easily dealt with.

------------------------------------------------

We will not give formal proofs for every possible pairing of isometries. Let us look at some other compositions, however, and decide what type of isometry should result.

Consider another composition from Activity 8—a reflection in one line followed by a reflection in a second line intersecting the first. Each reflection reverses the orientation, so the double reversal produces a direct isometry. Also, the point where the two mirror lines intersect is a fixed point, for it is unchanged by either reflection. Thus, this composition will be a rotation.

What happens when a translation is followed by another translation? The net effect is a translation whose vector is the sum of the vectors of the two original translations. Another easy composition to think about is a rotation followed by a rotation around the same center, which produces a rotation around this center through an angle equal to the (directed) sum of the given rotation angles.

What about two rotations with different centers? Activity 8 showed two examples of this. This movement will be a direct motion, so it must be either a rotation or a translation. Usually the composition of two rotations with different centers will be a rotation, but in the case where the sum of the two angles of rotation is an integer multiple of $360°$, the composition will be a translation. This is what happened in Activity 8c. (The matrix techniques that you will learn in Chapter 7 can be used to prove this.)

One composition warrants special mention. Suppose we compose a reflection with a translation parallel to the mirror line. The result will be an opposite isometry because of the reflection, and it will have no fixed points because of the translation. This is called a *glide reflection*. It fits the missing category in Table 6.1 (page 142).

We now have four different types of isometry in the Euclidean plane. As the following important theorem shows, these are the only possibilities.

**THEOREM 6.5**   In the Euclidean plane, there are only four types of isometry: translations, rotations, reflections, and glide reflections.

To prove this, we need a preliminary result (or lemma) (Coxeter, 1969, 40).

**LEMMA 6.6**   For four points $A$, $B$, $A'$, and $B'$, with $|AB| = |A'B'|$, there are exactly two isometries that give $f(A) = A'$ and $f(B) = B'$.

**Proof of Lemma 6.6**   Pick any point $C$ not collinear with $A$ and $B$. This creates a triangle, $\triangle ABC$. Because $|AB| = |A'B'|$, there are two ways to construct a triangle $A'B'C'$ that is congruent to $\triangle ABC$ (see Figure 6.4). Thus, there are only two isometries that give $f(A) = A'$ and $f(B) = B'$.

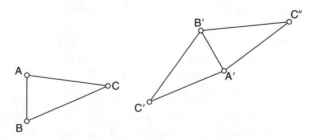

**FIGURE 6.4**
Only Two Possible Isometries

It is interesting to observe that the proof for Lemma 6.6 also shows that one of the two isometries will be direct and the other isometry will be opposite.

**Proof of Theorem 6.5** The approach for this proof is to examine all possible situations and describe two isometries that fit each situation. Because Lemma 6.6 says that there are only two possibilities, we will have described everything that can happen.

Suppose $f$ is an isometry and $A$ is a point for which $f(A) \neq A$. If there is no such point $A$, then every point is a fixed point for $f$. This is an isometry—rather dull but legitimate—that can be interpreted either as a translation by the zero vector or as a rotation through an angle of $0°$. Suppose there is an $A$ for which $f(A) \neq A$. We need to decide what happens to three points; let's choose points $B$ and $C$ so that $B = f(A)$ and $C = f(B)$.

**Case 1** Suppose $A = C$. Let $M$ be the midpoint of the segment $AB$, and let $\ell$ be the perpendicular bisector of this segment. The reflection in line $\ell$ takes $A$ to $B$ and $B$ to $C$. The $180°$ rotation around $M$ does as well. These two isometries are the only possibilities in this case.

**Case 2** Suppose $A \neq C$ and $A$, $B$, and $C$ are collinear on line $m$. Then a translation by the vector $AB$ works, as does a glide reflection in the line $m$ with vector $AB$.

**Case 3** Suppose $A$, $B$, and $C$ are three noncollinear points. Let $M$ be the midpoint of $AB$ and $N$ be the midpoint of $BC$. Then let $\ell$ be the line through $M$ and $N$, and let $O$ be the intersection point of the perpendicular bisectors of $AB$ and $BC$. Drop perpendiculars from $A$ and $B$ to $\ell$, and label the feet of $A$ and $B$ in $\ell$ as $D$ and $E$, respectively. (See Figure 6.5.) The isometry that does this is either a rotation around the point $O$ through $\angle AOB$ or a glide reflection in the line $\ell$ with vector $DE$. (In Figure 6.5, the image of $\triangle ABC$ under the rotation is $\triangle A'B'C_1$. The image of $\triangle ABC$ under the glide reflection is $\triangle A'B'C_2$.)

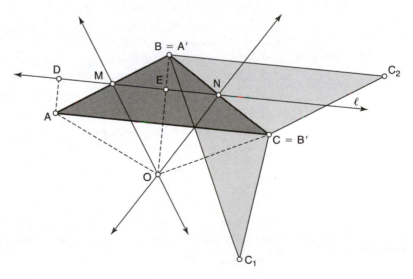

**FIGURE 6.5**
Case 3 of the Proof for Theorem 6.5

Thus, in every case, both possible isometries can be described by one of the four known types. Therefore, these are the only possible types of isometry.

## INVERSE ISOMETRIES

Activity 9 asks you to find an isometry that reverses a translation. If we think of translation as moving a geometric object a certain distance in a certain direction, the reverse motion is pretty obvious: Move the geometric object the same distance in the opposite direction. Thus, if $\vec{v}$ is the vector for the original translation, the inverse isometry is a translation by $-\vec{v}$. A rotation also has an easy inverse: rotate around the same center with the same angle, but in the opposite direction. Thus, a rotation through angle $\alpha$ would have an inverse rotation through angle $-\alpha$. (Equivalently, we could say that the new rotation is through the angle $360° - \alpha$).

Composing an isometry with its *inverse* undoes the original isometry and moves everything back to its starting point. We say that $f \circ f^{-1}$ produces the identity isometry. The *identity* is the isometry $i$ that leaves every point fixed, so that $i(X)$ always equals $X$. It can be thought of as a translation using the zero vector or as a rotation using an angle of $0°$. We can write $f \circ f^{-1} = i$.

The inverse of a reflection may be a surprise: It is the same reflection! In other words, any reflection composed with itself produces the identity. Now can you describe the inverse of a glide reflection?

The set of isometries in the Euclidean plane is an example of an important mathematical concept called a *group*.

**DEFINITION 6.5**    A *group* is a set with a binary operation that satisfies four properties: closure, associativity, identity, and inverses.

For the set of isometries, the operation is composition. This is a binary operation, because we combine two isometries at a time. Earlier, we started to verify closure when we showed that some compositions of two isometries produced another isometry. Rather than check every possible pairing, let us do a general proof.

**LEMMA 6.7**    The composition of any two isometries is an isometry. In other words, isometry is closed with respect to composition.

**Outline of a Proof**    If $f$ and $g$ are isometries, then they preserve distances and are one-to-one and onto. The composition $f \circ g$ is also one-to-one and onto. (You will have an opportunity to prove this in the exercises.) Because $|AB| = |g(A)g(B)|$ and $|f(g(A))f(g(B))| = |g(A)g(B)|$, we have $|AB| = |f(g(A))f(g(B))|$, showing that $f \circ g$ preserves distances.

Composition of functions is always associative. (What do we need to show to prove this?) The function $i$ that fixes all points is the identity, and we just explained what the inverse of each type of isometry will be. This gives us the following theorem.

**THEOREM 6.8**    The set of all isometries in the plane is a group.

Notice that the operation—composition—is not commutative. This means that it typically matters very much in what order you perform the isometries. For instance, the composition of two reflections in parallel lines will produce two different translations, depending on the order of the composition. Although some groups have a commutative operation, such as the group of integers with the addition operation, the group of isometries does not.

Suppose we focus our attention on the set of translations, which is a subset of the set of all isometries in the plane. The set of translations is also a group—the composition of two translations is a translation (closure), associativity still works, the identity is a translation by the zero vector, and the inverse of a translation is another translation. Thus, the set of translations forms a *subgroup* of the group of all isometries.

The set of rotations about a fixed center point is another subgroup of isometries. However, we cannot make a subgroup of reflections or of glide reflections. First of all, the identity isometry is not a reflection, because it is direct; thus, the identity property fails. Also, the closure property fails, because the composition of two opposite isometries will be direct and, hence, not a reflection.

Another way to form a subgroup of isometries is to restrict the domain and range to a particular geometric figure. For instance, suppose we use only the points of a given square. There are isometries that map points of a square to points of that same square. Certain rotations will use the points of the square as input and produce only that same set of points as output. Some reflections will also do this. However, no translations will, and no glide reflections will either. As we will see in more detail in Chapter 8, the small subset of isometries that map a square to itself forms a subgroup.

## USING ISOMETRIES IN PROOFS

Isometries can be useful as tools for proving various geometric facts. Recall that an isometry preserves congruence; therefore, finding an isometry between two objects is one way to verify the congruence of those objects. Sometimes a proof using isometries can provide a different insight than a more traditional proof can. For example, here is a transformational proof of a standard result—a proof that gives a visual sense of why the theorem is true.

**THEOREM 6.9**   **ASA Criterion for Congruent Triangles**   Suppose $\triangle ABC$ and $\triangle A'B'D$ satisfy $|AB| = |A'B'|$, $\angle ABC \cong \angle A'B'D$, and $\angle BAC \cong \angle B'A'D$. Then the two triangles are congruent.

**Proof**   Because $|AB| = |A'B'|$, there are exactly two isometries that map $A$ to $A'$ and $B$ to $B'$. Let $f_1$ be the isometry that sends $C$ to the point $C'$ on the side of $A'B'$ that is opposite $D$. (See Figure 6.6, next page.)

Now use an isometry, $r$, to reflect $\triangle A'B'C'$ across the line $A'B'$, and denote $r(C') = X$. Recall that isometries preserve congruence of angles. Thus,

$$\angle XA'B' \cong \angle C'A'B' \cong \angle CAB \cong \angle DA'B',$$

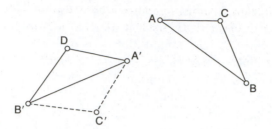

**FIGURE 6.6**
Proof of ASA

showing that point $X$ is on the line $DA'$. Similarly,

$$\angle XB'A' \cong \angle C'B'A' \cong \angle CBA \cong \angle DB'A',$$

showing that point $X$ is on the line $DB'$. Therefore, $X = D$. Because the composition $(r \circ f_1)$ is an isometry taking $\triangle ABC$ to $\triangle A'B'D$, the two triangles are congruent.

To complete this proof, we need to consider the other isometry, $f_2$, which sends $C$ to the point $C^*$ on the same side of $A'B'$ as $D$. Can you show that $C^* = D$? (You will have an opportunity to do this in the exercises.)

There are no simple rules about when to use isometries in a proof. Isometries are simply another way of thinking about geometric problems. These motions give us another tool that can be useful in understanding a particular situation.

## ISOMETRIES IN SPACE

The idea of isometry can be extended to three or even more dimensions by requiring the same properties we saw in two dimensions—a function that preserves distances and is one-to-one and onto. The distance between points $A$ and $B$ in $\Re^n$ can be calculated by using an extension of the Pythagorean Theorem:

$$\text{distance} = \sqrt{(a_1 - b_1)^2 + (a_2 - b_2)^2 + \cdots + (a_n - b_n)^2}.$$

With this formula, we can check whether a function preserves distances. For example, in $\Re^3$, define a function $f(x_1, x_2, x_3) = (x_2, x_3, x_1)$. Notice that this defines a function that takes a point in space, $(x_1, x_2, x_3)$, as input, and returns a point that has the same three coordinates in a different order. Will this be an isometry? It is not difficult to check the properties.

As you might expect, there are more possible types of isometry in 3-space than those we saw earlier. Translations, rotations, reflections, and glide reflections still occur in $\Re^3$, though they must be adapted to work with the additional dimension. However, there are also completely new types of isometry in three dimensions. We will not work with these new isometries; we merely list and briefly describe them here (Coxeter, 1969, 96–101).

**Translation**   Points are moved along a constant vector. This is a direct isometry without fixed points.

**Rotation**   Points are rotated around a given line. This is a direct isometry, and the axis of rotation is made up of fixed points.

**Reflection**   This isometry is like reflection in an ordinary mirror. The mirror is a plane that reflects a three-dimensional space. The points in the plane are fixed, and the isometry is opposite. (Orientation is a bit more complicated to define in $\Re^3$, but the general idea is the same.)

**Glide Reflection**   Points are reflected in a plane and then translated using a vector parallel to the plane. This is an opposite isometry with no fixed points.

**Screw**   This isometry is also called *twist*. It is a rotation followed by a translation along the axis of rotation. The name may make more sense if you imagine the rotation and translation happening simultaneously. This is a direct isometry, and though there are no fixed points, the axis line is transformed into itself.

**Rotary Reflection**   This is a reflection followed by a rotation around an axis perpendicular to the mirror plane. It is an opposite isometry, with a single fixed point where the axis intersects the mirror.

**Central Inversion**   Given a center point $O$, the image of a point $P$ is the point $P'$ on $\overrightarrow{PO}$ for which $O$ is the midpoint of the segment $PP'$. This isometry is opposite, and the center is the only fixed point. Central inversion is sometimes called *reflection in a point*.

Just as in $\Re^2$, isometries can be combined by composition. The identity isometry takes every point to itself, and each isometry has an inverse. Thus, the set of isometries in $\Re^3$ forms a group. There are many possible subgroups. Can you think of some examples?

## INVERSION IN A CIRCLE, REVISITED

Chapter 3 introduced the notion of inversion in a circle. We revisit the idea in this chapter because inversion is another example of a function that takes points in the plane as input and returns other points as output. In Activity 10, you were invited to explore the idea of inversion analytically, using coordinates and the methods of Chapter 4.

**DEFINITION 6.6**   The *inversion* of point $P = (x, y)$ with respect to the circle $x^2 + y^2 = 1$ is the point

$$P' = \left( \frac{x}{x^2 + y^2}, \frac{y}{x^2 + y^2} \right).$$

Notice that the circle of inversion is specified to be the standard unit circle around the origin. Any circle could be used, but the expressions for the function would have to be modified accordingly.

Does this analytic definition agree with the definition given in Chapter 3? We must verify that the points $P$ and $P'$ are collinear with the center $O$ of the inversion circle, and also that $|OP| \cdot |OP'| = r^2$. In this case, $O$ is the origin and $r = 1$. One way to verify collinearity is to calculate two slopes, $m_{OP}$ and $m_{OP'}$, and show that they are equal:

$$m_{OP} = \frac{y - 0}{x - 0} = \frac{y}{x}$$

$$m_{OP'} = \frac{\frac{y}{x^2+y^2} - 0}{\frac{x}{x^2+y^2} - 0} = \frac{y}{x}.$$

Calculating the product of the distances, we get

$$|OP| \cdot |OP'| = \sqrt{(x - 0)^2 + (y - 0)^2}$$
$$\cdot \sqrt{\left(\frac{x}{x^2 + y^2} - 0\right)^2 + \left(\frac{y}{x^2 + y^2} - 0\right)^2} = 1.$$

Notice that points on the inversion circle are fixed by this function, for if the coordinates of $P$ satisfy $x^2 + y^2 = 1$, then

$$\left(\frac{x}{x^2 + y^2}\right)^2 + \left(\frac{y}{x^2 + y^2}\right)^2 = 1.$$

Thus, the coordinates of $P'$ also satisfy this equation. As you worked on Activity 10, it probably appeared that $P' = P$. Can you show this analytically?

The function of inversion in a circle is not a transformation of the plane. There are two reasons for this. First, the origin is not part of the domain; there is no way to invert the point $(0, 0)$. We can say that the origin is inverted to a "point at infinity," which is a point that is not part of the plane. Second, the origin is not part of the range; no point of the plane inverts to the origin. In function terminology, this function is not defined on the entire plane, and it is not an onto function. It is true, however, that inversion is a one-to-one function, which can be proven analytically.

What happened in Activity 10c as you dragged line $\ell$ to various positions? When $\ell$ passed through the origin (the center of the inversion circle), its locus was $\ell$ itself. When $\ell$ did not pass through the origin, the locus was a circle that passed through the origin. This last fact was proved in Chapter 3. The reverse of this works as well: The inversion of a line gives a circle, and the inversion of a circle gives a line.

**THEOREM 6.10**  Given a circle $C$ with center $O$ and radius $r$, let $D$ be a circle that passes through $O$. The inversion of $D$ in $C$ will be a line that does not pass through point $O$.

**Proof**  We shall use the unit circle around the origin as $C$. A circle $D$ that passes through the origin must have the form $(x - a)^2 + (y - b)^2 = a^2 + b^2$.

How does the inversion affect this equation for the circle $D$? We have equations that convert a point $P = (x, y)$ to the new point $P'$:

$$x' = \frac{x}{x^2 + y^2} \quad \text{and} \quad y' = \frac{y}{x^2 + y^2}.$$

Solving this system of equations for $x$ and $y$ gives

$$x = \frac{x'}{(x')^2 + (y')^2} \quad \text{and} \quad y = \frac{y'}{(x')^2 + (y')^2}.$$

(Because inversion is its own inverse function, these equations should not be a surprise.) Using these to substitute into the circle $\mathcal{D}$ gives

$$\left( \frac{x'}{(x')^2 + (y')^2} - a \right)^2 + \left( \frac{y'}{(x')^2 + (y')^2} - b \right)^2 = a^2 + b^2.$$

After simplifying, we get

$$y' = \frac{1}{2b} - \frac{a}{b} x',$$

which is the equation of a straight line. Because the $y$-intercept, $\frac{1}{2b}$, cannot be 0, we know that this line does not pass through the origin.

In Activity 10d, you were asked if a circle $\mathcal{D}$ could equal its own locus under inversion. It turns out that this can occur, but only for certain special circles. Recall that two circles are said to be *orthogonal* if their tangents at their points of intersection are perpendicular.

**THEOREM 6.11**  If circle $\mathcal{D}$ is orthogonal to circle $\mathcal{C}$, then $\mathcal{D}$ is fixed by inversion in $\mathcal{C}$.

This theorem does *not* say that every point of $\mathcal{D}$ is fixed by the inversion; rather, it says that every point of $\mathcal{D}$ is mapped to another point of $\mathcal{D}$. There are only two fixed points—the two places where $\mathcal{C}$ and $\mathcal{D}$ intersect.

Theorem 6.11 is a corollary of a stronger fact that we will not prove here, namely, that angles are preserved by inversion. In other words, if two lines meet at an angle $\theta$, then their images under inversion will also meet at angle $\theta$. The images of these lines may be lines or circles, and in the latter case, we must think of tangent lines when measuring angle $\theta$. The circle of inversion, $\mathcal{C}$, is fixed, and the two intersection points are fixed; thus, the image of $\mathcal{D}$ must be the circle orthogonal to $\mathcal{C}$ at those two points. Circle $\mathcal{D}$ itself is the only option. For this reason, inversion is said to be a *conformal* function of the plane.

Inversion can be a surprisingly powerful proof tool. By inverting the points of the plane in a properly chosen circle, questions about circles become questions about lines, and vice versa. Theorems can be viewed in an entirely new light when their objects are inverted in this way.

As an example, let us use inversion to give another proof of Feuerbach's Theorem (first seen in Chapter 4). Here again is the statement of that theorem.

**THEOREM 6.12**  **Feuerbach's Theorem**  The nine-point circle of a triangle is tangent to the incircle and to all three excircles of the triangle.

The proof of this theorem is somewhat intricate, so here is a quick sketch of how we proceed: First, we consider the incircle of $\triangle ABC$ and only one of the excircles. Let us choose the excircle determined by the internal angle bisector at

$A$ and the external angle bisectors at $B$ and $C$. The midpoint $A'$ of the side $BC$ will be the center of an inversion circle with a carefully chosen radius. The incircle and the excircle are fixed by the inversion, and the nine-point circle of $\triangle ABC$ is inverted into a line. By showing what happens to the specific points $B'$ and $C'$ (the midpoints of the sides opposite angles $B$ and $C$, respectively), this line is shown to be a line tangent to both the incircle and this excircle. Therefore, the nine-point circle is tangent to the two circles. Because the same argument will work for the incircle and a different excircle, the tangency works for all three excircles (Coxeter and Greitzer, 1967).

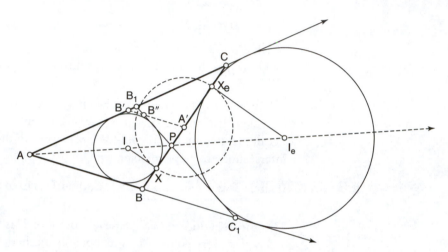

**FIGURE 6.7**
Proving Feuerbach's Theorem

**Proof of Theorem 6.12**    Figure 6.7 shows the situation we will consider: $\triangle ABC$ with its incircle and one excircle. Points $X$ and $X_e$ are where these circles are respectively tangent to side $BC$, and point $A'$ is the midpoint of $BC$. Find points $B_1$ on $AC$ and $C_1$ on $AB$ so that $AB_1$ is congruent to $AB$ and $AC_1$ is congruent to $AC$. Segment $B_1C_1$ is tangent to both circles, just as $BC$ is. (This segment can also be found by reflecting $BC$ in the angle bisector $AI$.)

Denote $a = |BC|$, $b = |CA|$, $c = |AB|$, and $s = (a + b + c)/2$. By Exercise 32, Chapter 3, we know that $|BX| = s - b$ and that $|CX_e| = s - b$. Then

$$|A'X| = \frac{1}{2}a - |BX| = \frac{b-c}{2} \quad \text{and} \quad |A'X_e| = \frac{1}{2}a - |CX_e| = \frac{b-c}{2}.$$

Construct a circle $\mathcal{C}$ centered at $A'$ with radius $|A'X|$. The incircle is orthogonal to $\mathcal{C}$ and so is the excircle; therefore, both of these circles are fixed by inversion in $\mathcal{C}$. The nine-point circle of $\triangle ABC$ passes through the center of $\mathcal{C}$, so it is inverted to a line—but which line? We will show that it is the line $B_1C_1$.

Let point $B'' = B_1C_1 \cap A'B'$. Point $B''$ turns out to be the inversion of $B'$, which is the midpoint of side $AC$. To see this, notice that $A'$, $B'$, and $B''$ are collinear by definition. Now we must check that the product $|A'B'| \cdot |A'B''| = |A'X|^2$.

First, note that $|A'B'| = \frac{c}{2}$ because $B'$ and $C'$ are the midpoints of their sides, and the segment $A'B'$ creates a triangle similar to $\triangle ABC$. Now define the point

$P = BC \cap B_1C_1$. The triangles $PA'B''$ and $PBC_1$ also will be similar. Thus,

$$\frac{|A'B''|}{|BC_1|} = \frac{|PA'|}{|PB|},$$

giving us

$$|A'B''| = \frac{|BC_1| \cdot |PA'|}{|PB|}.$$

Notice that $|BC_1| = |AC_1| - |AB| = b - c$ and that $|CP| + |PB| = a$. By Exercise 51, Chapter 2, we know that

$$\frac{|CP|}{|PB|} = \frac{b}{c}.$$

These last two facts imply that

$$|PB| = \frac{ac}{b+c}.$$

Further,

$$|PA'| = \frac{a}{2} - |PB| = \frac{a(b-c)}{2(b+c)}.$$

Therefore,

$$|A'B''| = \frac{(b-c) \cdot \frac{a(b-c)}{2(b+c)}}{\frac{ac}{b+c}} = \frac{(b-c)^2}{2c}.$$

The product $|A'B'| \cdot |A'B''|$ is thus

$$\frac{c}{2} \cdot \frac{(b-c)^2}{2c} = \left(\frac{b-c}{2}\right)^2 = |A'X|^2.$$

We have thus shown that $B''$ is the inversion of $B'$ with respect to this circle $\mathcal{C}$. Now define $C'' = B_1C_1 \cap A'C'$. Very similar calculations show that $C''$ is the inversion of $C'$ with respect to $\mathcal{C}$. (Recall that $C'$ is the midpoint of $AB$.) Therefore, the inversion of the nine-point circle with respect to $\mathcal{C}$ is the line $B''C'' = B_1C_1$. Because this line is tangent to both the incircle and this excircle, the nine-point circle is also tangent to these two circles.

The same argument works for either of the other two excircles, though each will use a different circle of inversion.

------------------------------------------

# 6.3 EXERCISES

Give clear and complete answers to the exercises, expressing your explanations in complete sentences. Include diagrams whenever appropriate.

1. Consider $\triangle ABC$ with vertices at $(3, 4)$, $(-1, 2)$, and $(2, -2)$. What happens to this triangle when it is put through each of the six functions in Activities 1 and 2? Either calculate the vertices of $\triangle A'B'C'$ or explain why they cannot be calculated.

2. Consider $\triangle RST$ with vertices at $(5, 0)$, $(0, 3)$, and the origin. What happens to this triangle when it is put through each of the six functions in Activities 1 and 2? Either calculate the vertices of $\triangle A'B'C'$ or explain why they cannot be calculated.

3. For each of the six functions in Activities 1 and 2, either prove that $f_n$ is an onto function or prove that it is not.

4. For each of the six functions in Activities 1 and 2, either prove that $f_n$ is a one-to-one function or prove that it is not.

5. For each of the six functions in Activities 1 and 2, prove or disprove that $f_n$ is an isometry.

6. Suppose that $f$ is an isometry. Prove that $\triangle ABC$ is congruent to $\triangle f(A)f(B)f(C)$.

7. Suppose that $f$ and $g$ are isometries.
   a. Prove that the composition $f \circ g$ is one-to-one.
   b. Prove that the composition $f \circ g$ is onto.

8. Suppose that $f$ and $g$ are isometries. Prove that the composition $f \circ g$ is also an isometry. (There are three things to prove.)

9. If a function $f : \Re^2 \longrightarrow \Re^2$ preserves distances, does $f$ have to be one-to-one? Does $f$ have to be onto? If you answer yes, give a proof; if you answer no, give a counterexample.

10. Prove that composition of isometries is associative. In other words, prove that
    $$(f \circ g) \circ h = f \circ (g \circ h).$$
    Did your proof address all of these concepts: function, onto, one-to-one, preserves distances?

11. Suppose that $f$ is an isometry.
    a. If $f$ is a translation, describe $f^{-1}$.
    b. If $f$ is a rotation, describe $f^{-1}$.
    c. If $f$ is a reflection, describe $f^{-1}$.
    d. Prove that $f^{-1}$ is also an isometry.

12. Suppose that $AB$ is a line segment.
    a. If $f$ is a translation, when will $f(AB)$ be parallel to $AB$? Justify your answer.
    b. If $f$ is a rotation, when will $f(AB)$ be parallel to $AB$? Justify your answer.
    c. If $f$ is a reflection, when will $f(AB)$ be parallel to $AB$? Justify your answer.

13. In Figure 6.1, one of the triangles is the original, and the other three are the result of a translation, a rotation, and a reflection. Write a paragraph explaining how you determined which triangle is which.

14. Investigate the following compositions of isometries. Decide what type of isometry is produced, and explain why you think so. Consider orientation and fixed points in your explanations.
    a. a reflection followed by a translation that is not parallel to the mirror
    b. a reflection followed by a rotation
    c. a rotation followed by a reflection
    d. a rotation followed by a translation

15. Write out a careful detailed proof of Theorem 6.1.

16. Give an example of two specific isometries to show that composition is not commutative. Give another example of two specific isometries that do commute.

17. A glide reflection can be thought of as a composition of a translation and a reflection. Is this pair of isometries commutative? In other words, does it matter whether you glide then reflect, or reflect then glide? Explain.

18. Describe the inverse of a glide reflection.

19. Central inversion through a point can also be done in the plane $\Re^2$. Draw a triangle and its image under this type of reflection. Is this an isometry? Is it direct or opposite? How else could you describe this isometry?

20. Consider $\triangle PQR$ with vertices at $(3, 5)$, $(-2, 1)$, and $(4, -2)$.
    a. Reflect this triangle through the origin, and calculate the vertices of $\triangle P'Q'R'$.
    b. Is this transformation an isometry? If so, which of the four types of isometry is it? If not, why not?

21. Consider $\triangle ABC$ with vertices at $(1, 5)$, $(3, 0)$, and $(-2, -2)$.
    a. Reflect this triangle through the origin, and calculate the vertices of $\triangle A'B'C'$.
    b. Is this transformation an isometry? If so, which of the four types of isometry is it? If not, why not?

22. Write a paragraph explaining why each part of Corollary 6.4 is true.

23. Sometimes a composition of two reflections has fixed points, and sometimes it doesn't. Write a paragraph in which you identify when it has fixed points, and explain why this happens.

24. Consider the set of rotations around one particular point. Prove that this set is a subgroup of the group of isometries in the plane.

25. Complete the transformational proof of Theorem 6.9.

26. Develop a transformational proof that the vertical angles formed by two intersecting lines are congruent.

27. Develop a transformational proof that the base angles of an isosceles triangle must be congruent.

28. Suppose that the diagonals of a quadrilateral are perpendicular bisectors of each other. Use isometries to prove that the quadrilateral must be a rhombus.

29. Let $P_1, \ldots, P_k$ be points on a circle with center $O$. Suppose that these points are evenly spaced in order around the circle. This means that the central angles

$$\angle P_1OP_2, \ \angle P_2OP_3, \ldots, \ \angle P_kOP_1$$

are all congruent. Prove that the polygon

$$P_1P_2 \ldots P_k$$

formed by these points is regular; that is, all sides of the polygon are congruent and all angles of the polygon are congruent. (It may be helpful to try some small examples first, such as $k = 3$ and $k = 4$.)

30. In Sketchpad, draw two points, $A$ and $B$, on the same side of a line, $\ell$, and let $X$ be a point on $\ell$. Locate $X$ so that the total distance $|AX| + |XB|$ is a minimum. Then find the reflection $B'$ of $B$ across $\ell$, and draw $AB'$.

    What conjecture can you make? Vary the points $A$ and $B$, and adjust the location of $X$. Does your conjecture still work? Prove your conjecture. (This is *Heron's Theorem.*)

31. Create a triangle, $\triangle ABC$, and measure its three angles. Translate the triangle by the vector $AB$. The two triangles should now be connected at point $B$. How do the three angles at point $B$ compare with the angles of $\triangle ABC$? Vary the triangle, and compare angles again.

    What is the sum of the three angles at $B$? What can you conclude about $\triangle ABC$? Prove your statement.

32. Let $H$ be the orthocenter of $\triangle ABC$. Show that when $H$ is reflected across the three sides of $\triangle ABC$, the three image points will lie on the circumcircle of the triangle (Yaglom, 1962, vol. I).

33. Suppose that $C$ is the unit circle around the origin. Let $(x - a)^2 + (y - b)^2 = c^2$ be any other circle $D$ not passing through the origin. Prove analytically that the inversion of the circle $D$ with respect to $C$ will still be a circle.

34. Construct a sketch of a triangle, its incircle and three excircles, and the nine-point circle. Confirm that the nine-point circle is tangent to the other four circles.

35. In the proof of the Feuerbach Theorem (Theorem 6.12), let $D$ be the point where the altitude from vertex $A$ meets the side $BC$. Show that the inversion of $D$ in circle $C$ is the point $P$ of Figure 6.7.

36. Prove that the set of integers under addition is a group. Is the set of integers under subtraction a group? Why or why not?

37. Consider points $A$ and $B$ in $\Re^3$. Let

$$A = (a_1, a_2, a_3) \quad \text{and} \quad B = (b_1, b_2, b_3).$$

Prove that

$$|AB| = \sqrt{(a_1 - b_1)^2 + (a_2 - b_2)^2 + (a_3 - b_3)^2}.$$

38. For each of the following functions in $\Re^3$, determine whether it is an isometry. If it is an isometry, prove it; if not, explain why not.

    a. $g_1(x, y, z) = (y, x, z)$
    b. $g_2(x, y, z) = (x, y, 3)$
    c. $g_3(x, y, z) = (2x, 2y, 2z)$
    d. $g_4(x, y, z) = (x, 2y, 3z)$
    e. $g_5(x, y, z) = (x + 2, y + 2, z + 2)$
    f. $g_6(x, y, z) = (x + y, 2y, z + y)$

39. Prove that the set of rotations about the origin is a group.

**40.** Prove or disprove that the set of rotations about the points $(1, -1)$ or $(-3, 3)$ is a group.

**41.** Prove or disprove that the set of translations in $\Re^3$ is a group.

**42.** Prove or disprove that the set of rotations about the $x$-axis in $\Re^3$ is a group.

**43.** Identify three different subgroups of the isometries in space that are listed on page 150–151.

**44.** In our discussion of the inversion of a point $P$ with respect to the standard unit circle $x^2 + y^2 = 1$, we observed that points of the unit circle get mapped to points of the unit circle, and we raised the question of whether $P'$ actually equals $P$ for each point $P$ on the standard unit circle. We know that points of the unit circle get mapped to points of the unit circle, but are these points each fixed points of the inversion? Resolve this question one way or the other.

**45.** Prove that inversion is a one-to-one function.

**46.** Work out the algebraic details in the proof of Theorem 6.10 to verify the results we claim in that proof.

Exercises 47 and 48 are especially for future teachers.

**47.** In the *Principles and Standards for School Mathematics,* the National Council of Teachers of Mathematics (NCTM) recommends that "Instructional programs from prekindergarten through grade 12 should enable all students to . . . apply transformations and use symmetry to analyze mathematical situations" (NCTM, 2000, 41). What does this mean for you and your future students?

  a. Study the Geometry Standard for one grade band (i.e., pre-K–2, 3–5, 6–8, or 9–12). What are the NCTM recommendations regarding transformational geometry?

  b. Find copies of school mathematics textbooks for these same grade levels. How are the NCTM standards implemented in those textbooks? Cite specific examples.

  c. Write a report in which you present and critique what you learn.

**48.** Design several classroom activities involving transformational geometry that would be appropriate for students in your future classroom. Write a short report explaining how the activities you design reflect both what you have learned in studying this chapter and the recommendations of the NCTM.

Reflect on what you have learned in this chapter.

**49.** Review the main ideas of this chapter. Describe, in your own words, the concepts you have studied and what you have learned about them. What are the important ideas? How do they fit together? Which concepts were easy for you? Which were hard?

**50.** Reflect on the learning environment for this course. Describe aspects of the learning environment that helped you understand the main ideas in this chapter. Which activities did you like? Dislike? Why?

## 6.4 CHAPTER OVERVIEW

In this chapter, we very likely stretched your understanding of *function* to include functions whose inputs and outputs are points in the coordinate plane. A function, $f$, whose domain and range are the set of points in the coordinate plane performs some kind of action on the plane; that is, $f$ transforms the plane.

Some transformations of the plane will stretch or shrink the plane. The function $f_1$ of Activity 1 pulled (or stretched) the plane away from the lines $y = \pm 1$. Some transformations of the plane project the entire plane onto a single line. The function $f_2$ in Activity 2 projected the entire plane onto the $y$-axis. Transformations of this type change the distances between points. Consequently, the distance between the original points $A$ and $B$ will not be the same as the distance between the images $f(A)$ and $f(B)$. Such transformations are interesting, but were not the primary focus of our study in this chapter.

In this chapter, we focused our attention on transformations of the plane that preserve distances and are one-to-one and onto. Transformations that meet all three of these criteria are called *isometries*. An isometry is classified according to whether it is direct or opposite and whether it has fixed points. *Rotations* and *translations* are direct isometries, while *reflections* and *glide reflections* are opposite isometries. A reflection has a whole line of fixed points, which forms the mirror line of the reflection. A rotation has a single fixed point, namely, the center of the rotation. Neither a translation nor a glide reflection has any fixed points. We have shown that every isometry of the plane is one of these four types.

Because an isometry is a distance-preserving function, it can be used as a powerful proof tool. If we can find an isometry between two figures, we know that they are congruent. We showed how to use isometries to develop another proof of the ASA criterion for triangle congruence.

An isometry is a function. Consequently, we can use composition to combine two (or more) isometries to construct a new isometry. We can also think about undoing the action of an isometry to get an inverse isometry. By considering a set of isometries and the operation of composition, we can ask whether the set is closed with respect to composition. We can check whether the identity isometry is in the set, and we can check whether the inverse of each isometry is also in the set. Function composition is associative. Therefore, if our set of isometries is closed, includes the identity, and includes the inverse of each isometry in the set, we have a *group* of isometries. This idea of a *group* is an important mathematical concept, one that is typically studied at greater depth in a course like Abstract Algebra.

In this chapter, we focused on isometries of the plane. All of these concepts can be extended to three-, four-, and even higher-dimensional space. We will leave an exploration of these ideas to another book in another course!

Another important function that uses points for inputs and outputs is *inversion with respect to a given circle*. Inversion sets up interesting relationships between lines and circles, fixing some and radically changing others. Under inversion, some lines get turned into circles, and some circles get turned into lines. As a proof tool, inversion gives a way to restate questions about circles as questions about lines. We used this strategy in proving Feuerbach's Theorem.

We have developed a number of theorems in this chapter. They are listed here again for your convenience.

**Theorem 6.1**    In the Euclidean plane, the images of three noncollinear points completely determine an isometry. In other words, if we know the outputs for three noncollinear points $A$, $B$, $C$, we can figure out what the isometry does to any point $X$.

**Theorem 6.2**    An isometry preserves collinearity. In other words, three points that are collinear will still be collinear after going through an isometry.

**Theorem 6.3**  An isometry preserves betweenness. In other words, if point $B$ is between points $A$ and $C$ along a line, then point $f(B)$ will be between points $f(A)$ and $f(C)$ along their line.

**Corollary 6.4**  Under an isometry, the image of

$$\left.\begin{array}{l} \text{a line segment} \\ \text{a triangle} \\ \text{an angle} \\ \text{a circle} \end{array}\right\} \text{ is a congruent } \left\{\begin{array}{l} \text{line segment,} \\ \text{triangle,} \\ \text{angle,} \\ \text{circle.} \end{array}\right.$$

**Theorem 6.5**  In the Euclidean plane, there are only four types of isometry: translations, rotations, reflections, and glide reflections.

**Lemma 6.6**  For four points $A$, $B$, $A'$, and $B'$, with $|AB| = |A'B'|$, there are exactly two isometries that give $f(A) = A'$ and $f(B) = B'$.

**Lemma 6.7**  The composition of any two isometries is an isometry. In other words, isometry is closed with respect to composition.

**Theorem 6.8**  The set of all isometries in the plane is a group.

**Theorem 6.9 ASA Criterion for Congruent Triangles**  Suppose $\triangle ABC$ and $\triangle A'B'D$ satisfy $|AB| = |A'B'|$, $\angle ABC \cong \angle A'B'D$, and $\angle BAC \cong \angle B'A'D$. Then the two triangles are congruent.

**Theorem 6.10**  Given a circle $\mathcal{C}$ with center $O$ and radius $r$, let $\mathcal{D}$ be a circle that passes through $O$. The inversion of $\mathcal{D}$ in $\mathcal{C}$ will be a line that does not pass through point $O$.

**Theorem 6.11**  If circle $\mathcal{D}$ is orthogonal to circle $\mathcal{C}$, then $\mathcal{D}$ is fixed by inversion in $\mathcal{C}$.

**Theorem 6.12 Feuerbach's Theorem**  The nine-point circle of a triangle is tangent to the incircle and to all three excircles of the triangle.

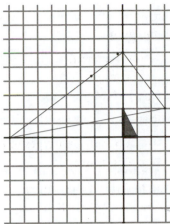

## CHAPTER SEVEN

T his chapter examines how to combine the ideas of coordinates and isometry, thus bringing together the analytic and transformational approaches to the geometry of the plane. To do this, we must first describe in algebraic expressions what an isometry does to the $x$-coordinate and $y$-coordinate of a point. We can do this using a system of equations. Matrix notation is a convenient way to represent systems of equations and thus to describe isometries. Facts about matrices provide additional insights into isometries.

# Isometries and Matrices

Do the following activities, writing your explanations clearly in complete sentences. Include diagrams whenever appropriate. Save your work for each activity, as later work sometimes builds on earlier work. You will find it helpful to read ahead into the chapter as you work on these activities.

1. In this chapter, we represent points $(x, y)$ as column vectors $\begin{bmatrix} x \\ y \end{bmatrix}$.

   a. In Sketchpad, create a coordinate system. Draw a vector from the origin to point $\begin{bmatrix} -2 \\ 3 \end{bmatrix}$.

      SKETCHPAD TIP   The **Appearance Tools.gsp** document contains tools for drawing arrows. You can find this document in **Sketchpad | Samples | Custom Tools**.

   b. Create an arbitrary point, $P$, and translate it by the vector of part a. Measure the coordinates of $P$ and of its image $P'$. Move $P$, and observe how these coordinates change.

   c. Complete the following equation to describe this situation:

   $$P' = \begin{bmatrix} x' \\ y' \end{bmatrix} = \begin{bmatrix} x \\ y \end{bmatrix} + \begin{bmatrix} ? \\ ? \end{bmatrix}.$$

2. In Sketchpad, create a coordinate system.

   a. Construct the standard unit circle. (Recall that this is the circle centered at the origin with radius 1.) Construct an arbitrary point $P$ on this circle.

   b. Select the origin. Select **Center** from the Transform menu to mark the origin as a center of rotation. Then select point $P$, and rotate it by 60° to get the new point $P'$.

   c. Measure the coordinates of both $P$ and $P'$. Calculate the values of $\sin(60°)$ and $\cos(60°)$. Drag $P$, and observe the coordinates of $P$ and $P'$. It will be particularly helpful to observe what happens at the locations $\begin{bmatrix} 1 \\ 0 \end{bmatrix}$, $\begin{bmatrix} 0 \\ 1 \end{bmatrix}$, $\begin{bmatrix} -1 \\ 0 \end{bmatrix}$, and $\begin{bmatrix} 0 \\ -1 \end{bmatrix}$.

   d. Describe the relationship between $P$ and $P'$ by a matrix equation in the following pattern:

   $$\begin{bmatrix} x' \\ y' \end{bmatrix} = \begin{bmatrix} a & b \\ c & d \end{bmatrix} \begin{bmatrix} x \\ y \end{bmatrix}.$$

   In other words, find values for $a$, $b$, $c$, and $d$ that describe this relationship.

   e. Select the point $P$ and the unit circle, then choose **Edit | Split Point From Circle**. Vary $P$, and observe $P'$ and their coordinates. Does your matrix equation from part d still describe this rotation correctly?

3. Describe in words the inverse of the rotation in Activity 2. Then write a matrix equation to describe this inverse.

4. In a new sketch, create a coordinate system and an arbitrary point, $P$.
   a. Mark the vertical axis as a mirror and reflect $P$ across this mirror.
   b. Measure the coordinates of $P$ and of $P'$. Drag $P$ and observe these coordinates.
   c. Write a matrix equation to describe this reflection.

5. Repeat Activity 4, using the horizontal axis as the mirror.

6. In a new sketch, create a coordinate system and an arbitrary point, $P$.
   a. Construct a line, $\ell$, through the origin $O$. Measure the line's angle of inclination.
   b. Reflect point $P$ across $\ell$ to get point $P_1$. Measure the coordinates of $P$ and of $P_1$. Drag $P$ and observe these coordinates. What relationship do you see between the coordinates of $P$ and those of $P_1$?

   SKETCHPAD TIP    To label point $P_1$, type P[1].

   c. Now reflect $P$ across the horizontal axis to get point $P_2$. Measure $\angle P_1OP_2$. It may be helpful to construct segments showing this angle. What do you observe? How is this angle related to the angle of inclination? Drag $P$. Do your observations still hold?
   d. Now vary line $\ell$ and drag $P$ again. Do your observations hold in this new situation? How could you explain this?

7. The general pattern of the matrix equations in these activities is

$$\begin{bmatrix} x' \\ y' \end{bmatrix} = \begin{bmatrix} a & b \\ c & d \end{bmatrix} \begin{bmatrix} x \\ y \end{bmatrix} + \begin{bmatrix} e \\ f \end{bmatrix}.$$

Suppose we know three specific input points and their images, as follows:

$$\begin{bmatrix} 0 \\ 0 \end{bmatrix} \longrightarrow \begin{bmatrix} -2.4 \\ 0.7 \end{bmatrix}, \qquad \begin{bmatrix} 2.5 \\ 0 \end{bmatrix} \longrightarrow \begin{bmatrix} 0 \\ 0 \end{bmatrix}, \qquad \begin{bmatrix} 3 \\ 1 \end{bmatrix} \longrightarrow \begin{bmatrix} 0.2 \\ -1.1 \end{bmatrix}.$$

   a. Use this information to find values for $a$, $b$, $c$, $d$, $e$, and $f$ in the matrix equation.
   b. In a new sketch, create a coordinate system and an arbitrary point, $P$. Measure the abscissa and the ordinate of point $P$.
   c. Using the $x$- and $y$-coordinates of $P$, calculate in Sketchpad the value of $x'$ according to your equation in part a. Also calculate $y'$.
   d. Select the $x'$ and $y'$ values, *in that order*, then choose **Graph | Plot as (x,y)**. Label this image as $P'$.
   e. Drag point $P$ to

$$\begin{bmatrix} 0 \\ 0 \end{bmatrix}, \qquad \begin{bmatrix} 2.5 \\ 0 \end{bmatrix}, \qquad \text{and} \qquad \begin{bmatrix} 3 \\ 1 \end{bmatrix}$$

   to verify that your calculations in part a are correct.
   f. Which of the four types of isometry is this? Explain why you think so.

8. In the activities so far, there have been many square matrices. Find the determinant of each matrix. What pattern do you see?

**9.** Following is the equation for a transformation that does not fit the pattern of Activities 1–8:

$$\begin{bmatrix} x' \\ y' \end{bmatrix} = \begin{bmatrix} 3 & -4 \\ -4 & -3 \end{bmatrix} \begin{bmatrix} x \\ y \end{bmatrix} + \begin{bmatrix} 0 \\ 6 \end{bmatrix}.$$

   a. In a new sketch, create a coordinate system and an arbitrary point, $P$. Measure the abscissa and the ordinate of $P$. Calculate the coordinates of the image $P'$ under this transformation, and plot the point. (This is similar to what you did in Activity 7.)

   b. Construct a triangle and its interior, using the following three points as the vertices:

$$A = \begin{bmatrix} 0 \\ 0 \end{bmatrix}, \qquad B = \begin{bmatrix} 1 \\ 0 \end{bmatrix}, \qquad C = \begin{bmatrix} 1 \\ 2 \end{bmatrix}.$$

   c. Select point $P$ and the interior of $\triangle ABC$. Choose **Edit | Merge Point to Triangle**.

   d. Select point $P'$ (only) and trace it. Then select point $P$ (only) and animate it. Describe what happens.

   e. Is this transformation an isometry? Is it direct or opposite?

   f. What is the ratio of the areas for the triangles? What is the determinant of the matrix?

## 7.2 DISCUSSION

In this chapter, we examine how to work with isometries analytically, that is, by using coordinates and algebra. In Chapter 4, you explored how to set up coordinates, and you learned (or reviewed) some basic facts about equations and their relation to geometric objects. Now we will explore how to use matrix equations to represent geometric functions (transformations). The algebra of vectors and matrices will allow us to better understand isometries and their properties. We will work only in $\Re^2$, although this same approach can be applied in any number of dimensions.

## USING VECTORS TO REPRESENT TRANSLATIONS

Recall what a translation does: Each input point is moved a constant distance in a constant direction to its new location. This idea of constant distance and constant direction can be nicely described by a vector. We could write the vector in polar coordinates, for polar coordinates are based on exactly these two quantities— distance and direction. However, because it is more common to work in Cartesian coordinates, we will use the Cartesian system in this chapter.

    Your Sketchpad diagram in Activity 1 shows both the beginning and ending points of the translation. The input point and its image point are always the same distance apart and are always in the same relative position to each other. The vector for this translation is $\begin{bmatrix} -2 \\ 3 \end{bmatrix}$. This vector says to start at the input point

$\begin{bmatrix} x \\ y \end{bmatrix}$, go left 2 units and then go up 3 units. Your diagram in Activity 1a should show this vector beginning at the origin. However, a vector can begin anywhere; it is the distance and direction of the vector that are critical, not its location. If the beginning point for the vector is $\begin{bmatrix} x \\ y \end{bmatrix}$, where is the ending point? The answer is the image point of the translation.

You have probably noticed that we are using the same notation for points as for vectors. We can interpret the notation $\begin{bmatrix} x \\ y \end{bmatrix}$ either as the vector or as the point $(x, y)$. This ambiguity is deliberate. Using the same notation for both meanings allows us to write expressions such as

$$\begin{bmatrix} x' \\ y' \end{bmatrix} = \begin{bmatrix} x \\ y \end{bmatrix} + \begin{bmatrix} -2 \\ 3 \end{bmatrix} = \begin{bmatrix} x - 2 \\ y + 3 \end{bmatrix}.$$

In this expression, $\begin{bmatrix} x \\ y \end{bmatrix}$ is a point and $\begin{bmatrix} -2 \\ 3 \end{bmatrix}$ is the translation vector. The combined expression describes how this translation works.

## USING MATRICES TO REPRESENT ROTATIONS

To perform a rotation, we must know two things: the point to use as the center of the rotation and the angle $\theta$ through which to turn. The angle $\theta$ can be positive or negative, small or huge; it can even exceed $360°$. The center of rotation can be any point in the plane. In Activity 2, you examined the situation with the rotation center at the origin. As you might expect, this situation is simpler than the general rotation.

Because rotations involve angles, it is convenient, for the moment, to work in polar coordinates. The starting point is $(r, \alpha)$, which says that the input is distance $r$ from the rotation center and at an inclination of $\alpha$ from the horizontal axis. The image point is the same distance, $r$, from the center but the angle has changed to $(\alpha + \theta)$. Thus, the image has polar coordinates $(r, \alpha + \theta)$. See Figure 7.1. How do

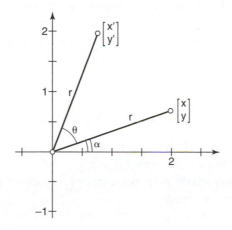

**FIGURE 7.1**
Rotation Around the Origin

we express this in rectangular coordinates? We can write the input point as

$$\begin{bmatrix} x \\ y \end{bmatrix} = \begin{bmatrix} r\cos(\alpha) \\ r\sin(\alpha) \end{bmatrix}.$$

You should examine Figure 7.1 and convince yourself that this is correct. Similarly, we can write the image point as

$$\begin{bmatrix} x' \\ y' \end{bmatrix} = \begin{bmatrix} r\cos(\alpha + \theta) \\ r\sin(\alpha + \theta) \end{bmatrix}.$$

Our challenge is to figure out how the right side of this equation relates to the input point $\begin{bmatrix} x \\ y \end{bmatrix}$. By using an identity from trigonometry, we can rewrite the upper coordinate as

$$\begin{aligned} r\cos(\alpha + \theta) &= r(\cos(\alpha)\cos(\theta) - \sin(\alpha)\sin(\theta)) \\ &= \cos(\theta)r\cos(\alpha) - \sin(\theta)r\sin(\alpha) \\ &= \cos(\theta)x - \sin(\theta)y. \end{aligned}$$

Because the angle of rotation $\theta$ is a constant, the factors $\cos(\theta)$ and $-\sin(\theta)$ are also constants. These are the values for the first row of the matrix in Activity 2d. A similar calculation with $r\sin(\alpha + \theta)$ will produce the values for the second row of the matrix.

Once you have completed the matrix for the rotation through angle $\theta$, finding the matrix for the inverse rotation is easy. There are (at least) two approaches to solving this problem: We can use trigonometry, or we can use linear algebra. The inverse rotation uses the same center, namely, the origin, but the direction of rotation is reversed; that is, the angle of rotation is $-\theta$. The trigonometric approach to finding the matrix for rotation in the opposite direction is simply to replace $\theta$ by $-\theta$ everywhere in the matrix and use even/odd identities to simplify the results. The linear algebra approach is to find the inverse matrix.

Of course, any point can be the center of a rotation. We could find the more general matrix pattern by setting the point $(h, k)$ as the center and doing similar work with trigonometric identities; however, it is a messy calculation. An alternative approach is to think of this rotation in three steps: Translate the rotation center $(h, k)$ to the origin, perform a rotation there, then translate the origin back to $(h, k)$. The composition of these three steps will perform the rotation around $(h, k)$, and the composition of the three matrix functions will calculate this rotation. We will examine this in more detail when we discuss composition of isometries later in this chapter.

## USING MATRICES TO REPRESENT REFLECTIONS

The action of a reflection uses a line. Points are reflected across this mirror line to their counterparts on the other side in a very specific way. Recall from Chapter 6 that the mirror is the perpendicular bisector of the line segment between any point and its reflected image.

Let us use this idea to examine what is going on in Activity 4. Suppose we have a point $\begin{bmatrix} x \\ y \end{bmatrix}$. To begin with, let us suppose that $x > 0$. A positive value for $x$ puts the point in the right side of the plane. With the $y$-axis as the mirror, the segment between this point and its image must be horizontal. Furthermore, the image is the same distance to the left of the $y$-axis as the original point is to the right. In coordinates, this means that the image has the same $y$-coordinate as the input point and that its $x$-coordinate has the same magnitude but the opposite sign. In vector notation,

$$\begin{bmatrix} x' \\ y' \end{bmatrix} = \begin{bmatrix} -x \\ y \end{bmatrix}.$$

Does this vector equation still work for points of the form $\begin{bmatrix} 0 \\ y \end{bmatrix}$, that is, for points that lie on the reflection line? Does the equation still work for points that have $x < 0$?

We can write this vector equation as a matrix equation, similar to the equations you wrote for rotations. We need to have

$$\begin{bmatrix} -x \\ y \end{bmatrix} = \begin{bmatrix} a & b \\ c & d \end{bmatrix} \begin{bmatrix} x \\ y \end{bmatrix}.$$

By multiplying out the right side and comparing coordinates, it is easy to see that reflection across the vertical axis is represented by the matrix equation

$$\begin{bmatrix} x' \\ y' \end{bmatrix} = \begin{bmatrix} -1 & 0 \\ 0 & 1 \end{bmatrix} \begin{bmatrix} x \\ y \end{bmatrix}.$$

In Activity 5, you examined reflections across the $x$-axis. The pattern is similar to that in Activity 4, but here the segment between point $P$ and its image $P'$ must be vertical, and the distance from $P$ to the horizontal axis is the same as the distance from $P'$ to the horizontal axis. It is not difficult to write this reflection as a matrix equation.

Reflections in other lines are more complicated to represent. In Activity 6, you examined the situation where the mirror line $\ell$ passes through the origin. The point $P_1$ is the reflection of point $P$ across $\ell$. What did you observe about the relationship between the coordinates of $P$ and those of $P_1$? It can be difficult to describe this relationship—unless you were very lucky in how you picked line $\ell$. To find a matrix description of this relationship, it is helpful to consider an alternative way of performing this reflection. You should be able to prove that the reflection across any line $\ell$ that passes through the origin can be broken into two steps: first, reflect across the horizontal axis, then rotate around the origin. (Examine your sketch for Activity 6 and look for congruent triangles.) Activity 6c suggests that $\angle P_1 O P_2$ is twice the angle of inclination for $\ell$. Your proof should show that the angle of rotation will be twice the angle of inclination for the mirror line (Wallace and West, 1992, 229).

Now we can think of the reflection across $\ell$ as two steps. We know how to represent each step as a matrix equation. The first step, the reflection across the horizontal axis, is

$$\begin{bmatrix} x' \\ y' \end{bmatrix} = \begin{bmatrix} 1 & 0 \\ 0 & -1 \end{bmatrix} \begin{bmatrix} x \\ y \end{bmatrix}.$$

The second step, rotating around the origin through an angle of $2\alpha$, is

$$\begin{bmatrix} x'' \\ y'' \end{bmatrix} = \begin{bmatrix} \cos(2\alpha) & -\sin(2\alpha) \\ \sin(2\alpha) & \cos(2\alpha) \end{bmatrix} \begin{bmatrix} x' \\ y' \end{bmatrix}.$$

The challenge is how to combine these steps into a single matrix. We can accomplish this combination by a substitution—replace the intermediate point $\begin{bmatrix} x' \\ y' \end{bmatrix}$ in the second equation by $\begin{bmatrix} 1 & 0 \\ 0 & -1 \end{bmatrix} \begin{bmatrix} x \\ y \end{bmatrix}$. This creates the following somewhat messy expression:

$$\begin{bmatrix} x'' \\ y'' \end{bmatrix} = \begin{bmatrix} \cos(2\alpha) & -\sin(2\alpha) \\ \sin(2\alpha) & \cos(2\alpha) \end{bmatrix} \left( \begin{bmatrix} 1 & 0 \\ 0 & -1 \end{bmatrix} \begin{bmatrix} x \\ y \end{bmatrix} \right)$$

By carrying out the matrix multiplication, we can simplify this expression. Bear in mind that multiplication of matrices is associative.

## COMPOSITION OF ISOMETRIES

Isometries are functions that use points as input and output. Combining isometries, then, means combining functions. There are many ways to combine functions, and you have seen many combinations of functions in other courses. However, some of the standard ways to combine functions do not always produce an isometry. For instance, the sum of two nonzero translations is no longer an isometry. (Can you verify this?) In Chapter 6, we saw that the composition of two isometries will always produce another isometry. Remember that in composing isometries, we must perform the isometries in sequence, with the output of the first being used as the input for the next. The usual function notation for composition of functions underscores this fact:

$$(f \circ g)(P) = f(g(P)).$$

The right side of this equation tells us how to proceed: The input point $P$ is put into the function $g$, then the result of that process is used as input for the next function $f$.

The discussion of Activity 6 dealt with a two-step composition, the goal of which was to find a single matrix that represents reflection across the line $y = mx$. This reflection was broken into two steps: reflection across the $x$-axis followed by rotation around the origin. Each step was represented by a matrix. We calculated the first step, reflection across the $x$-axis, by multiplying a matrix times a vector:

$$\begin{bmatrix} 1 & 0 \\ 0 & -1 \end{bmatrix} \begin{bmatrix} x \\ y \end{bmatrix}.$$

This product then replaced the input for the second multiplication, which is the rotation. Doing the substitution gave us the rather complicated expression shown earlier. Fortunately, multiplication of matrices and vectors is associative. Thus, the expression equals

$$\begin{bmatrix} x' \\ y' \end{bmatrix} = \left( \begin{bmatrix} \cos(2\alpha) & -\sin(2\alpha) \\ \sin(2\alpha) & \cos(2\alpha) \end{bmatrix} \begin{bmatrix} 1 & 0 \\ 0 & -1 \end{bmatrix} \right) \begin{bmatrix} x \\ y \end{bmatrix}.$$

By multiplying the two matrices, we get a much simpler expression for reflection across $y = mx$.

The important idea here is that we can calculate compositions by multiplying their matrices. Therefore, knowing the matrix patterns for some basic isometries can help us create the matrices for much more complicated motions.

You may recall from linear algebra that multiplication of matrices is not commutative, except in a few special circumstances. This tells us that composition of isometries is not commutative in general. We already know that from Chapter 6, and the implementation of isometries using matrices confirms this fact in another way. So, it is important to write the matrices in the correct order—right to left, in the way compositions are always written.

Here is a specific example of a composition: Suppose we want to perform the reflection across the $x$-axis followed by the reflection across the line $y = x$. What is the resulting composition? First, we need to write the matrix for each step:

$$\text{reflection across the } x\text{-axis: } \begin{bmatrix} 1 & 0 \\ 0 & -1 \end{bmatrix}$$

$$\text{reflection across } y = x: \begin{bmatrix} 0 & 1 \\ 1 & 0 \end{bmatrix}$$

The matrix for the composition is the product of these two matrices:

$$\begin{bmatrix} 0 & 1 \\ 1 & 0 \end{bmatrix} \begin{bmatrix} 1 & 0 \\ 0 & -1 \end{bmatrix} = \begin{bmatrix} 0 & -1 \\ 1 & 0 \end{bmatrix}.$$

Notice the order of the multiplication—the matrix for the first step is on the right.

Look carefully at the result. The upper-left and lower-right entries are the same, and the lower-left and upper-right are negatives of each other. This is like the matrix pattern for a rotation around the origin, discussed earlier. With an angle of 90°, we get precisely this matrix. (You should compare this with the product in the reverse order. The reversed product will still be a rotation around the origin, but not the same one.)

We can also use matrix algebra to find equations for compositions using translations. Suppose we want an equation that describes a rotation around a point other than the origin, such as around the point $(h, k)$. We can do this in three steps: Translate $(h, k)$ to $(0, 0)$, do the rotation around the origin, then translate the origin back to $(h, k)$. Here is an expression for this process:

$$\begin{bmatrix} x' \\ y' \end{bmatrix} = \left( \begin{bmatrix} \cos(\theta) & -\sin(\theta) \\ \sin(\theta) & \cos(\theta) \end{bmatrix} \left( \begin{bmatrix} x \\ y \end{bmatrix} - \begin{bmatrix} h \\ k \end{bmatrix} \right) \right) + \begin{bmatrix} h \\ k \end{bmatrix}.$$

Do you see how each step is represented in this expression? Matrix multiplication is distributive; thus, with some work, we can simplify this expression.

We have not discussed glide reflections yet, precisely because a glide reflection is defined as a composition, specifically as a reflection followed by a translation. You should have some idea how this works by now:

image = reflection matrix · input + translation vector

Suppose we want to do a glide reflection using the diagonal line $y = x$ and a translation of 2 units in the northeast direction. You should convince yourself that this is

$$\begin{bmatrix} x' \\ y' \end{bmatrix} = \begin{bmatrix} 0 & 1 \\ 1 & 0 \end{bmatrix} \begin{bmatrix} x \\ y \end{bmatrix} + \begin{bmatrix} \sqrt{2} \\ \sqrt{2} \end{bmatrix}.$$

## THE GENERAL FORM OF A MATRIX REPRESENTATION

Activity 7 gives three inputs and their images under some isometry. Because the three input points are not collinear, Theorem 6.1 tells us that this is enough information to completely determine the isometry. In other words, knowing what happens to these three points is enough information to figure out what the isometry does to any other point. Similarly, the coordinates of these three points and their images are enough information to figure out the matrix equation for the isometry. It just takes a little algebra.

Even though the equation given in Activity 7 contains six parameters—the values $a$, $b$, $c$, $d$, $e$, and $f$—you should find that only four numbers are actually involved: $-2.4$, $0.7$, $0.96$, and $0.28$ (with a few negative signs tossed in). In particular, the matrix has an important pattern. The upper-left and lower-right entries are the same value (except for perhaps a negative sign), and the lower-left and upper-right entries are the same (again, except for perhaps a negative sign). Look back at the other matrices from Activities 1–6; all of them should have this same pattern.

Another important fact comes from the determinant of the isometry matrix. In Activities 2–7, there were five square matrices. Every determinant was either $+1$ or $-1$. If we write the translation of Activity 1 as a matrix equation, it is

$$\begin{bmatrix} x' \\ y' \end{bmatrix} = \begin{bmatrix} 1 & 0 \\ 0 & 1 \end{bmatrix} \begin{bmatrix} x \\ y \end{bmatrix} + \begin{bmatrix} -2 \\ 3 \end{bmatrix}.$$

Here, the matrix is the identity matrix, for the translation is done by adding the vector. The determinant is again $+1$.

But there is more; think about which type of isometry produced each value:

$$\text{translation} \longrightarrow \text{determinant} = 1,$$
$$\text{rotation} \longrightarrow \text{determinant} = 1,$$
$$\text{reflection} \longrightarrow \text{determinant} = -1,$$
$$\text{glide reflection} \longrightarrow \text{determinant} = -1.$$

(The isometry in Activity 7 was a glide reflection.) The direct isometries have a determinant of $+1$, while the opposite isometries have a determinant of $-1$. To prove this, we will use a tool from multivariable calculus.

**THEOREM 7.1**   Suppose an isometry is represented by the equation

$$\begin{bmatrix} x' \\ y' \end{bmatrix} = \begin{bmatrix} a & b \\ c & d \end{bmatrix} \begin{bmatrix} x \\ y \end{bmatrix} + \begin{bmatrix} e \\ f \end{bmatrix}.$$

Then the determinant of this matrix, $ad - bc$, equals $+1$ if the isometry is direct and $-1$ if the isometry is opposite.

**Proof**   First, recall the *cross product* and some basic facts about it.

- Let $\vec{u}$ and $\vec{v}$ be vectors in three-dimensional space $\mathfrak{R}^3$. Then the three vectors $\vec{u}$, $\vec{v}$, and $\vec{u} \times \vec{v}$, in that order, form a right-handed triple.
- If $A$, $B$, and $C$ are three points, then the area of $\triangle ABC = \frac{1}{2}|\overrightarrow{AB} \times \overrightarrow{AC}|$.
- $\vec{v} \times \vec{u} = -(\vec{u} \times \vec{v})$.

Next, consider what this isometry does to three special points:

$$A = \begin{bmatrix} 0 \\ 0 \end{bmatrix} \longrightarrow \begin{bmatrix} e \\ f \end{bmatrix} = A'$$

$$B = \begin{bmatrix} 1 \\ 0 \end{bmatrix} \longrightarrow \begin{bmatrix} a + e \\ c + f \end{bmatrix} = B'$$

$$C = \begin{bmatrix} 0 \\ 1 \end{bmatrix} \longrightarrow \begin{bmatrix} b + e \\ d + f \end{bmatrix} = C'.$$

Because the cross product can only be used in three-dimensional space, think of points $A'$, $B'$, and $C'$ as lying in the plane $z = 0$. Thus, their full coordinates are

$$A' = \begin{bmatrix} e \\ f \\ 0 \end{bmatrix}, \qquad B' = \begin{bmatrix} a + e \\ c + f \\ 0 \end{bmatrix}, \qquad C' = \begin{bmatrix} b + e \\ d + f \\ 0 \end{bmatrix}.$$

We are interested in the vectors $A'B'$ and $A'C'$, which are

$$\begin{bmatrix} a \\ c \\ 0 \end{bmatrix} \quad \text{and} \quad \begin{bmatrix} b \\ d \\ 0 \end{bmatrix},$$

respectively. The cross product of these two vectors is

$$\overrightarrow{A'B'} \times \overrightarrow{A'C'} = \begin{vmatrix} \vec{i} & \vec{j} & \vec{k} \\ a & c & 0 \\ b & d & 0 \end{vmatrix}$$

$$= (ad - bc)\vec{k}.$$

Thus, this cross product will point either upward or downward, depending on the sign of $ad - bc$.

The area of $\triangle A'B'C'$ is $\frac{1}{2}|\overrightarrow{A'B'} \times \overrightarrow{A'C'}| = \frac{1}{2}(ad - bc)$. Because we are working with an isometry, this area equals the area of $\triangle ABC$, which is $\frac{1}{2}$. Thus, $|ad - bc| = 1$.

How does this computation tell us if the orientation is direct or opposite?

Notice that points $A$, $B$, and $C$ were labeled in counterclockwise order around their triangle. Thus, the angle from $\overrightarrow{AB}$ to $\overrightarrow{AC}$ is positive. Let us consider what happens with the three vectors $\overrightarrow{A'B'}$, $\overrightarrow{A'C'}$, and $\overrightarrow{A'B'} \times \overrightarrow{A'C'}$. If the angle from $\overrightarrow{A'B'}$ to $\overrightarrow{A'C'}$ is positive, then this is a right-handed triple in three dimensions, and $\overrightarrow{A'B'} \times \overrightarrow{A'C'}$ points upward. Thus, $ad - bc > 0$, making the determinant equal to $+1$. The orientation of $\triangle A'B'C'$ is, therefore, the same as the orientation of $\triangle ABC$, confirming that the isometry is direct.

On the other hand, if the angle from $\overrightarrow{A'B'}$ to $\overrightarrow{A'C'}$ is negative, then the right-handed triple will be $\overrightarrow{A'C'}$, $\overrightarrow{A'B'}$, and $\overrightarrow{A'C'} \times \overrightarrow{A'B'}$, with the cross product $\overrightarrow{A'C'} \times \overrightarrow{A'B'}$ pointing upward. Therefore, $\overrightarrow{A'B'} \times \overrightarrow{A'C'}$ points downward. This means $ad - bc < 0$; so, the determinant equals $-1$, confirming that the orientation of $\triangle A'B'C'$ is opposite that of $\triangle ABC$.

Here are the two possible patterns for the equation of an isometry:

$$\begin{bmatrix} x' \\ y' \end{bmatrix} = \begin{bmatrix} a & -b \\ b & a \end{bmatrix} \begin{bmatrix} x \\ y \end{bmatrix} + \begin{bmatrix} e \\ f \end{bmatrix}$$

for direct isometries and

$$\begin{bmatrix} x' \\ y' \end{bmatrix} = \begin{bmatrix} a & -b \\ -b & -a \end{bmatrix} \begin{bmatrix} x \\ y \end{bmatrix} + \begin{bmatrix} e \\ f \end{bmatrix}$$

for opposite isometries. In either situation, we must have $a^2 + b^2 = 1$, so that distances are preserved by the transformation. You should also verify that the sign of each determinant matches its orientation.

## USING MATRICES IN PROOFS

How can we be certain that the patterns just mentioned actually represent isometries? The isometries in the activities seem to fit these equations, but will every such matrix equation represent an isometry? Recall the definition of isometry from Chapter 6.

**DEFINITION 7.1**    A function $f$ on the plane is an *isometry* if $f$ is one-to-one, $f$ is onto, and $f$ preserves distances.

Using techniques from linear algebra, you should be able to prove that matrix functions such as those in our patterns are indeed one-to-one and onto. This is what we mean by calling them *transformations* of the plane. However, we still have to resolve the question of preserving distances. Let us start by looking at the first pattern, the one for direct isometries.

For two points $\begin{bmatrix} x_1 \\ y_1 \end{bmatrix}$ and $\begin{bmatrix} x_2 \\ y_2 \end{bmatrix}$, the distance between them is, of course,

$$\sqrt{(x_1 - x_2)^2 + (y_1 - y_2)^2}\,.$$

We need to calculate the image of each point and compute the distance between their image points. The first point is transformed by the matrix equation in this way:

$$\begin{bmatrix} a & -b \\ b & a \end{bmatrix} \begin{bmatrix} x_1 \\ y_1 \end{bmatrix} + \begin{bmatrix} e \\ f \end{bmatrix} = \begin{bmatrix} ax_1 - by_1 + e \\ bx_1 + ay_1 + f \end{bmatrix}.$$

Similarly, the image of the second point is $\begin{bmatrix} ax_2 - by_2 + e \\ bx_2 + ay_2 + f \end{bmatrix}$.

Next, we calculate the distance between these images:

$$\sqrt{((ax_1 - by_1 + e) - (ax_2 - by_2 + e))^2 + ((bx_1 + ay_1 + f) - (bx_2 + ay_2 + f))^2}.$$

There is more than one way to do the algebra; we will leave the details for the exercises. It is important that $a^2 + b^2 = 1$. Eventually, this simplifies to the same value for distance as calculated earlier.

In Chapter 6, we found that one of the characteristics of an isometry is whether it has any fixed points. We observed that a rotation has a fixed point at the center of rotation, and a reflection has fixed points on its mirror line. Translations and glide reflections have no fixed points. With matrix equations, we can now prove these statements.

Consider a nonzero rotation around the origin. The matrix equation of this transformation is

$$\begin{bmatrix} x' \\ y' \end{bmatrix} = \begin{bmatrix} \cos(\theta) & -\sin(\theta) \\ \sin(\theta) & \cos(\theta) \end{bmatrix} \begin{bmatrix} x \\ y \end{bmatrix}.$$

For a fixed point, we have $\begin{bmatrix} x' \\ y' \end{bmatrix} = \begin{bmatrix} x \\ y \end{bmatrix}$. So, we need to solve

$$\begin{bmatrix} x \\ y \end{bmatrix} = \begin{bmatrix} \cos(\theta) & -\sin(\theta) \\ \sin(\theta) & \cos(\theta) \end{bmatrix} \begin{bmatrix} x \\ y \end{bmatrix}.$$

This equation is equivalent to the following system, which we must solve for the variable $\theta$:

$$\begin{cases} x = x \cos(\theta) - y \sin(\theta) \\ y = x \sin(\theta) + y \cos(\theta) \end{cases}.$$

Solving the first equation for $y$ and substituting into the second equation leads us to the statement $(2 - 2\cos(\theta))x = 0$. If the first factor, $(2 - 2\cos(\theta))$, equals 0, then $\theta = 0$, and we have the identity isometry for which every point is fixed. If $\theta \neq 0$, however, $x = 0$, and the equations say that $y$ must be 0 also. Hence, the origin, the center of the rotation, is the only fixed point. Other questions about fixed points will appear in the exercises.

There are many propositions about composition of isometries that we can prove by matrix calculations. For instance, what sort of isometry is produced by a composition of two reflections? This depends on whether the mirror lines intersect; the answers can be proved by matrix calculations. With some facts from linear algebra, we can prove a much more general theorem about composition.

**THEOREM 7.2** The composition of two direct isometries is direct. The composition of two opposite isometries is also direct. However, the composition of a direct and an opposite isometry is opposite.

**Proof** Suppose two isometries are represented by the expressions

$$\vec{x}' = M_1 \vec{x} + \vec{v}_1 \quad \text{and} \quad \vec{x}' = M_2 \vec{x} + \vec{v}_2$$

where $M_1$ and $M_2$ are the matrices, and $\vec{v}_1$ and $\vec{v}_2$ are the vectors. The composition of these isometries is

$$\vec{x}' = M_2(M_1 \vec{x} + \vec{v}_1) + \vec{v}_2$$
$$= M_2 M_1 \vec{x} + (M_2 \vec{v}_1 + \vec{v}_2).$$

The quantity $(M_2\vec{v}_1 + \vec{v}_2)$ represents the translation component of the composition. The direct/opposite decision comes solely from the product $M_2 M_1$. Because $det(M_2 M_1) = det(M_2)det(M_1)$, the theorem follows by comparing $+$ and $-$ signs.

--------------------------------------------

## SIMILARITY TRANSFORMATIONS

It should be obvious from your diagram in Activity 9 that the transformation is *not* an isometry. It is a legitimate transformation, for it is both one-to-one and onto; however, distances are not preserved. For example, the distance from $A = \begin{bmatrix} 0 \\ 0 \end{bmatrix}$ to $B = \begin{bmatrix} 1 \\ 0 \end{bmatrix}$ is 1, while the distance between their image points, $A' = \begin{bmatrix} 0 \\ 6 \end{bmatrix}$ and $B' = \begin{bmatrix} 3 \\ 2 \end{bmatrix}$, is 5. When we compare other distances in Activity 9, we see this number 5 again: $BC = 2$ and $B'C' = 10$; $CA = \sqrt{5}$ and $C'A' = 5\sqrt{5}$. The transformation increases the length of each of these line segments by a factor of 5.

As you no doubt saw in your diagram, $\triangle ABC$ is similar to $\triangle A'B'C'$. The image is the same shape as $\triangle ABC$ but is a different size and in a different position. The matrix function of Activity 9 shows the precise relation between the two triangles. This function, which is called a *similarity*, behaves much like an isometry. There are direct similarities and opposite similarities. Some similarities have fixed points, while others do not. The important difference between isometries and similarities is that similarities have a factor that gives the size change.

The pattern for a matrix equation of a similarity is very much like that of an isometry:

$$\begin{bmatrix} x' \\ y' \end{bmatrix} = \begin{bmatrix} a & -b \\ b & a \end{bmatrix} \begin{bmatrix} x \\ y \end{bmatrix} + \begin{bmatrix} e \\ f \end{bmatrix}$$

for direct similarities and

$$\begin{bmatrix} x' \\ y' \end{bmatrix} = \begin{bmatrix} a & -b \\ -b & -a \end{bmatrix} \begin{bmatrix} x \\ y \end{bmatrix} + \begin{bmatrix} e \\ f \end{bmatrix}$$

for opposite similarities. What is different from isometries is that in a similarity, $a^2 + b^2$ does not have to equal 1.

The equation in Activity 9 fits the pattern for an opposite similarity. We saw that 5 is an important number for this similarity. Suppose we separate a 5 from the matrix, as follows:

$$\begin{bmatrix} x' \\ y' \end{bmatrix} = 5 \begin{bmatrix} \frac{3}{5} & -\frac{4}{5} \\ -\frac{4}{5} & -\frac{3}{5} \end{bmatrix} \begin{bmatrix} x \\ y \end{bmatrix} + \begin{bmatrix} 0 \\ 6 \end{bmatrix}.$$

This gives us another way to write the similarity:

$$\begin{bmatrix} x' \\ y' \end{bmatrix} = \begin{bmatrix} \frac{3}{5} & -\frac{4}{5} \\ -\frac{4}{5} & -\frac{3}{5} \end{bmatrix} \left( 5 \begin{bmatrix} x \\ y \end{bmatrix} \right) + \begin{bmatrix} 0 \\ 6 \end{bmatrix}.$$

Written this way, the expression shows the two steps of the similarity. The new matrix is an isometry matrix, with its determinant equal to 1. The factor of 5 is the size change. We can understand this similarity as first increasing the sides of $\triangle ABC$ to five times their original size, then performing a glide reflection (that is, a reflection and a translation).

Here is another interesting observation about this similarity matrix: its determinant is $-25 = -5^2$. The negative sign, as before, tells us that this is an opposite transformation, reversing the orientation. The reappearance of 5 is not a coincidence; the size factor affects both dimensions of the plane, and thus it causes area to change by a factor of 25. The factor of 5 affects both rows of the matrix, so it multiplies the determinant twice and causes a change by a factor of 25. (The linear algebra is more technical than that, but this is the general idea. See Lay, 1997, 201–204 for the details.) For any similarity in the plane, the determinant of its matrix tells the square of the size change.

## 7.3 EXERCISES

Give clear and complete answers to the exercises, expressing your explanations in complete sentences. Include diagrams whenever appropriate.

1. Every isometry for the Euclidean plane can be written in one of the following two patterns:
$$\begin{bmatrix} x' \\ y' \end{bmatrix} = \begin{bmatrix} a & -b \\ b & a \end{bmatrix} \begin{bmatrix} x \\ y \end{bmatrix} + \begin{bmatrix} c \\ d \end{bmatrix}$$
or
$$\begin{bmatrix} x' \\ y' \end{bmatrix} = \begin{bmatrix} a & -b \\ -b & -a \end{bmatrix} \begin{bmatrix} x \\ y \end{bmatrix} + \begin{bmatrix} c \\ d \end{bmatrix},$$
with the requirement that $a^2 + b^2 = 1$. Prove algebraically that the second pattern keeps distances invariant.

2. Write the general matrix equation for a rotation through angle $\theta$ around the origin.
   a. What is this matrix if $\theta = 90°$? If $\theta = 180°$?
   b. Find the general matrix equation for the inverse of this rotation.

3. Write the matrix equation for the identity isometry. Is it direct or opposite? Is it a translation? Is it a rotation? Is it a reflection? Is it a glide reflection? Explain why or why not for each answer.

4. Verify that reflection across the line $y = x$ is represented by the equation
$$\begin{bmatrix} x' \\ y' \end{bmatrix} = \begin{bmatrix} 0 & 1 \\ 1 & 0 \end{bmatrix} \begin{bmatrix} x \\ y \end{bmatrix}.$$

5. Here are equations for two translations:
$$T_1\left(\begin{bmatrix} x \\ y \end{bmatrix}\right) = \begin{bmatrix} x \\ y \end{bmatrix} + \begin{bmatrix} a \\ b \end{bmatrix},$$
$$T_2\left(\begin{bmatrix} x \\ y \end{bmatrix}\right) = \begin{bmatrix} x \\ y \end{bmatrix} + \begin{bmatrix} c \\ d \end{bmatrix}.$$
Show that the transformation $(T_1 + T_2)\left(\begin{bmatrix} x \\ y \end{bmatrix}\right)$ is not an isometry.

6. Find matrix equations for the following.
   a. a reflection in the line $x = h$
   b. a reflection in the line $y = k$

7. Suppose the line $y = mx$ has angle of inclination $\alpha$. Prove that a reflection in this line is equivalent to a reflection across the $x$-axis followed by a rotation around the origin through an angle of $2\alpha$.

8. a. Prove analytically that two reflections in parallel lines create a translation. (You may assume that the lines are vertical.)
   b. Prove analytically that two reflections in intersecting lines create a rotation. (You may assume that the lines intersect at the origin. The $x$-axis and an arbitrary line $y = mx$ are good choices.) What is the angle of rotation in this situation?

9. Find a matrix that represents a reflection across the $y$-axis followed by reflection across the

*x*-axis. Verify that this composition is a rotation about the origin. What is the angle of this rotation?

10. Verify that the matrix equation

$$\begin{bmatrix} x' \\ y' \end{bmatrix} = \begin{bmatrix} 0 & 1 \\ 1 & 0 \end{bmatrix} \begin{bmatrix} x \\ y \end{bmatrix} + \begin{bmatrix} \sqrt{2} \\ \sqrt{2} \end{bmatrix}$$

represents a glide reflection along the diagonal line $y = x$, with a translation of 2 units in the northeast direction.

11. Find a matrix representation for a glide reflection across the line $y = -x$, with a translation by 3 units in the direction parallel to this midline.

12. An alternative way to understand a reflection across the line $y = mx$ is by the following three steps: Rotate the line to the horizontal axis, reflect across the horizontal axis, and then rotate the line back to its original position. Use this idea to find a matrix equation for reflection across $y = mx$. Simplify your answer.

13. Triangle *A* has vertices at $(-1, 2)$, $(3, 4)$, and $(-2, 4)$. Triangle *B* has vertices at $(3, 1)$, $(7, 3)$, and $(4, -1)$. Create accurate drawings of these triangles. Find an isometry to prove that they are congruent, and give the matrix equation for your isometry. Identify the kind of isometry you use.

14. If an isometry has fixed points, they will satisfy the equation $\begin{bmatrix} x' \\ y' \end{bmatrix} = \begin{bmatrix} x \\ y \end{bmatrix}$.
    a. Prove analytically that a nonzero translation has no fixed points.
    b. Prove analytically that a reflection has a line of fixed points. (*Hint:* Use a line through the origin. You may need double-angle formulas from trigonometry.)

15. A nonzero translation will not have fixed points. However, a line parallel to a translation vector will be invariant, meaning that the image of any point on this line will also be on the line. Prove this analytically.

16. Give a simplified expression for the rotation through angle $\theta$ around point $(h, k)$. (Follow the suggestions on page 166.) What does this simplify to if $\theta = 90°$? If $\theta = 180°$?

17. Pick a point, *P*, on the *x*-axis and a point, *Q*, on the *y*-axis. Consider the following operation: Rotate 90° around *P*, then rotate 90° around *Q*. Find a matrix equation for this composite operation. Then find any fixed points, and tell what kind of isometry this is.

18. For two lines $\ell$ and $m$, let $Ref(\ell)$ and $Ref(m)$ stand for the reflections across the lines. We can form two compositions from these: $Ref(\ell) \circ Ref(m)$ and $Ref(m) \circ Ref(\ell)$. When are these compositions equal? Experiment with various lines to find a conjecture, then prove your conjecture analytically.

19. Pick three points *P*, *Q*, and *R*. Define the isometry $H_P$ as the half turn on *P*, that is, a rotation of 180° around *P*. The isometries $H_Q$ and $H_R$ are defined similarly. How does the composition $H_P \circ H_Q \circ H_R$ compare with $H_R \circ H_Q \circ H_P$? Prove your answer analytically.

20. Write a matrix equation for reflection in the line $y - 1 = -2(x - 3)$.

21. It takes three points and their images to completely determine an isometry. Suppose we know that an isometry *g* does the following:

$$\begin{bmatrix} 2 \\ 0 \end{bmatrix} \longrightarrow \begin{bmatrix} 1 \\ 2 \end{bmatrix}, \qquad \begin{bmatrix} 0 \\ -1 \end{bmatrix} \longrightarrow \begin{bmatrix} -1 \\ 1 \end{bmatrix}.$$

There are two possibilities for this isometry. Find the matrix equation for each, and identify what type of isometry each is.

22. Give an example of two isometries that do not commute. Find the matrix representations of these isometries. Then compute both possible compositions, thus analytically showing that these isometries do not commute.

23. Consider the similarity of Activity 9.
    a. Does this similarity have any fixed points? If so, find them; if not, explain why not.
    b. Find a matrix equation for the inverse of this similarity. Calculate the determinant of the matrix for this inverse, and compare this with the determinant you calculated in Activity 9.

24. Suppose a similarity has two fixed points. Prove that it must be an isometry.

25. Let $M$ represent the matrix of a similarity.
    a. Give a geometric reason to explain why
    $$det(M^{-1}) = \frac{1}{det(M)}.$$
    b. Now give a linear algebra explanation for this.

26. Consider the following equation:
    $$\begin{bmatrix} x' \\ y' \\ 1 \end{bmatrix} = \begin{bmatrix} a & b & e \\ c & d & f \\ 0 & 0 & 1 \end{bmatrix} \begin{bmatrix} x \\ y \\ 1 \end{bmatrix}.$$
    Explain how all four types of isometry can be represented by this pattern. Discuss the determinant in your explanation. Give a specific example for each type.

Exercises 27–28 are especially for future teachers.

27. In the *Principles and Standards for School Mathematics,* the National Council of Teachers of Mathematics (NCTM) recommends that "Instructional programs from prekindergarten through grade 12 should enable all students to ... apply transformations and use symmetry to analyze mathematical situations." In particular, students in grades 9–12 should be able to "understand and represent translations, reflections, rotations, and dilations of objects in the plane by using sketches, coordinates, vectors, function notation, and matrices" (NCTM, 2000, 308). Expectations for high school students build directly on experiences these students have in the middle grades. What does this mean for you and your future students?
    a. Study the Geometry Standard for Grades 9–12. What prerequisite content knowledge and skills do students need to acquire in the middle grades in order to continue their study of transformational geometry in high school? Cite specific examples.
    b. Study the Geometry Standard for Grades 6–8. What content knowledge and skills are students expected to learn in the middle grades? Find some mathematics textbooks for middle school grade levels. How are the NCTM recommendations regarding transformational geometry implemented in these textbooks? Cite specific examples.
    c. Find some high school mathematics textbooks. How are the NCTM recommendations regarding vector and matrix representations for transformations of objects in the plane implemented in these textbooks? Cite specific examples.
    d. Write a report in which you present and critique what you learn. Your report should include your responses to parts a–c.

28. Study the Geometry Standard for one grade band. Design several classroom activities involving transformations of objects in the plane that would be appropriate for students in that grade band. These activities should engage students in using a variety of representations of geometric motions. Write a short report explaining how the activities you design reflect both the NCTM recommendations and what you are learning in this course.

Reflect on what you have learned in this chapter.

29. Review the main ideas of this chapter. Describe, in your own words, the concepts you have studied and what you have learned about them. What are the important ideas? How do they fit together? Which concepts were easy for you? Which were hard?

30. Reflect on the learning environment for this course. Describe aspects of the learning environment that helped you understand the main ideas in this chapter. Which activities did you like? Dislike? Why?

In this chapter, we combined the notion of coordinates with that of isometry. This allowed us to represent geometric functions in algebraic form—specifically, in matrix notation. Matrix tools provide new ways to understand isometry and similarity.

Translation is represented by adding a vector. Rotation around the origin is represented by multiplying by a certain pattern of matrix. Reflection in a line through the origin is also represented by matrix multiplication. Thus, we can use matrix notation to represent isometries as functions.

We can also accomplish composition of isometries through matrix operations. Multiplying matrices is an important part of this process. Composition allows us to find matrix expressions for more general rotations and reflections, as well as for glide reflections.

After a variety of examples, we saw two general patterns for isometry in the plane:

$$\begin{bmatrix} x' \\ y' \end{bmatrix} = \begin{bmatrix} a & -b \\ b & a \end{bmatrix} \begin{bmatrix} x \\ y \end{bmatrix} + \begin{bmatrix} c \\ d \end{bmatrix}$$

or

$$\begin{bmatrix} x' \\ y' \end{bmatrix} = \begin{bmatrix} a & -b \\ -b & -a \end{bmatrix} \begin{bmatrix} x \\ y \end{bmatrix} + \begin{bmatrix} c \\ d \end{bmatrix},$$

with the requirement that $a^2 + b^2 = 1$. The first pattern is for direct isometries, and the second pattern is for opposite isometries.

Given a matrix expression for an isometry, the sign of the determinant of the matrix tells whether the isometry is direct or opposite. A determinant of $+1$ indicates a direct isometry, while $-1$ indicates an opposite isometry.

Matrix expressions offer powerful tools for proofs that involve isometries, such as proofs verifying that two triangles are congruent or proofs finding the result of a composition of isometries.

A *similarity* is a transformation in two steps: a size change followed by an isometry. Similarities have many features in common with isometries, including the distinction between direct and opposite. Matrix representations are useful for studying geometric similarity.

## CHAPTER EIGHT

# Symmetry in the Plane

In Chapter 6, we studied congruence by looking at ways to move one figure on top of another. Congruent figures are identical in the measure of their sides, the arrangement of their vertices, the measure of their angles, and so on. If two figures are congruent, the only difference between them is their position. Using isometries, we can change the position of one figure to make it correspond to its congruent partner; in other words, an isometry can move one congruent figure on top of its partner.

We saw that there are four types of isometries in the plane: translations, rotations, reflections, and glide reflections. We also saw that the composition of two isometries will again be an isometry. Recall that translations and rotations are direct isometries, and reflections and glide reflections are opposite isometries.

The identity isometry leaves every point exactly where it is. Because of the identity isometry, any geometric figure is congruent to itself. However, many geometric figures are congruent to themselves in more interesting ways. The purpose of this chapter is to study the ways in which a figure can be self-congruent and to use these self-congruences to classify figures. As we do so, we will again encounter the concept of a mathematical group.

Do the following activities, writing your explanations clearly in complete sentences. Include diagrams whenever appropriate. Save your work for each activity, as later work sometimes builds on earlier work. You will find it helpful to read ahead into the chapter as you work on these activities.

1. Construct a square. The objective of this activity is to find all the ways in which this square can be congruent to itself.

   a. Construct a line segment near your square. Select the endpoints of this segment, and mark it as a translation vector. Then translate the square by this vector. (It may be easier to keep track of things if you label the vertices or if you make the image a different color from the original square.) Vary the vector until the image of the square lies on top of the original square. Find all ways in which this can be done.

   b. Construct the center of your square, and mark it as a center of rotation. Create an angle near the square. Measure your angle, then choose **Edit | Preferences | Units** to set the angle units to **directed degrees**. Select three points for this angle (as if you planned to measure it again), then choose **Transform | Mark Angle**. Use this marked angle to rotate the square around its center. Vary the angle so that the image of the square lies on top of the original square. Find all ways in which this can be done.

   c. Draw a line near your square. Construct the reflection of the square in this line. Then vary the position and direction of the mirror line so that the image of the square lies on top of the original square. Find all the ways this can be done.

   d. Look back at your results in this activity. Using diagrams, symbols, or words, summarize all the ways in which a square can be congruent to itself.

2. A *figure* is a set of points in the plane. Typically, a figure is a familiar object, such as a square or a triangle, but any collection of points will do.

   A *symmetry* of a particular figure, $S$, is an isometry $f$ for which $f(S) = S.$ This means that the image of the figure lies on top of the original figure. It does not mean that each point is in the same place; the points can shift while the overall figure looks the same, as you saw in Activity 1.

   List the symmetries for each figure in Figure 8.1. How are your two lists similar and how are they different?

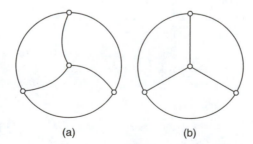

**FIGURE 8.1**
Figures for Activity 2

(a)          (b)

3. Each of the symmetries you listed for Figure 8.1b is an isometry; thus, each has an inverse isometry. Are these inverses also symmetries for this figure? Explain why or why not.

4. The capital letters of the English alphabet are listed below in a *sans serif* font. Sort these letters into sets so that every element in the same set has the same symmetries.

A, B, C, D, E, F, G, H, I, J, K, L, M,<br>N, O, P, Q, R, S, T, U, V, W, X, Y, Z

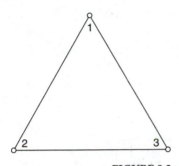

**FIGURE 8.2**
An Equilateral Triangle

5. Figure 8.2 shows an equilateral triangle. For this activity, you may find it helpful to cut out a paper triangle, label its vertices as in Figure 8.2, and use it to perform the isometries. Be careful to label the back side of the triangle correctly.

   The triangle in Figure 8.2 has six symmetries:
   - $R_0$     a rotation through $0°$
   - $R_{120}$  a rotation through $120°$
   - $R_{240}$  a rotation through $240°$
   - $V$      a reflection across a vertical line through the uppermost vertex
   - $L$      a reflection across a diagonal line through the lower-left vertex
   - $R$      a reflection across a diagonal line through the lower-right vertex

   a. Sketch the triangle, showing the lines of reflection and the center of rotation.

   b. These symmetries can be combined by composition. For instance, $V \circ R_{120} = L$. Complete the following table. (Recall that composition is not commutative!)

|          | $R_0$ | $R_{120}$ | $R_{240}$ | $V$ | $L$ | $R$ |
|----------|-------|-----------|-----------|-----|-----|-----|
| $R_0$    |       |           |           |     |     |     |
| $R_{120}$|       |           |           |     |     |     |
| $R_{240}$|       |           |           |     |     |     |
| $V$      |       |           |           | $L$ |     |     |
| $L$      |       |           |           |     |     |     |
| $R$      |       |           |           |     |     |     |

   c. Based on your table, which symmetry is the identity? What is the inverse of each symmetry?

6. A *frieze* is a pattern that is infinite along one line. Wallpaper borders are friezes for they have a pattern that repeats as the border goes around the room. Here is another example of a frieze:

...P P P P P P P P P P P P P P P P P P P P P P...

This particular frieze has a translation as its only symmetry.

   Create at least three more examples of friezes so that each has a different set of symmetries. Mark the initial pattern used to create each frieze, and tell what symmetries each example has. (*Hint:* It may help to use a Cartesian coordinate system and a square grid.)

**FIGURE 8.3**
A Wallpaper Pattern

7. A pattern that is infinite in two directions is often called a *wallpaper pattern*. Figure 8.3 shows a portion of a wallpaper pattern based on reflections in both the vertical and the horizontal directions. Notice the darker scalene triangle, which is the basis for the pattern.

   Using a Cartesian coordinate system and a square grid in Sketchpad, construct vertical and horizontal lines to divide the plane into squares, as was done in Figure 8.3. The **Snap Points** option will be helpful for this, but be sure to turn it off once your lines have been drawn. In one of the squares, create a nonsymmetric figure. (This figure does not have to be a triangle.)
   a. Using only translations, create a wallpaper pattern based on your figure.
   b. Using reflections in one direction and translations in another, create a wallpaper pattern based on your figure.

8. A *tiling* of the plane is a finite collection of polygon patterns that can be assembled to fill the plane completely. Sometimes this collection can consist of a single shape repeated over and over.

   Draw a quadrilateral. Use this to begin a tiling of the plane. Show at least 16 copies of the quadrilateral in your picture, and mark the original quadrilateral. Print your tiling. Vary the original quadrilateral to create a new tiling, and print that one also. (Coloring the regions can enhance your pictures.)

9. A polygon is called *regular* if all its sides are the same length and all its angles are congruent. The equilateral triangle and the square are familiar examples of regular polygons.

   Can a plane be tiled by equilateral triangles? By squares? By regular pentagons? By regular hexagons? In each case, either draw an example or explain why it cannot be done.

The term *figure* is not a precise one in geometry. For our purposes, a figure can be any collection of points in the plane: finite or infinite, bounded or unbounded, solid or hollow, and so on. For instance, Figure 8.4 shows three figures, all based on a familiar constellation.

**FIGURE 8.4**
The Constellation
Orion the Hunter

The first figure of Orion is a finite collection of points, arranged in a way that suggests the shape of a hunter. The second figure consists of points and line segments, and the shape can be seen more clearly. (Do you see the belt and the sword?) The third figure shows the hunter and how the stars fit within his shape. Any of these figures could be studied geometrically.

The wording in Activity 6 is another example of imprecise language: The phrase "infinite along one line" is intended to suggest a pattern that continues indefinitely in one particular direction, without ever coming to an end. However, *infinite* is the wrong term to use. After all, the second figure of Orion contains an infinite number of points on its line segments. The proper term for a pattern that extends indefinitely is *unbounded*, which means there is no boundary to stop the pattern. The figures for Orion are *bounded*, for we can draw a circle around any of them to act as a boundary.

## SYMMETRIES

Activity 1 asks you to find various ways in which a square can be congruent to itself. In a trivial sense, the identity isometry does this; $i(square) = square$ because $i(P) = P$ for every point $P$. No doubt you discovered some more interesting ways to have $f(square) = square$. One way is to use a rotation of 180° around the center of the square. This isometry will produce an image identical to the original square, not only in shape but also in location. This does not, however, mean that each point is in its original position. For instance, this rotation moves the upper-right vertex to the lower-left corner. Overall, however, the rotated image of the figure uses precisely the same points as the original figure. This 180° rotation is one way in which the square is self-congruent. There are two other rotations, as well as four reflections, that show self-congruence. Thus, a square is self-congruent in eight ways, which demonstrates that a square is a very symmetric figure.

**DEFINITION 8.1**    A *symmetry* of a figure $S$ is an isometry $f$ for which $f(S) = S$.

If $S$ is the domain of an isometry, the notation $f(S) = S$ means that the range $f(S)$ is exactly the same set of points as the domain. Individual points do not have to be fixed, but the overall set will be.

Figure 8.1a and 8.1b share some symmetries: the identity (of course), a rotation of 120°, and a rotation of 240°. Figure 8.1b has three additional symmetries: reflections across certain lines. Thus, Figure 8.1b, with its straight segments, has more symmetry than Figure 8.1a, with its curved segments.

The regular polygons are very symmetric figures (see Figure 8.5). Each regular polygon has rotations and reflections as its symmetries. How many symmetries of each type are there for a regular *n*-gon?

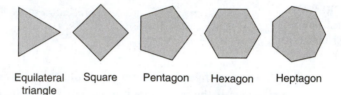

**FIGURE 8.5**
Some Regular Polygons

Equilateral    Square    Pentagon    Hexagon    Heptagon
triangle

A circle is an extremely symmetric figure. Which rotations are symmetries for the circle? Which reflections are symmetries for the circle?

## GROUPS OF SYMMETRIES

In Chapter 6, we saw that the set of isometries on the plane, together with the operation of composition, forms a mathematical structure called a *group*. This structure requires that four properties must be satisfied by the set and its operation: closure, associativity, identity, and inverse.

**DEFINITION 8.2**    A *group* is a set, $G$, together with a binary operation, $\circ$, such that

- for any two elements $x$, $y$ of $G$, $x \circ y$ is in $G$ (the *closure* property);
- for any elements $x$, $y$, $z$ of $G$, $x \circ (y \circ z) = (x \circ y) \circ z$ (the *associative* property);
- there is an element $i$ of $G$ so that $i \circ x = x \circ i = x$ (the *identity* property); and
- for any element $x$ of $G$, there is an element $x^{-1}$ also in $G$ so that $x \circ x^{-1} = x^{-1} \circ x = i$ (the *inverse* property).

The operation for a group can be addition, multiplication, or some other operation. The operation must be *binary*, meaning that the operation combines only two elements at a time. The difficulty is knowing what to do with expressions such as $x \circ y \circ z$. Should it be computed as $x \circ (y \circ z)$ or as $(x \circ y) \circ z$? This is why the associative property is important; it says that we can pair the elements in either way and still get the same result. The $\circ$ symbol in the definition was chosen to remind you of composition of functions, for that is the operation we use for isometries in general and for symmetries in particular.

An important fact about symmetries is that the set of symmetries for any figure will be a group. Activity 3 is a small step toward proving this. The full proof is not difficult, and it will appear in the exercises.

Notice that the definition of *group* says nothing about commutativity. Composition of functions is seldom commutative. A few special combinations of functions commute, but the huge majority do not. This is true for isometries and symmetries as well. In Activity 5, for instance, you were told that $V \circ R_{120} = L$. You should have found that $R_{120} \circ V = R$.

Calculating these compositions can be confusing. It may help to label the three vertices on the equilateral triangle, as in Figure 8.2, and then to write each symmetry in its *cycle notation*. Here is an example:

The rotation of $120°$ can be represented by $R_{120} = (1\,2\,3)$. The notation $(1\,2\,3)$ denotes the cycle $1 \longrightarrow 2 \longrightarrow 3 \longrightarrow 1$ and tells us that the function $R_{120}$ carries point 1 to point 2, point 2 to point 3, and point 3 to point 1. All other points of the figure are carried along also, but these three points are the critical ones.

The reflection $V$ can be represented by $V = (1)(2\,3)$. This says that point 1 is carried to itself while points 2 and 3 change places. For the composition, we can write

$$V \circ R_{120} = (1)(2\,3) \circ (1\,2\,3).$$

Observe that the transformation we applied first, $R_{120}$, appears as the *rightmost* transformation in this expression. Recall that $V$ and $R_{120}$ are functions being applied to points in the plane. Thus, $V \circ R_{120}$ is a composition of functions.

If $P$ is a point in the plane, we can calculate $V \circ R_{120}(P)$. This expression can also be written as $V(R_{120}(P))$, showing that we work from the inside out. In other words, point $P$ is the input to the transformation $R_{120}$, and the output of this transformation (a point) is the input to the transformation $V$. The final output (again a point) is the result of these two steps: first applying $R_{120}$ to the point $P$, and then applying the transformation $V$ to its result.

To simplify a composition given in cycle notation, start at the right and see what happens to each point. For example, in the composition $(1)(2\,3) \circ (1\,2\,3)$, point 1 goes to point 2, which then goes to point 3. We can begin to write this as $(1\,3\,\ldots)$. Now, what happens to point 3? It is carried to point 1 and then left alone. So, we close off the cycle: $(1\,3)$. Point 2 remains; it is carried to point 3, which goes back to point 2. The simplified composition is therefore written as $(1\,3)(2)$. From Figure 8.2, we can see that this result says to interchange points 1 and 3 while keeping point 2 fixed. This is the reflection $L$.

In reverse order, the composition is

$$R_{120} \circ V = (1\,2\,3) \circ (1)(2\,3) = (1\,2)(3) = R.$$

Look at the group table you made for the symmetries of an equilateral triangle in Activity 5. Six symmetries are involved; which one is the identity? It is somewhat traditional to place the identity in the first row and the first column of a group table, and this row and column are easy to complete. Now look at the second row of your table; how many different symmetries appear in this row? Is this true in the other rows? What about the columns?

We hope you observed in Activity 5 that every row of the group table contains all six symmetries, and so does every column. In other words, each element of a finite group will appear exactly once in each row and once in each column. This is a very useful fact that is not difficult to prove. Suppose that two entries in the row for element $a$ are equal, specifically, the entries in the columns for $b$ and for $c$. This means that

$$a \circ b = a \circ c$$

for certain elements $a$, $b$, $c$. Because $a$ has an inverse, we can calculate

$$a^{-1} \circ a \circ b = a^{-1} \circ a \circ c,$$

and we see that $b = c$. Thus, the two equal entries are, in fact, in the same column and must be the same entry! This implies that the elements in a row are all different, and because the number of entries equals the number of elements, every element must be used. A similar proof shows that each element of a group will appear exactly once in each column. These facts can greatly reduce your work when completing a group table.

## CLASSIFYING FIGURES BY THEIR SYMMETRIES

In Activity 5, you completed a group table for the symmetries of an equilateral triangle. Each figure you worked with in Activity 2 has its own set of symmetries. The group table for one of these symmetries is exactly the same as the group table for an equilateral triangle. (Which one?) When two sets have the same group table, we say that the groups are *isomorphic*.

A set of symmetries for a figure with the operation of composition forms the figure's *symmetry group*. This is true for the equilateral triangle, for Figures 8.1a and 8.1b, and for the letters of the alphabet in Activity 4. In Activity 4, each letter has a corresponding symmetry group. However, these 26 letters represent only four symmetry groups: There is the group consisting of just the identity isometry, represented by F, G, and so on. There is the group containing the identity and one other rotation, represented by N, S, and Z. The group with the identity and one reflection goes with A, B, and lots of other letters. The letters H, I, X have the largest symmetry group and, hence, have the most symmetry. What symmetries does this group contain?

Though the symmetry groups for the alphabet are small groups, they illustrate the two basic types of symmetry groups. One type of symmetry group is the *cyclic group* $C_n$, which contains only rotations. The name comes from the fact that every member of the group is a power of the smallest rotation in the group, and these powers occur in a cycle. For instance, Figure 8.1a has $R_{120}$ as its smallest (nontrivial) rotation. If we take powers of this rotation, we get

$$R_{120}^1 = R_{120},$$
$$R_{120}^2 = R_{240},$$
$$R_{120}^3 = R_{360} = R_0,$$
$$R_{120}^4 = R_{480} = R_{120},$$

and so on. Continuing the list of powers will not create any new rotations. The cycle is of length 3 and the symmetry group of this figure is $C_3$. (In this situation, the powers of $R_{240}$ also create a cycle of length 3; so, we could use $R_{240}$ to generate the same symmetry group. This is not true in general.)

Figure 8.1b has three rotations, and it also has three reflections. Its symmetry group is called $D_3$, a *dihedral group*. The term *di - hedral* means "two sides" and refers to the symmetry across a reflection line. Notice that the dihedral group $D_3$ has six elements, half of which are rotations and the other half of which are reflections. Thus, $D_3$ contains $C_3$ as a subgroup. This is true in general; $D_n$ has twice as many elements as $C_n$, and $D_n$ contains $C_n$ as a subgroup. (This statement needs a proof, of course, which you will be asked to develop in the exercises.)

A major theorem about symmetry groups is credited to Leonardo da Vinci, who discovered it during his architectural work (Bix, 1994, 134).

**THEOREM 8.1**  **Leonardo's Theorem**  A finite symmetry group for a figure in the plane must be either the cyclic group $C_n$ or the dihedral group $D_n$.

This theorem says that the size of the symmetry group can vary but that there are only two options for its type: symmetry based solely on rotations or symmetry with both rotations and reflections. The symmetry groups for Figures 8.1a and 8.1b are typical examples of these two types.

To prove Leonardo's Theorem, we need a preliminary result (Sibley, 1998, 185–186).

**LEMMA 8.2**  A finite symmetry group has a point that is fixed for every one of its symmetries.

**Proof**  We must be clear about the meaning of this statement. The lemma says that every symmetry of the figure fixes the same point $P$. We must locate $P$, and then, no matter which symmetry we choose, we must prove that $f(P) = P$.

The proof will use coordinates. Suppose the finite symmetry group is $\{ f_1, f_2, \ldots, f_n \}$, and for convenience, let $f_1$ be the identity. Pick any point $P_1$, and calculate

$$f_1(P_1) = P_1$$
$$f_2(P_1) = P_2$$
$$f_3(P_1) = P_3$$
$$\vdots$$
$$f_n(P_1) = P_n.$$

Form the new point

$$P = \frac{P_1 + P_2 + \cdots + P_n}{n}.$$

Point $P$ is called the *center of gravity* for this collection of points.

Now, $f_1(P) = P$ because $f_1$ is the identity isometry. What happens for $f_2(P)$? First, think about what $f_2$ does to the list of points.

$$f_2(P_1) = f_2(f_1(P_1)) = (f_2 \circ f_1)(P_1)$$
$$f_2(P_2) = f_2(f_2(P_1)) = (f_2 \circ f_2)(P_1)$$
$$f_2(P_3) = f_2(f_3(P_1)) = (f_2 \circ f_3)(P_1)$$
$$\vdots$$
$$f_2(P_n) = f_2(f_n(P_1)) = (f_2 \circ f_n)(P_1).$$

Earlier we proved that the compositions $(f_2 \circ f_1), (f_2 \circ f_2), \ldots, (f_2 \circ f_n)$ are all different from each other. This list of compositions fills the $f_2$ row of the group table with $n$ different results. Therefore, the list $f_2(P_1), f_2(P_2), \ldots, f_2(P_n)$ contains all of the original $n$ points, though in a different order. Calculating the center of gravity for the points $f_2(P_1), f_2(P_2), \ldots, f_2(P_n)$ is thus the same as calculating the original center of gravity $P$. So, the center of gravity remains fixed for $f_2$, that is, $f_2(P) = P$.

The same reasoning works for $f_3, \ldots, f_n$, showing that this particular point $P$ is fixed for every symmetry in this group.

------------------------------------------------

Because the members of a finite symmetry group have at least one fixed point, they must be either rotations or reflections. Even the identity falls into these categories, for the identity isometry can be interpreted as a rotation of $0°$. Later, we will examine symmetry groups that include translations and glide reflections. Because of Lemma 8.2, we know that these more complicated symmetry groups will be infinite groups.

**Proof of Theorem 8.1**   Let us start small and work upward.

**Case 1**   Suppose the finite symmetry group has a single rotation—which must be the identity—and has no reflections. This is the cyclic group $C_1$.

**Case 2**   Suppose the finite symmetry group has a single rotation and has one reflection. This is the dihedral group $D_1$.

**Case 3**   Suppose the finite symmetry group has a single rotation and has more than one reflection. Let's focus our attention on two of the reflections. Because these two reflections share a fixed point, their mirror lines intersect. Recall that the composition of two reflections in intersecting lines is a rotation. Because this cannot be the identity rotation (why not?), this case is impossible.

**Case 4** Suppose the finite symmetry group has more than one rotation and has no reflections. Let $R_\alpha$ be the rotation with the smallest angle, modulo $360°$. It turns out that every other rotation in this group can be generated by a power of $R_\alpha$.

To prove this claim, assume that $R_\theta$ is a rotation in the symmetry group but that $R_\theta$ is not a power of $R_\alpha$. This means that $\theta$ is not a multiple of $\alpha$. Consequently, $\theta$ lies between two consecutive multiples of $\alpha$. Expressed algebraically, there is a nonnegative integer $k$ so that $k\alpha < \theta < (k+1)\alpha$. Then, $0 < (\theta - k\alpha) < \alpha$. However, having $R_\theta$ in the symmetry group means that the rotation $R_{(\theta - k\alpha)}$ is in the group, because $R_{(\theta - k\alpha)} = R_\theta \circ (R_\alpha^{-1})^k$. This contradicts the choice of $\alpha$ as the smallest positive angle for a rotation and shows that $\theta$ must be a multiple of $\alpha$.

Let $n$ be the smallest power of $R_\alpha$ that equals the identity. This symmetry group is the cyclic group $C_n$.

**Case 5** Suppose the finite symmetry group has more than one rotation and has at least one reflection. Call this reflection $F_1$, and once again let $R_\alpha$ be the rotation with the smallest angle, with $R_\alpha^n = identity$. Notice that $R_\alpha^j \circ F_1$ is an opposite isometry, so it must be a reflection. Every choice of $j$ produces a reflection, and every choice produces a different reflection. (Why is this true? Think about a column in a group table.)

Are these all of the reflections? Assume $F_2$ is a reflection not equal to $F_1$. The reflection $F_2$ has the same fixed point as $F_1$, and their mirror lines intersect there; thus, the composition $F_2 \circ F_1$ is a rotation, say, $R_{k\alpha}$. Then, $F_2 = R_{k\alpha} \circ F_1$, which is already accounted for.

**FIGURE 8.6**
The Pentagon in Washington, D.C., an Example of $D_5$ Symmetry

Thus, there are $n$ reflections in addition to the $n$ rotations. This is the dihedral group $D_n$.

Finite symmetry groups are very common. As da Vinci noticed, architects often use small dihedral groups, such as $D_1$ and $D_2$, when designing a building. Can you find some examples on or near your campus? Some very interesting buildings use other symmetry groups. The Pentagon in Washington, D.C., uses $D_5$. The famous Opera House in Sydney, Australia, has $C_1$ as its symmetry group when considered as a whole, though each individual component of the building uses $D_1$.

**FIGURE 8.7**
The Opera House in Sydney, Australia, an Example of $D_1$ Symmetry

Examples of symmetry appear frequently in nature. The leaves of trees, such as maple trees, usually have $D_1$ as their symmetry group. The leaves of the poplar tree are strikingly symmetric. The sassafras tree is an exception however; its asymmetric leaves have $C_1$ as their symmetry group. Flowers may have $D_4$, $D_5$, or even $D_6$ as a symmetry group. Snowflakes are famous for having $D_6$ symmetry.

**FIGURE 8.8**
Logo of Key College Publishing, Another Example of $D_1$ Symmetry

Commercial logos often exhibit symmetry. The logo of Key College Publishing, shown in Figure 8.8, has $D_1$ as its symmetry group. It is fun to look through the Yellow Pages and classify the logos by their symmetries. Hub caps on cars are another interesting source of finite symmetry groups.

## FRIEZES AND SYMMETRY

The symmetry groups discussed so far have been finite groups. Furthermore, the figures they describe have been *bounded*, meaning that these figures do not continue indefinitely. (A more technical definition for *bounded* says that the figure can be surrounded by a circle.) Also, these finite symmetry groups have used only rotations and reflections. There is an easy explanation for why translations have not been used: If a figure has a translation $T$ as one of its symmetries, then $T^2$,

$T^3, \ldots$, must also all be symmetries of the figure. (The inverse $T^{-1}$ and all its powers must be symmetries, too.) Thus, the figure must be unbounded in the direction of the translation, and the symmetry group will be infinite.

Let's look more closely at this situation. Suppose a pattern has translations as symmetries and all its translation symmetries use parallel vectors. This sort of pattern is called a *frieze*. A frieze will be infinite along a single line parallel to the vectors of its translations. Frieze patterns are often seen in architecture as ornamentation along the top of a building.

Activity 6 gave a simple example of a frieze that uses the letter P as its base pattern and has translations as its only symmetries. Other letters suggest other types of patterns, such as the following:

$$\ldots\, \text{N N N N N N N N N N N N N N N N N N N N N N N N}\, \ldots$$

$$\ldots\, \text{W W W W W W W W W W W W W W W W W W W W W W}\, \ldots$$

$$\ldots\, \text{X X X X X X X X X X X X X X X X X X X X X X X}\, \ldots$$

The N pattern has both translations and 180° rotations in its set of symmetries. The W pattern has translations and reflections across vertical lines. The X pattern is a busy one: It has translations, 180° rotations, reflections in vertical lines, and a reflection in a horizontal line. The X pattern also has glide reflections, formed by the composition of the horizontal reflection and the translations.

Figure 8.9 shows some more examples of friezes. Try to identify the symmetries of each. (These figures are given in color on the CD accompanying this book.)

Notice that a frieze pattern follows a line, which we call the *midline* of the pattern. For convenience, assume that the midline is horizontal.

**FIGURE 8.9**
Examples of Friezes

**THEOREM 8.3**   The only possible symmetries for a frieze pattern are horizontal translations along the midline, rotations of 180° around points on the midline, reflections in vertical lines perpendicular to the midline, a reflection in the horizontal midline, and glide reflections using the midline.

**Proof**   The midline must be fixed by any symmetry of the frieze. Consider what various isometries would do to the midline. Translations that are not parallel to this midline change the position of the midline. Rotations that are not 180° or 0° change the inclination of the midline. Rotations of 180° that are not centered on the midline produce a line parallel to, but different from, the midline. Reflections in a line that is not parallel or perpendicular to the midline change the inclination of the line. Reflections in a horizontal line other than the midline change the position of the midline. The same is true for glide reflections in other horizontal lines. Thus, we have eliminated all possibilities except those listed in the theorem (Sibley, 1998, 191).

--------------------------------------------

By definition, any frieze will have an infinite number of translational symmetries along its midline. We can use the other possible symmetries to classify and count the frieze groups.

**THEOREM 8.4**   There exist exactly seven symmetry groups for friezes.

**Proof**   Because all friezes have translations, let's consider the other types of symmetries and the various combinations of symmetries. We will see that many of these combinations cannot occur.

We will use the following abbreviations:

- $H$ is the reflection in the horizontal midline.
- $V$ is a reflection in a vertical line.
- $R$ is a rotation of 180° about a center on the midline.
- $G$ is a glide reflection using the midline.

The following table lists all possible combinations:

|     | H   | V   | R   | G   | Result       |
| --- | --- | --- | --- | --- | ------------ |
| 1.  | Yes | Yes | Yes | Yes | X pattern    |
| 2.  | Yes | Yes | Yes | No  | not possible |
| 3.  | Yes | Yes | No  | Yes | not possible |
| 4.  | Yes | Yes | No  | No  | not possible |
| 5.  | Yes | No  | Yes | Yes | not possible |
| 6.  | Yes | No  | Yes | No  | not possible |
| 7.  | Yes | No  | No  | Yes | E pattern    |
| 8.  | Yes | No  | No  | No  | not possible |
| 9.  | No  | Yes | Yes | Yes | ∨∧ pattern   |
| 10. | No  | Yes | Yes | No  | not possible |
| 11. | No  | Yes | No  | Yes | not possible |
| 12. | No  | Yes | No  | No  | W pattern    |
| 13. | No  | No  | Yes | Yes | not possible |
| 14. | No  | No  | Yes | No  | Z pattern    |
| 15. | No  | No  | No  | Yes | Γ L pattern  |
| 16. | No  | No  | No  | No  | P pattern    |

In this table, we have shown seven frieze patterns, each having a different combination of symmetries. Why are there no others? Think about some of the possible compositions. If a frieze has the horizontal reflection $H$ and we compose that with a translation along the midline, the result is $G$, a glide reflection. This eliminates options 2, 4, 6, and 8. If a frieze has both horizontal and vertical reflections, the composition $H \circ V$ is a 180° rotation (because the composition of two such reflections is direct and is not the identity). This eliminates option 3 (and 4 again). The composition $G \circ R$ is opposite, so it is either $H$ or $V$. However, the upper half of the pattern will end up on top again, so the composition must equal $V$. This eliminates options 5 and 13. Composing $V \circ R$ equals either $H$ or $G$, depending on whether the mirror for $V$ passes through the center for $R$. Either result eliminates option 10. The composition $G \circ V$ is direct, so it equals either $R$ or a translation. However, the upper half of the pattern ends up on the bottom this time, so the composition must equal $R$. This eliminates option 11 (and 3 again).

Therefore, we are left with options 1, 7, 9, 12, 14, 15, and 16, and we have shown friezes to match each of these possibilities.

## WALLPAPER SYMMETRY

Allowing translations as symmetries creates patterns that must be unbounded. The friezes have translational symmetries along one line, with all the translation vectors in the same direction. Activity 7 introduces the possibility of translational symmetry in more than one direction. Such a pattern will, of course, use the entire plane. For obvious reasons, these are called *wallpaper patterns*. Translations are not the only possible types of symmetry for wallpaper, however. Rotations, reflections, and glide reflections can occur as well.

If you have ever looked at a book of wallpaper samples or at the patterns on wrapping paper, you may have noticed that 90° angles are not the only patterns used. It is quite common to see patterns using 60° and 120° rotations. The following theorem, which we will not prove, states an important fact about these patterns.

**THEOREM 8.5**  **The Crystallographic Restriction**  The minimal angle of rotation for a wallpaper symmetry is 60°, 90°, 120°, 180°, or 360°. All other rotation angles for a symmetry must be multiples of the minimal angle for that pattern (Sibley, 1998, 194).

This theorem arose in the study of crystals. It can also be seen in the network of cracks in drying mud; the cracks tend to occur at angles of 90° or 120° (Shorlin et al., 2000).

How many symmetry groups are there for wallpaper patterns? The proof is more complicated than that of the frieze groups (see *Topics in Geometry* [Bix, 1994] for a full treatment, including the classification scheme used by crystallographers). We will merely give the answer.

**THEOREM 8.6**  There are exactly 17 wallpaper groups.

The Alhambra is a Moorish palace built in Spain during the thirteenth and fourteenth centuries. It is well known among mathematicians for the elaborate tiling patterns on its walls. All 17 wallpaper groups are represented in the Alhambra.

The study of crystals is concerned with symmetry in three dimensions, not just two. The three-dimensional *crystallographic groups* describe the possible symmetries in three dimensions. There are 230 groups for unbounded symmetry in three dimensions. It is possible to study symmetry in four, five, and even higher dimensions by using the same ideas of isometry and symmetry group. For instance, it has been proven that in four dimensions, there are 4783 possible symmetry groups for unbounded patterns (Bix, 1994, 134).

## TILINGS

One method for creating symmetric patterns in the plane is to make a *tiling* of the plane. A tiling is a collection of nonoverlapping polygons, laid edge to edge to cover the entire plane. *Edge to edge* means that an edge of one polygon must also be an edge of the adjacent polygon. The usual way of stacking bricks, staggered for greater stability, is not a tiling. (It is, rather, a *tessellation.*) The hexagonal pattern in a honeycomb, however, is a tiling.

When you ask Sketchpad for a coordinate system and it displays the grid, this collection of squares or rectangles is a tiling. There are a great variety of tilings. Figure 8.10 shows a tiling based on the Greek cross (Wells, 1991, 89).

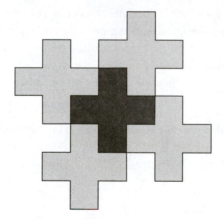

**FIGURE 8.10**
The Greek Cross Tiling

The artist M. C. Escher is famous for his tilings. He created elaborate interlocking figures that fill the plane: horsemen, fish, birds, and so on. In his *Circle Limit* works, Escher developed several tilings of the Poincaré disk. To the viewer of these striking works, it appears that the figures shrink to microscopic size as they get nearer to the boundary circle. However, using the unusual method of measuring distance in the Poincaré disk, all the figures in the series of *Circle Limit* illustrations are exactly the same size; that is, they are congruent figures. This will make more sense after you explore hyperbolic geometry in the Poincaré disk. (See Chapter 9.)

**FIGURE 8.11**
*Circle Limit IV*, by M. C. Escher,
a Tiling of the Poincaré Disk

Elementary and middle school mathematics classrooms often have a set of pattern blocks. These plastic polygons can be assembled to form multicolored patterns. It is intriguing to watch young children work with these blocks. They often create highly symmetric patterns, without being aware of doing so. These patterns are often the beginnings of tilings.

In an *elementary* tiling, all the regions are congruent to one basic shape (Kay, 1994, 246). The Greek Cross tiling is elementary, as is the collection of squares in a grid. In Activity 8, you were asked to create an elementary tiling using a quadrilateral. By rotating and translating your quadrilateral, you should have been able to fill the computer screen completely—and, in theory, completely fill the plane.

**THEOREM 8.7**   Any quadrilateral can be used to create an elementary tiling of the plane.

**Proof**   We will use the notation of Figure 8.12 (next page), where $ABCD$ is the initial quadrilateral. Let $M_1$ be the midpoint of side $BC$. Rotate $ABCD$ through $180°$ around $M_1$. This creates the quadrilateral $A'B'C'D'$, where $B' = C$ and $C' = B$. Now let $M_2$ be the midpoint of $C'D'$ and rotate $A'B'C'D'$ through $180°$ around $M_2$ to create the quadrilateral $A''B''C''D''$, where $C'' = D'$ and $D'' = C' = B$. Lastly, let $M_3$ be the midpoint of $D''A''$ and rotate $A''B''C''D''$ through $180°$ around $M_2$ to create the quadrilateral $A'''B'''C'''D'''$, where $A''' = D'' = C' = B$ and $D''' = A''$. Notice that the four angles at $B$ are congruent to the four different angles of the original quadrilateral. These angles sum to $360°$, so the four quadrilaterals fit together at $B$ without overlapping.

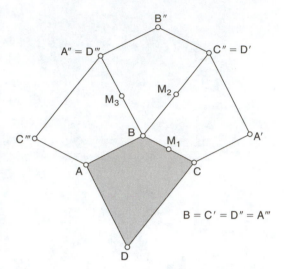

**FIGURE 8.12**
Tiling by a Quadrilateral

Now, $\angle C'''AD$ is congruent to $\angle B''D'A'$. Thus, we can translate this entire picture by vector $AD'$, and the image will fit exactly onto the existing copy. This is also true for the inverse translation. Also, $\angle C'''A''B''$ is congruent to $\angle DCA'$. So, we can translate the entire picture by vector $CA''$ or by its inverse, and the image will fit exactly onto the existing copy. Therefore, this picture, which is composed of four copies of the original quadrilateral, will tile the plane by translations, showing that the original quadrilateral tiles the plane.

------------------------------------------------

With Theorem 8.7, we can prove a similar fact about triangles.

**COROLLARY 8.8**    Any triangle can be used to tile the plane.

**Proof**    Rotate the triangle $180°$ about the midpoint of one of its sides. The resulting figure of two congruent triangles sharing an edge is a parallelogram. Tile the plane with this parallelogram in the obvious way, then divide each parallelogram to create congruent copies of the original triangle.

------------------------------------------------

Corollary 8.8 holds for any triangle: scalene, isosceles, or equilateral. Thus, we have seen that any triangle and any quadrilateral can be used to make a tiling. A natural question arises: Can a pentagon be used to tile the plane? What about a hexagon, or a polygon with even more sides?

Let's try to answer a more specific question: Which *regular* polygons can we use to tile the plane? Recall that a polygon is regular if all its sides are congruent and all its angles are congruent. The equilateral triangle and the square are familiar examples. A tiling based on a regular polygon is called a *regular* tiling. In Activity 9, you were asked to think about these elementary tilings with regular polygons. Can you fit equilateral triangles together around a vertex without overlapping? What about regular pentagons?

To answer these questions, it is helpful to know the measure of the congruent angles of a regular $n$-gon. It is easy to derive a formula for this. Pick any point $P$

in the interior of the $n$-gon, and construct segments from $P$ to the vertices. This creates $n$ triangles, each with angle sum of $180°$. The total angle sum for the $n$ triangles would be $180°n$. However, one angle of each triangle is interior to the polygon; these angles surround point $P$. Because the angle sum of these interior angles should not be included in the total for the $n$-gon, we need to subtract $360°$. Thus, the angle sum for a regular $n$-gon is $(n-2)180°$. To find the size of each angle, divide by $n$. The measure of one angle of a regular $n$-gon is

$$\frac{(n-2)180°}{n}.$$

Now think about putting $k$ regular $n$-gons together around a vertex of a tiling. If they are to fit without overlapping, we must have

$$k\frac{(n-2)180°}{n} = 360°.$$

Solving for $k$ gives

$$k = \frac{2n}{n-2}.$$

The value $k$ must be a natural number and cannot be less than 3. (Do you see why?) By trying some values, this leaves only $n = 3, 4, 6$ as solutions. Therefore, the only regular tilings are those using equilateral triangles, squares, and regular hexagons.

We can create some fascinating patterns using two or more regular polygons for the tiling. If every vertex in a tiling such as this is identical, it is called a *semiregular* tiling. Figure 8.13 shows an example of semiregular tiling with octagons and squares. There is a well-developed theory of semiregular tilings; we encourage you to read further on this topic.

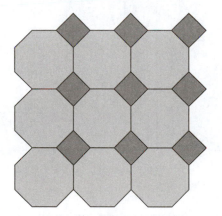

**FIGURE 8.13**
A Semiregular Tiling

The *Penrose tiles* are a curious phenomenon. These tiles are named for the physicist Roger Penrose, who discovered them. A pair of Penrose tiles are constructed from a rhombus by dividing it into two quadrilaterals, called a *kite* and a *dart*, as shown in Figure 8.14 (next page). In this figure, $\phi$ represents the *golden ratio* $\frac{1}{2}(1 + \sqrt{5})$. With these shapes, we can tile the plane in a nonperiodic way. In other words, a Penrose tiling does not have translational symmetry. This is quite different from regular and semiregular tilings, which have many translations as symmetries.

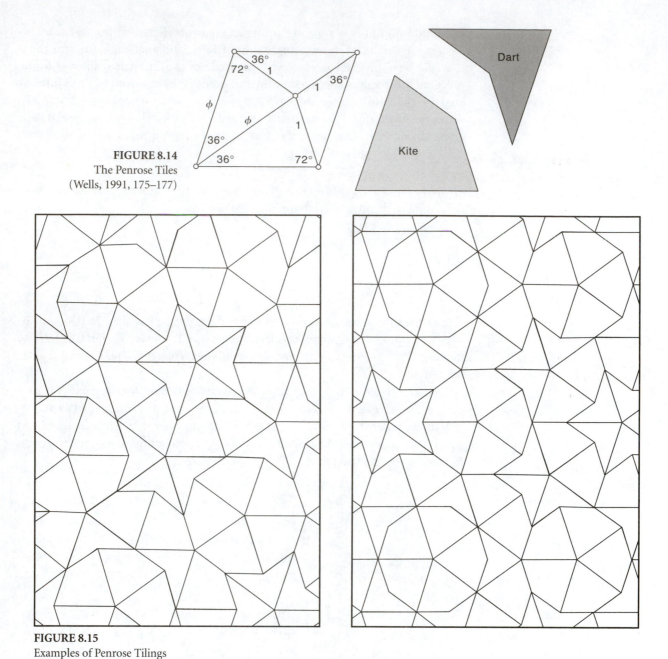

**FIGURE 8.14**
The Penrose Tiles
(Wells, 1991, 175–177)

**FIGURE 8.15**
Examples of Penrose Tilings

## 8.3 EXERCISES

Give clear and complete answers to the following exercises, expressing your explanations in complete sentences. Include diagrams whenever appropriate.

1. For each set of letters you found in Activity 4, tell what the symmetry group is.

2. Label the vertices of a square as 1, 2, 3, and 4, going counterclockwise from the upper-right corner. Listed below are some cycles. Decide which are legitimate symmetries for your square.

For those that are symmetries, tell specifically what each is.

   a. $(1\,3)\,(2\,4)$
   b. $(1\,2\,3)\,(4)$
   c. $(1\,3)\,(2)\,(4)$
   d. $(1\,4\,3\,2)$
   e. $(1\,2)\,(3)\,(4)$

3. Following are some compositions of symmetries for a square (labeled as in Exercise 2). Compute these compositions. Then identify each symmetry in the composition and the symmetry of the result.

   a. $(1\,4)\,(2\,3) \circ (1\,2\,3\,4)$
   b. $(1\,2)\,(3\,4) \circ (1\,3)\,(2\,4)$
   c. $(1)\,(2\,4)\,(3) \circ (1\,4\,3\,2)$

4. Describe in detail all symmetries of a regular $n$-gon for $n = 3, 4, 5, 6$. Then generalize, describing the symmetries of a regular $n$-gon.

5. What are the symmetries of a circle? Of a line? Give an example of a symmetry that a circle has that a line does not, and vice versa.

6. Look in the Yellow Pages for company logos. Find examples of at least eight different symmetry groups, and label each example with its symmetry group.

7. Identify the symmetries of each frieze in Figure 8.9.

8. Find examples of each of the seven types of friezes. (Try to find examples that are more interesting than the letter patterns given in the text. If you have difficulty finding these, you may create your own in Sketchpad.) Mark the basic pattern used to generate each frieze. Tell what symmetries each of your examples has.

9. Create a figure that has the symmetry group $C_5$. Do the same for $C_6$, $D_5$, and $D_6$. Label your figures with their symmetry group.

10. The groups $C_2$ and $D_1$ each contain two isometries. From the perspective of group theory, the two groups are identical (isomorphic). However, they are geometrically different. Show this by finding a figure that has $C_2$, but not $D_1$, as its symmetry group and another figure that has $D_1$, but not $C_2$, as its symmetry group.

11. There are two different groups of order four, shown in the following tables. Create a geometric figure whose symmetry group is isomorphic to the group represented by each table. (Two groups are *isomorphic* if their group tables are identical except for the symbols used.) Explain how you create these figures to match the tables.

a.

| $*_1$ | $a$ | $b$ | $c$ | $d$ |
| --- | --- | --- | --- | --- |
| $a$ | $a$ | $b$ | $c$ | $d$ |
| $b$ | $b$ | $c$ | $d$ | $a$ |
| $c$ | $c$ | $d$ | $a$ | $b$ |
| $d$ | $d$ | $a$ | $b$ | $c$ |

b.

| $*_2$ | $w$ | $x$ | $y$ | $z$ |
| --- | --- | --- | --- | --- |
| $w$ | $w$ | $x$ | $y$ | $z$ |
| $x$ | $x$ | $w$ | $z$ | $y$ |
| $y$ | $y$ | $z$ | $w$ | $x$ |
| $z$ | $z$ | $y$ | $x$ | $w$ |

12. A group you are familiar with is the "clock arithmetic" group, which uses the integers from 1 to 12. In this group, $6 + 9 = 3$, because 9 hours after 6 o'clock is 3 o'clock. Create a geometric figure whose symmetry group is isomorphic to the clock arithmetic group. (*Hint:* Think about commutativity.)

13. Suppose that $S$ is the finite set of symmetries for a planar figure. By verifying the four axioms of a group, prove that $S$ with composition forms a finite group.

14. In the operation table of a finite group, prove that any column contains all the elements of the group.

15. The property given in Exercise 14 is also true for infinite groups. Find a proof for the infinite case.

16. Consider the cyclic group $C_3$ and the dihedral group $D_3$.

   a. Prove that $C_3$ is a group.
   b. Prove that $D_3$ is a group.
   c. Show that all the elements of $C_3$ are contained in the group $D_3$, thus confirming that $C_3$ is a *subgroup* of $D_3$.

17. Consider the cyclic group $C_n$ and the dihedral group $D_n$.
    a. Prove that $C_n$ is a group.
    b. Prove that $D_n$ is a group.
    c. Show that all the elements of $C_n$ are contained in the group $D_n$, thus confirming that $C_n$ is a *subgroup* of $D_n$.

18. Prove that the rotations of $D_n$ account for half of its symmetries. Thus, $D_n$ has twice as many elements as $C_n$ has.

19. Using Sketchpad, construct the regular tilings for $n = 3$, for $n = 4$, and for $n = 6$.

20. Construct an arbitrary triangle. Use this triangle to create a tiling of the plane.

21. Create an example of a semiregular tiling different from that in Figure 8.13. (*A suggestion:* There are at least two that can be made with hexagons and triangles.)

22. Make a set of Penrose tiles, and explore ways to tile the plane with them. Construct at least one Penrose tiling that does not have translational symmetry. Prove that your tiling does not have translational symmetry.

Exercises 23 and 24 are especially for future teachers.

23. In the *Principles and Standards for School Mathematics,* the National Council of Teachers of Mathematics (NCTM) recommends that "Instructional programs from prekindergarten through grade 12 should enable all students to . . . apply transformations and use symmetry to analyze mathematical situations" (NCTM, 2000, 41). What does this mean for you and your future students?
    a. Study the Geometry Standard for one grade band (i.e., pre-K–2, 3–5, 6–8, or 9–12). What are the NCTM recommendations regarding transformations such as slides, flips, and turns and combinations of these motions?
    b. Find copies of school mathematics textbooks for these same grade levels. How are the NCTM standards reflected in those textbooks? Cite specific examples.
    c. Write a report in which you present and critique what you learn. Your report should include your responses to parts a and b.

24. Design several classroom activities involving symmetries and combinations of symmetries that would be appropriate for students in your future classroom. Write a short report explaining how the activities you design reflect both what you have learned in studying this chapter and the NCTM recommendations.

Reflect on what you have learned in this chapter.

25. Review the main ideas of this chapter. Describe, in your own words, the concepts you have studied and what you have learned about them. What are the important ideas? How do they fit together? Which concepts were easy for you? Which were hard?

26. Reflect on the learning environment for this course. Describe aspects of the learning environment that helped you understand the main ideas in this chapter. Which activities did you like? Dislike? Why?

## 8.4 CHAPTER OVERVIEW

In this chapter, we explored the various ways in which a geometric figure can be congruent to itself. These are the *symmetries* of the figure. We also discussed in greater depth the mathematical structure known as a *group*.

A *group* is a set, $G$, together with a binary operation, $\circ$, such that

- for any two elements $x$, $y$ of $G$, $x \circ y$ is in $G$ (the *closure* property);
- for any elements $x$, $y$, $z$ of $G$, $x \circ (y \circ z) = (x \circ y) \circ z$ (the *associative* property);

- there is an element $i$ of $G$ so that $i \circ x = x \circ i = x$ (the *identity* property); and

- for any element $x$ of $G$, there is an element $x^{-1}$ also in $G$ so that $x \circ x^{-1} = x^{-1} \circ x = i$ (the *inverse* property).

Sets of symmetries with the operation of composition are classic examples of groups. Cycle notation for symmetries gives a convenient way to calculate compositions of symmetries.

We saw a variety of examples of figures with finite symmetry groups. Such figures must be bounded. The finite symmetry groups can be *cyclic,* which means they are made up solely of rotations, or they can be *dihedral,* which means they contain both rotations and reflections. A major theorem, credited to Leonardo da Vinci, says that any finite symmetry group for a planar figure must be one of these two types.

**Leonardo's Theorem**　A finite symmetry group for a figure in the plane must be either the cyclic group $C_n$ or the dihedral group $D_n$.

Symmetry groups that contain translations must belong to unbounded figures. If all translations in the group are parallel to each other, the figure is called a *frieze.* We proved that there are seven possible types of friezes. However, the translations might use more than one direction, which gives rise to the 17 *wallpaper patterns.* The *Crystallographic Restriction* says that only certain angles can be rotational symmetries for a wallpaper pattern:

**The Crystallographic Restriction**　The minimal angle of rotation for a wallpaper symmetry is 60°, 90°, 120°, 180°, or 360°. All other rotation angles for a symmetry must be multiples of the minimal angle for that pattern.

An idea related to wallpaper symmetry is *tiling* the plane, that is, filling the plane with congruent copies of a polygon. We saw that we can use any quadrilateral and any triangle to tile the plane. Triangles, squares, and regular hexagons can be used to tile the plane; when such a tiling uses just one kind of regular polygon, it is called a *regular tiling. Semiregular tilings* use two regular polygonal shapes. The surprising Penrose tiles can be used to produce tilings that fill the plane without any translational symmetry.

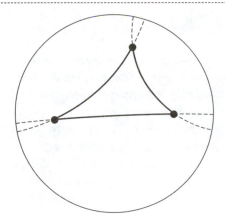

## CHAPTER NINE

# Hyperbolic Geometry

How important are axioms or postulates in a mathematical system? In Chapter 1, we presented Euclid's postulates, and in Chapter 2, we took a lot of time to carefully develop the negation of Playfair's Postulate. What if we used one of the two possible negations of Playfair's Postulate in place of the usual Euclidean parallel postulate? How much difference could negating one of Euclid's postulates really make?

The Poincaré disk is a model of the hyperbolic plane. It is a geometric situation that satisfies all of the axioms for the hyperbolic plane. The hyperbolic plane is a world in which a small change has been made in Euclid's Fifth Postulate, the parallel postulate. We invite you to explore the not-so-small ramifications of this change.

This chapter has a slightly different format from most of the previous chapters. We divided the activities and the discussion into two sections. In Part I, we invite you to do some preliminary explorations in the Poincaré disk. You are invited to explore what lines, triangles, and circles look like in this universe. Then, in Part II, we take a deeper look at the parallel postulate and begin to explain what is going on in this strange new universe. Your investigations of the hyperbolic parallel axiom—one of the two possible negations of Euclid's Parallel Postulate—will lead to being able to explain why the things you observe in the Poincaré disk are happening.

# PART I:   EXPLORING A NEW UNIVERSE

## 9.1  ACTIVITIES

To do the activities in this chapter, you will need the **Poincare Disk** tools, which are available in the **Poincare Disk.gsp** document found in **Samples | Sketches | Investigations**. The original document should be locked. Copy it onto your own disk so that you can work with it. If you save one copy of **Poincare Disk.gsp** on your own disk and then open it and save it under a different name for each activity, you can save yourself the step of going back to the original locked document each time.

Do the following activities, writing your explanations clearly in complete sentences. Include diagrams whenever appropriate. Save your work for each activity, as later work sometimes builds on earlier work. You will find it helpful to read ahead into the chapter as you work on these activities.

1. Open **Poincare Disk.gsp,** and save it on your own disk under a different name.
   a. Adjust the scroll bars on the side and bottom of the window so that you can see most of the white disk, as well as the *Disk Controls* button, in the worksheet. We refer to the boundary of this white disk as the *fundamental circle* of the hyperbolic plane.

      Press the *Disk Controls* button several times, and observe what happens. As you work on these activities, you can move this button out of your way, if necessary, by dragging the solid bar on the left side of this button.

      Find the blue dot marked *Radius.* As you work on the activities in this chapter, you can drag this dot to make the interior of the Poincaré disk larger or smaller. Your work on these activities will take place in the interior of the Poincaré disk.

   b. Choose the **Hyperbolic Segment** tool. (It is on the bottom icon in the Toolbox.) Construct several hyperbolic segments within the space of the white disk. How are these segments similar to ordinary Euclidean segments? How are these hyperbolic segments different from ordinary Euclidean segments?

      SKETCHPAD TIP    You may notice that the special hyperbolic tools work a little differently than the usual Euclidean tools. For example, to use the **Hyperbolic Segment** tool, first choose the tool, then click on the endpoints of the line segment you want to construct. If the point does not already exist, it will be constructed as you construct the segment. In general, when you use the hyperbolic tools, first choose the tool, then click on the objects that tool needs in order to do its job.

   c. After you have several hyperbolic segments in the Poincaré disk, construct some Euclidean segments with the same endpoints. Construct at least one example showing Euclidean and hyperbolic segments that appear to coincide. Construct at least one example showing Euclidean and hyperbolic segments sharing the same endpoints that clearly do not coincide. Press the *Disk Controls* button to see if you can observe what is special about the hyperbolic segments that appear to coincide with Euclidean segments.

d. Choose the **Hyperbolic Line** tool, and construct several hyperbolic lines. Construct some hyperbolic lines that contain some of the hyperbolic segments you have already constructed. What do you notice about the way the hyperbolic lines meet the fundamental circle? How could you test your conjecture about what you observe?

e. What happens to a hyperbolic segment (or a hyperbolic line) if you drag one of its points across the fundamental circle into the blue space? Does the same thing happen when you drag the endpoint of a Euclidean segment across the fundamental circle?

2. Open **Poincare Disk.gsp,** and save it on your own disk under a different name. Adjust the scroll bars on the side and bottom of the window so that you can see the entire boundary of the fundamental circle.

a. Construct a hyperbolic triangle.

b. Use the **Hyperbolic Angle** tool to measure the angles of this triangle.

SHORTCUT    You can use the **Hyperbolic Angle** tool to construct and measure an angle in one step.

c. Calculate the angle sum of this triangle.

d. Drag one of the vertices of this triangle, and observe how the angle sum changes. Does there appear to be a maximum and/or a minimum value for this sum? If so, what are these maximum or minimum values?

e. Describe the size, shape, and position of a hyperbolic triangle whose angle sum seems to be as large as possible. What factors seem to contribute to making the angle sum larger? Describe the size, shape, and position of a hyperbolic triangle whose angle sum seems to be as small as you can make it. What factors seem to contribute to making the angle sum smaller?

3. Open **Poincare Disk.gsp,** and save it on your own disk under a different name.

a. Construct a hyperbolic triangle. Use the **Hyperbolic P. Bisector** tool to construct the hyperbolic perpendicular bisector of each side of the triangle. Construct the intersection of two of these perpendicular bisectors, and label this point $O$. In Chapter 2, we saw that the point of concurrence of the three perpendicular bisectors of a triangle in Euclidean space is the circumcenter of the triangle.

SKETCHPAD TIP    To use the **Hyperbolic P. Bisector** tool, first choose the tool, then click on the endpoints of the line segment for which you want the perpendicular bisector.

SHORTCUT    You can use the **Hyperbolic P. Bisector** tool to construct a line segment and its perpendicular bisector in one step: If the segment you want to bisect has not already been constructed, choose the tool and click two places in the Poincaré disk. This action will construct a segment with those endpoints as it constructs its perpendicular bisector.

b. Are the three hyperbolic perpendicular bisectors concurrent? Drag one of the vertices of the hyperbolic triangle. Does your observation continue to hold? Always? Sometimes? Can you make a conjecture?

c. If the three perpendicular bisectors are concurrent at point $O$, is $O$ the center of a circumcircle of $\triangle ABC$? Is this circumcircle—when it exists—unique? What would you need to know to prove your conjecture?

d. In earlier chapters, we assumed that if two lines, $\ell$ and $m$, are perpendicular to two different sides of $\triangle ABC$, then $\ell$ and $m$ must intersect each other. Can we assume this in the Poincaré disk? Why or why not?

4. Open **Poincare Disk.gsp,** and save it on your own disk under a different name.

a. Construct a hyperbolic triangle, $\triangle ABC$. Use the **Hyperbolic A. Bisector** tool to construct the hyperbolic angle bisector of each angle of the triangle. Construct the point of intersection of two of these angle bisectors, and label this point $I$.

b. Are the three angle bisectors concurrent? Always? Sometimes? Can you make a conjecture?

c. If the three angle bisectors are concurrent at $I$, is $I$ the incenter of an incircle of $\triangle ABC$? Is the incircle of $\triangle ABC$—if it exists—unique? What would you need to know to prove your conjecture?

SKETCHPAD TIP   As you investigate this situation, you might want to use the **Hyperbolic Perpendicular** tool. To use this tool, click on point $P$, then click on any two distinct points of line $\ell$. This gives you the unique perpendicular from $P$ to $\ell$. (For some versions of Sketchpad, you will have to do these steps in the reverse order.)

5. Open **Poincare Disk.gsp,** and save it on your own disk under a different name.

a. Use one of the hyperbolic circle tools to construct a hyperbolic circle, $c_1$, in the Poincaré disk.

SKETCHPAD TIP   Using the **Hyperbolic Circle by CP** tool is similar to using the ordinary Euclidean **Circle by Center+Point** tool. For the **Hyperbolic Circle by CR** tool, select the desired center point of the circle, then select the endpoints of a line segment that designates the radius of the circle.

b. Construct a point, $P$, on $c_1$, and measure the hyperbolic distance from the center of the circle to point $P$. Move $P$ around the circle, and notice how the distance from $P$ to the center of the circle changes.

c. Move $c_1$ around the Poincaré disk, changing both its center and its radius. Are your observations consistent with what you already know about circles? Can you explain what is going on?

d. How are hyperbolic circles similar to Euclidean circles? How are they different?

SKETCHPAD TIP   You can use the **Circle** tool in the Toolbox or on the Construct menu to construct some Euclidean circles if this makes it easier to compare the two kinds of circles.

## HYPERBOLIC LINES AND SEGMENTS

As you worked on the activities that introduced this chapter, you observed that hyperbolic lines are somewhat different from Euclidean lines. In the Poincaré disk model of the hyperbolic plane, a line is a circular arc that meets the fundamental circle orthogonally. Two circles are *orthogonal* if the tangents constructed at their points of intersection are perpendicular. Consider the fundamental circle and a hyperbolic line, $\ell$, in the Poincaré disk. Let $X$ be a point where $\ell$ meets the fundamental circle. If $t_1$ is the tangent at $X$ to the fundamental circle and $t_2$ is the tangent at $X$ to the circular arc that represents $\ell$, then $t_1$ is perpendicular to $t_2$.

As you worked on Activity 1, you probably observed that some hyperbolic lines appear to be less curved than other lines. You may have noticed that those hyperbolic lines that pass through the center point of the fundamental circle appear to coincide exactly with Euclidean lines, and that the farther the hyperbolic line is from the center of the fundamental circle, the more curved it appears. This apparent curvature is due to the distortion caused by our attempt to represent the infinite hyperbolic plane in a bounded disk on a flat surface. Hyperbolic lines are straight lines in the world of the hyperbolic plane!

The diameter of a Poincaré disk should also be included as a hyperbolic line. We can think of a diameter of the fundamental circle as an arc of a circle of infinite radius. Because a diameter is perpendicular to the tangent lines constructed at its endpoints, any diameter of the fundamental circle is a hyperbolic line in the Poincaré disk model.

## THE POINCARÉ DISK MODEL OF THE HYPERBOLIC PLANE

The Poincaré disk is a model of a geometric world in which a different set of rules apply. In Chapter 1, we introduced Euclid's postulates; since then, we have been exploring geometric structures of Euclidean space using the model of a flat plane as represented in a Sketchpad diagram. Working with the Poincaré disk model will help us become familiar with geometric structures that are possible in the hyperbolic plane.

Whenever we set up a model of a geometric world, we must identify what we are using to represent points and lines in the model. Then we must verify that the axioms we are using are accurately represented in that model. In the Poincaré disk model, the *points* of the hyperbolic plane are the points interior to the fundamental circle. The points that lie on or are exterior to the fundamental circle are not points in the hyperbolic plane. The points that lie *on* the fundamental circle play an important role, even though they do not lie in the hyperbolic plane. (The points exterior to the fundamental circle are of no concern to us in this model.) *Lines* in the Poincaré disk are the circular arcs that are orthogonal to the fundamental circle.

Hyperbolic lines, like their Euclidean counterparts, extend as far as possible in the plane. We could say that any line "extends to infinity." This idea of points at infinity requires a special name.

**DEFINITION 9.1**     The points where a hyperbolic line meets the fundamental circle are *ideal points*. These points are said to lie at infinity for the hyperbolic plane. We denote the set of points at infinity with the Greek letter $\Omega$ (read as "omega").

Euclid's first four postulates should hold in this model, because the only axiom we are changing is the parallel postulate. Let's see how the first four postulates work in the Poincaré disk model.

1. *Given two distinct points A and B in the hyperbolic plane, there is a unique line that passes through A and B.* When you are working in the Poincaré disk, you can draw or construct points anywhere within the fundamental circle. Using the special hyperbolic tools, you can construct hyperbolic lines and hyperbolic segments through any pair of these points. These hyperbolic lines and segments are the portions of circular arcs that lie within the fundamental circle and that are orthogonal to it.

   The construction we are describing is a Euclidean construction. Given two points $A$ and $B$, we can always treat these as the endpoints of a chord of an ordinary (Euclidean) circle. In Chapter 3, we learned that we can construct infinitely many (Euclidean) circles with chord $AB$. So the question here is: For any two points in the hyperbolic plane, is *at least one* of these circles orthogonal to the fundamental circle? Is it always so? Is *exactly one* circle through points $A$ and $B$ orthogonal to the fundamental circle? From your investigations in the activities, you may suspect that the answers to these questions are "yes," "yes," and "yes." But this needs to be proven.

2. *Any line segment can be extended indefinitely.* The points of the hyperbolic plane are the points interior to the fundamental circle and do not include the points on the fundamental circle itself. In a certain sense, the fundamental circle is the boundary of the hyperbolic plane. However, the hyperbolic plane does not include the points on its boundary. Thus, a hyperbolic line does not have endpoints.

   A hyperbolic *segment,* like a Euclidean segment, does have endpoints. We say that a segment is *closed* because it includes its endpoints. A hyperbolic *line,* like its Euclidean counterpart, does not include its endpoints, and so it is said to be *open.* A hyperbolic line is open, even though it is bounded by the fundamental circle. So you can always extend a hyperbolic line segment—even if only a little bit. When you use a Sketchpad worksheet with the Poincaré disk model, you may be limited by the pixels on the computer screen in actually carrying this out, but there is always room in the hyperbolic plane to extend a line segment, even if you cannot draw this using Sketchpad.

3. *Given two distinct points A and B, a circle centered at A with radius AB can be drawn.* The Poincaré disk model does have tools for constructing hyperbolic

circles—either by CP (center and point) or by CR (center and radius). As you worked on Activity 5, you constructed several examples of hyperbolic circles, and you probably observed that these hyperbolic circles look a lot like Euclidean circles, except that the center of the hyperbolic circle may have seemed to be a bit off center.

When you measured the hyperbolic distance from the center to a point on a hyperbolic circle, you should have observed that this distance is constant, as it should be. But it probably didn't appear to your eye to be constant—at least not for all your examples of hyperbolic circles. This has to do with the way distance is measured in the hyperbolic plane. We will discuss the hyperbolic metric—the distance measure—a little later. For the moment, we can say that yes, this postulate, too, is satisfied in the hyperbolic plane.

4. *Any two right angles are congruent.* The Poincaré disk model has a tool for constructing the hyperbolic perpendicular to any line. The underlying construction—a circular arc that is orthogonal to the given hyperbolic line—is a Euclidean construction. For the hyperbolic lines to be perpendicular, the Euclidean circular arcs representing those lines must be orthogonal. Hyperbolic lines are perpendicular if the corresponding Euclidean tangent lines are perpendicular. Thus, just as all right angles are congruent in Euclidean space, all hyperbolic right angles are congruent in the Poincaré disk.

So, Euclid's first four postulates do hold in the Poincaré disk. These four postulates are also postulates or axioms in the hyperbolic plane. In Part II of this chapter, we will investigate the hyperbolic parallel postulate. Once we have verified that all five hyperbolic axioms hold in the Poincaré disk, we will be able to say that the Poincaré disk is a model of the hyperbolic plane.

The Poincaré disk is just one model of the hyperbolic plane. There are other models. One of these is the upper-half-plane model, which appears in the exercises.

## HYPERBOLIC TRIANGLES

As you worked on Activity 2, what did you observe as the angle sum of a hyperbolic triangle? Recall that when we proved that the angles of a triangle add up to the sum of two right angles (see Theorem 3.1), we used Euclid's Fifth Postulate. So, in a geometric world where Euclid's Fifth Postulate does not hold, the angle sum of a triangle might not be 180°.

Were you able to construct a triangle in the Poincaré disk whose angle sum was almost 180°? What did this triangle look like? Compare the size of the triangle with the size of the Poincaré disk. You may have noticed that the location of the triangle in the Poincaré disk makes a difference. There is a visual distortion in the Poincaré disk that is less apparent near the center of the disk. This distortion is greater near the edge of the disk. Relatively large triangles and triangles close to the boundary of the Poincaré disk will appear more distorted than relatively small triangles close to the center of the Poincaré disk. This distortion is due to the way that distance is measured in this model.

Points lying on the fundamental circle—the $\Omega$-points—are not considered to be points in the hyperbolic plane. If the lines forming the sides of a triangle intersect at an $\Omega$-point, we say that these lines do not intersect; that is, if the lines do not intersect in the world inside the Poincaré disk model, they do not intersect.

**DEFINITION 9.2**    Lines that do not intersect are said to be *parallel lines.*

If a triangle could have one or more of its vertices on the fundamental circle, such a triangle would have parallel sides and would not be a triangle at all!

The idea that the angle sum of a hyperbolic triangle is less than 180° is so startling that we define the difference between 180° (the expected angle sum) and the actual angle sum of a hyperbolic triangle as the *defect of the triangle.*

**DEFINITION 9.3**    The *defect of* $\triangle ABC$ is the difference between 180° and the angle sum of that triangle:

$$\text{defect}(\triangle ABC) = 180° - (m\angle A + m\angle B + m\angle C).$$

In the hyperbolic plane, every triangle has a positive defect. (You will have an opportunity to prove this in the exercises.)

A geometric figure with three vertices, some of which lie on the fundamental circle, is called an *asymptotic triangle.* An asymptotic triangle has one, two, or all three of its vertices at infinity. These vertices are not really in the hyperbolic plane (rather they lie, in some sense, just beyond the edge of hyperbolic space). In an asymptotic triangle, the sides that meet at infinity do not intersect; that is, some (or all) of the sides of an asymptotic triangle are parallel.

An asymptotic angle, such as $\angle PQR$ with its vertex $Q$ actually on the fundamental circle, would have an (hyperbolic) angle measure of 0° at $Q$. Can you prove this? Think about the hyperbolic lines $PQ$ and $QR$ and how these lines meet the fundamental circle. If $\overleftrightarrow{PQ}$ and $\overleftrightarrow{QR}$ are both orthogonal to the fundamental circle at $Q$, what can you conclude about the measure of $\angle PQR$?

If all three vertices of $\triangle PQR$ lie on the fundamental circle—that is, if $\triangle PQR$ is triply asymptotic—what is the defect of this triangle? Can you prove your conjecture?

To find the angle sum of a quadrilateral, a pentagon, or any *n*-gon, an effective strategy is *to triangulate the figure*—that is, construct diagonals to break up the figure into triangles. In the Euclidean plane, for example, we can divide any quadrilateral into two triangles by drawing one diagonal. The angle sum of a quadrilateral is equal to the angle sum of two triangles, which is 360° for *Euclidean* quadrilaterals.

We can do the same thing in the hyperbolic plane. We triangulate a hyperbolic polygon by constructing an appropriate number of diagonals. The fact that hyperbolic triangles have angle sum less than 180° has a profound impact on the angle sum for quadrilaterals and other polygons in the hyperbolic plane.

## HYPERBOLIC CIRCLES

A *circle* is a set of points that are equidistant from a fixed center point. The *radius* of a circle is the distance from the center to a point on the circle. As you saw from your work on Activity 5, there are circles in the hyperbolic plane. Hyperbolic circles look like Euclidean circles. They appear to be round, just as ordinary Euclidean circles are, but the center appears to be a bit off center. However, the distance from the center of the circle to any point on the circle is, in fact, constant, just as in Euclidean space.

In the Euclidean plane, as the radius $r$ of a circle increases toward infinity, the curvature of the circle decreases. In other words, the arc of the circle gets straighter and straighter. Eventually, in the limit as $r \longrightarrow \infty$, the circle would be a straight line. Working in the Poincaré disk, we find that we are not allowed to construct hyperbolic circles *on* or *outside of* the fundamental circle. However, we can imagine the limiting situation where the hyperbolic circle has its center on the fundamental circle and has infinite radius. Such an object is a *horocycle* (Baragar, 2001; Coxeter, 1969; Greenberg, 1980).

## MEASURING DISTANCE IN THE POINCARÉ DISK MODEL

Measuring the length of a segment in the Poincaré disk is a bit tricky. The idea is that the hyperbolic plane, like the Euclidean plane, is infinite. Postulate 2 requires that any line segment can be extended indefinitely. This means that our method for measuring distance must give lines of infinite length, even though the lines are bounded by the fundamental circle.

Recall that a rule for measuring distance is called a *metric*. You are familiar with the usual distance formula for measuring distance in the Euclidean plane, which is based on the Pythagorean Theorem. If point $A$ has coordinates $(a_1, a_2)$ and point $B$ has coordinates $(b_1, b_2)$, the Euclidean distance between $A$ and $B$, denoted $d(A, B)$, is given by the formula

$$d(A, B) = \sqrt{(a_1 - b_1)^2 + (a_2 - b_2)^2}.$$

Chapter 5 introduced a different metric for the Euclidean plane, the taxicab metric, based on horizontal and vertical distances. Both the taxicab metric and the Pythagorean metric satisfy three basic axioms; a metric for the Poincaré disk must also satisfy these same axioms. Let us briefly restate these axioms.

Suppose $A$ and $B$ are points in the space. The distance from $A$ to $B$ must satisfy the following requirements:

**Metric Axiom 1**  If $A$ and $B$ are points, then $d(A, B) \geq 0$, and $d(A, B) = 0$ if and only if $A = B$.

**Metric Axiom 2**  If $A$ and $B$ are points, then $d(A, B) = d(B, A)$.

**Metric Axiom 3**  (the *triangle inequality*) If $A$, $B$, and $C$ are points, then $d(A, B) + d(B, C) \geq d(A, C)$.

The formula for measuring the distance between points in the Poincaré disk must meet these three requirements, as well as allow the length of a line to be infinite, even though the hyperbolic plane is bounded by the fundamental circle. (We told you this would be a bit tricky!) Poincaré developed the following formula for the distance between two points:

$$d(A, B) = \left| \ln \left( \frac{AM/AN}{BM/BN} \right) \right| = \left| \ln \left( \frac{AM \cdot BN}{AN \cdot BM} \right) \right|.$$

To use this formula, you must imagine the Poincaré line segment $AB$ extended until it reaches the boundary of the hyperbolic plane (that is, until it reaches the fundamental circle). The two points where it meets the boundary are $M$ and $N$ (it does not matter which is which). The calculation then uses four *Euclidean distances*, $AM$, $AN$, $BM$, and $BN$.

It is necessary to verify that this distance formula meets the requirements given for being a metric in the hyperbolic plane.

1. The first axiom requires that $d(A, B) \geq 0$; the absolute value takes care of this for us. But what about the requirement that $d(A, B) = 0$ if and only if $A = B$? How can the natural logarithm, $\ln(x)$, be equal to 0? Will this happen if and only if $A$ and $B$ are the same point?

2. The second axiom requires that $d(A, B) = d(B, A)$. In other words, we must show that

$$\left| \ln \left( \frac{AM \cdot BN}{AN \cdot BM} \right) \right| = \left| \ln \left( \frac{BM \cdot AN}{BN \cdot AM} \right) \right|.$$

The expressions $\frac{AM \cdot BN}{AN \cdot BM}$ and $\frac{BM \cdot AN}{BN \cdot AM}$ are reciprocals of each other. What is going on in the hyperbolic distance formula that allows $d(A, B)$ to be the same as $d(B, A)$?

3. The third axiom, the triangle inequality, can be proved by integration in the complex plane. This requires mathematical ideas that go well beyond the scope of this course.

In the exercises, you will have an opportunity to prove that the formula developed by Poincaré does indeed meet the necessary requirements for a metric in the hyperbolic plane.

Axiom 2 requires that any hyperbolic segment can be extended indefinitely. Is there a maximum length for line segments in the hyperbolic plane? Suppose segment $AB$ lies in line $\ell$ in the Poincaré disk. For the length of segment $AB$ to increase, points $A$ and $B$ must each move closer to the edge of the hyperbolic plane. One way this could happen would be for $A$ to move closer to $N$ and $B$ closer to $M$. Thus, the Euclidean lengths of $AN$ and $BM$ would both decrease and would even approach 0. Meanwhile, the Euclidean lengths of $AM$ and $BN$ would be getting longer (possibly up to the maximum diameter of the Poincaré disk). In other words,

$$\left| \ln \left( \frac{AM \cdot BN}{AN \cdot BM} \right) \right| \longrightarrow \left| \ln \left( \frac{MN \cdot MN}{0 \cdot 0} \right) \right|.$$

Thus, even though hyperbolic lines are bounded by the Poincaré disk, they may be infinitely long. (There is another case to consider: the case where $A$ moves closer to $M$ and $B$ closer to $N$. You will have an opportunity to complete this proof in the exercises.)

## CIRCUMCIRCLES AND INCIRCLES OF HYPERBOLIC TRIANGLES

In Chapter 2, we saw that the perpendicular bisectors of the three sides of Euclidean $\triangle ABC$ are concurrent. The point where these lines intersect, which we have been calling $O$, is equidistant from the three vertices of the triangle. Because the distances $OA$, $OB$, and $OC$ are equal, the point $O$ is the center of a circle through the three vertices $A$, $B$, and $C$. Because this circle surrounds $\triangle ABC$, it is called the circumcircle of $\triangle ABC$.

**Euclidean Theorem**    The three perpendicular bisectors of the sides of a triangle are concurrent. The point where they intersect, called the *circumcenter*, is often denoted as $O$. (See Theorem 2.7.)

What is the situation in the hyperbolic plane? As you worked on Activity 3, you probably observed that when the three perpendicular bisectors of the sides of the triangle did intersect, they were concurrent (that is, they intersected in the same point). In this case, the point of concurrency appears to be the center of the circumcircle of the triangle. (This needs to be proved.) But there is the disturbing fact that sometimes the three perpendicular bisectors of the sides of the triangle did not intersect! Somehow these lines, which were constructed perpendicular to the three sides of a triangle, turned out to be parallel.

**CONJECTURE 9.1**    If the three perpendicular bisectors of the sides of a triangle in the Poincaré disk are concurrent at a point $O$, then the circle with center $O$ and radius $OA$ also contains the points $B$ and $C$.

Your proof for this conjecture will be similar to your proof of the corresponding Euclidean theorem, but you must now account for the bewildering case where the three perpendicular bisectors do not intersect. Our explorations of the parallel postulate in Part II of this chapter will give you the tools you need to explain what is happening.

Another of the concurrence properties for Euclidean triangles, which we explored in Chapter 2, stated that the angle bisectors of a triangle are concurrent.

**Euclidean Theorem**    The three angle bisectors of a triangle are concurrent. The point where they intersect, called the *incenter*, is often denoted as $I$. (See Theorem 2.6.)

As you worked on Activity 4, it should have appeared that the three angle bisectors are concurrent. A proof of this conjecture will follow an outline similar to the one given in Chapter 2 for Theorem 2.6.

**CONJECTURE 9.2**   The three angle bisectors of a triangle in the Poincaré disk are concurrent. The point where they intersect, called the *incenter,* is often denoted as *I.*

The first step in the proof of this conjecture is to drop perpendiculars from *I* to each side of the triangle; call the feet of these perpendiculars *W*, *X*, and *Y*. As you work through the details of this proof, you should be able to prove that $IW \cong IX \cong IY$, so that *I* is the center of a circle through points *W*, *X*, and *Y*. Is this circle an incircle of the hyperbolic $\triangle ABC$? If so, is this incircle unique? Of course, your answers to these questions will require proofs.

To answer these and other questions about what is going on in the hyperbolic plane, we need to develop a deeper understanding of the parallel postulate as it pertains to the hyperbolic plane.

## CONGRUENCE OF TRIANGLES IN THE HYPERBOLIC PLANE

Your proofs of Conjectures 9.1 and 9.2 require you to determine whether two triangles in the hyperbolic plane are congruent. Because distances are visually distorted in the Poincaré disk, we cannot rely on visual inspection; we must approach the question of congruence of triangles by using the axioms and theorems of the hyperbolic plane. Because Euclid's first four postulates hold in the hyperbolic plane, any theorems that we proved using only these first four postulates will also hold in the hyperbolic plane.

In Chapter 2, we agreed to accept the SAS criterion for congruence of triangles as an axiom. Using SAS with Euclid's first four postulates, we were able to prove the ASA, SSS, and AAS criteria for triangle congruence *without using Euclid's Fifth Postulate.* (See Exercises 22–24, Chapter 2.) Thus, we can continue to use these criteria to prove congruence of triangles in the hyperbolic plane.

An important difference between the Euclidean plane and the hyperbolic plane is the AAA criterion. In the Euclidean plane, two triangles that have three pairs of congruent angles will be similar but not necessarily congruent. In the hyperbolic plane, however, two triangles with three pairs of congruent angles must be congruent. After learning some facts about quadrilaterals in the hyperbolic plane, you will be asked to prove this in the exercises.

# PART II: THE PARALLEL POSTULATE IN THE POINCARÉ DISK

Playfair's Postulate, which is equivalent to Euclid's Fifth Postulate, says:

Given any line $\ell$ and any point $P$ not on $\ell$, there is exactly one line on $P$ that is parallel to $\ell$.

In the Euclidean plane, two lines either intersect or they are parallel. In fact, we use this postulate to define what we mean when we say two lines are parallel.

**DEFINITION 9.4** Two lines, $\ell$ and $m$, are *parallel* if they do not intersect.

We will continue to use this definition for parallel lines in the hyperbolic plane. If two hyperbolic lines do not intersect, they are parallel. However, in the Poincaré disk, there are two different situations that give nonintersecting hyperbolic lines: There are hyperbolic lines that intersect (or appear to intersect) on the fundamental circle, and there are hyperbolic lines that do not intersect at all. We will investigate both of these kinds of parallel lines.

## 9.3 ACTIVITIES

Do the following activities, writing your explanations clearly in complete sentences. Include diagrams whenever appropriate. Save your work for each activity, as later work sometimes builds on earlier work. You will find it helpful to read ahead into the chapter as you work on these activities.

6. Open a copy of **Poincare Disk.gsp,** and save it on your own disk under a different name.
    a. Construct a hyperbolic line, $a_1$, in the Poincaré disk. Construct two additional points on $a_1$, and label these $O_1$ and $O_2$. (In Sketchpad, you can type the labels as O[1] and O[2] to get the subscripts.) Drag $O_1$ and $O_2$ to opposite ends of $a_1$—actually drag $O_1$ and $O_2$ *all the way* to the ends of $a_1$. By dragging these points as far to the ends of $a_1$ as Sketchpad allows, you are, in effect, dragging them to infinity. When you have dragged $O_1$ and $O_2$ as far as possible to the ends of $a_1$, they represent the $\Omega$-points of $a_1$.
    b. Construct a point, $P$, not on $a_1$.
    c. Construct a hyperbolic line through $P$ that is parallel to (that is, does not intersect) $a_1$. Label this line $a_2$.
    d. With $O_1$ and $O_2$ dragged off to the ends of $a_1$ so that they represent the $\Omega$-points of $a_1$, construct hyperbolic lines $PO_1$ and $PO_2$. Construct a point, $R$, on the ray $PO_1$ and a point, $S$, on the ray $PO_2$. The rays $PR$ and $PS$ are called the *limiting parallel rays* from $P$ to the line $a_1$.
        Are $\overrightarrow{PR}$ and $\overrightarrow{PS}$, as constructed here, the last pair of rays through $P$ that intersect $a_1$ or the first pair of rays through $P$ that do not intersect $a_1$? Explain.

e. Look again at line $a_2$, and observe its position relative to the limiting parallel rays $PR$ and $PS$. How many ways could you reposition line $a_2$ so that it continues to be parallel to $a_1$? If you reposition line $a_2$ so that it is no longer parallel to $a_1$, what happens to its position relative to the limiting parallel rays $PR$ and $PS$?

7. Open a copy of **Poincare Disk.gsp**.
   a. Construct a hyperbolic line, $\ell$, in the Poincaré disk.
   b. Construct a point, $P$, that is not on $\ell$. Using the **Hyperbolic Perpendicular** tool, construct a perpendicular from $P$ to $\ell$. Label the foot of the perpendicular from $P$ to $\ell$ as $Q$.
   c. Construct the limiting parallel rays $PR$ and $PS$, as you did in Activity 6d. Measure $\angle QPR$ and $\angle QPS$. What do you observe?
   d. Move point $P$, and adjust $\overrightarrow{PR}$ and $\overrightarrow{PS}$ (if necessary) so that they are again limiting parallel rays from $P$ to $\ell$. Does your observation about the measures of $\angle QPR$ and $\angle QPS$ continue to hold? Can you make a conjecture?

8. Open a copy of **Poincare Disk.gsp**.
   a. Construct a hyperbolic line, $a_1$, in the Poincaré disk.
   b. Construct a point, $P$, that is not on $a_1$. Drop a perpendicular from $P$ to line $a_1$. Label the foot of the perpendicular from $P$ to $a_1$ as $Q$.
   c. Construct the hyperbolic perpendicular bisector of segment $PQ$, and call this line $a_2$. Mark the intersection of $PQ$ and $a_2$ as $R$. Measure the hyperbolic distance $QR$.
   d. Construct a point, $X$, on $a_1$ and a point, $Y$, on $a_2$. Measure the hyperbolic distance from $X$ to $Y$. Move points $X$ and $Y$ to make this distance as small as possible. What do you observe?
   e. Move point $P$ so that lines $a_1$ and $a_2$ are changed. Adjust points $X$ and $Y$ so that the hyperbolic distance from $X$ to $Y$ is again as small as possible. Does your conjecture continue to hold? What would you need to know to prove your conjecture?

9. A quadrilateral, $ABCD$, with right angles at $A$ and $B$ and congruent sides $AD$ and $BC$ is called a *Saccheri quadrilateral*. Construct an example of this in the Poincaré disk. Use the **Poincare Disk** tools to construct right angles at $A$ and $B$. Construct points $C$ and $D$ so that sides $AD$ and $BC$ are the same length. Measure the angles at $C$ and $D$. What do you observe? Can you make a conjecture?

10. A quadrilateral with three right angles is called a *Lambert quadrilateral*. Construct an example of this in the Poincaré disk. What do you observe about the fourth angle of this quadrilateral? Make a conjecture.

11. Construct a quadrilateral with four right angles in the Poincaré disk. What do you observe? Make a conjecture. What would you need to know to prove your conjecture?

# THE HYPERBOLIC AND ELLIPTIC PARALLEL POSTULATES

In Chapter 2, we discussed how to develop the negation of Playfair's Postulate. We found that the negation of Playfair's Postulate is as follows:

There is a line $\ell$ and there is a point $P$ not on $\ell$ such that either there are no lines through $P$ parallel to $\ell$ or there is more than one line through $P$ parallel to $\ell$.

This negation of Playfair's Postulate says that either one thing happens (there are no lines through $P$ parallel to $\ell$) or another thing happens (there is more than one line through $P$ parallel to $\ell$). Let us make two separate statements from this negation of Playfair's Postulate. In doing this, we will get two different non-Euclidean postulates—one for elliptic space and another for hyperbolic space.

## The Elliptic Parallel Postulate

If we take the first part of the negation of Playfair's Postulate, we get the statement that there is a line $\ell$ and there is a point $P$ not on $\ell$ such that there are no lines through $P$ parallel to $\ell$. Because lines are either parallel or intersecting, we can state this a little more simply as follows:

**Elliptic Parallel Postulate** There is a line $\ell$ and there is a point $P$ not on $\ell$ such that every line through $P$ intersects $\ell$.

Spherical geometry is one possible model of an *elliptic space.* Consider the geometry of our planet Earth, as an airplane pilot must: A straight line from any city (or airport) to another must follow some kind of curve. (Otherwise, the airplane would be on a trajectory into outer space!) The straightest line, or *geodesic,* on a sphere is a great circle. A *great circle* on a sphere is a circle whose center coincides with the center of the sphere itself. Thus, the diameter of a great circle is the same as the diameter of the sphere. The great circles that we are most familiar with on Earth are the longitude lines and the equator. Of course, there are many other great circles. If you look at the routes traveled by airlines from one city to another, you will see that many of these routes follow the path of a great circle; you might have to look at the cities on a globe rather than on a flat map to see this. Any flat map of our world necessarily has some distortion. Our world is not flat, and to represent it on a flat piece of paper requires that we stretch it or tear it in some way to fit it to the paper.

The elliptic parallel postulate, as stated above, is an existential statement, claiming that there is at least one situation in which this property of points and lines is true. (Playfair's Postulate is a universal statement, which states that its property holds in every situation. The negation of a universal statement is always an existential statement. See page 40 to review the process of negating a quantified statement.) In elliptic space, it is possible to prove a slightly stronger statement than the elliptic parallel postulate. Because this is not the focus of our study in this chapter, we will simply state this theorem here.

**THEOREM 9.3**    **Elliptic Parallel Theorem**    Given any line $\ell$ and any point $P$ not on $\ell$, every line through $P$ intersects $\ell$.

For an example of this, consider Earth (assuming it has a perfect spherical shape) as a model of spherical space. The *points* in this model are the points on Earth's surface, and the *lines*—called *great circles* of the sphere—are the circles whose center coincides with Earth's center. Thus, the equator and all the lines of longitude are among the lines in this model. Consider the equator as the line $\ell$ and any point $P$ not on the equator. There is a longitudinal line through $P$ that intersects the equator twice. In fact, there are many lines (great circles) through $P$ other than that longitudinal line; every line (every great circle) through $P$ will intersect the equator twice at diametrically opposite points.

## The Hyperbolic Parallel Postulate

In this chapter, we are focusing on the hyperbolic parallel postulate. We want to investigate the consequences of accepting this postulate in place of the Euclidean parallel postulate. In Activity 6, you began to investigate a world in which it is possible to have a line $\ell$ and a point $P$ not on $\ell$ with more than one line through $P$ that does not intersect $\ell$. In the Poincaré disk model, we have a world in which the hyperbolic parallel postulate holds.

**Hyperbolic Parallel Postulate**    There is a line $\ell$ and there is a point $P$ not on $\ell$ such that more than one line through $P$ is parallel to $\ell$.

The Poincaré disk is one possible model of the *hyperbolic plane*. As with elliptic space, any attempt to represent the hyperbolic plane on a flat surface, such as a sketch on paper or an interactive diagram on a computer screen, introduces some distortion. Nevertheless, we can learn a lot about the hyperbolic plane by studying images in the Poincaré disk model. An important consequence of the hyperbolic parallel postulate is the existence of at least one triangle whose angle sum is less than 180°.

**THEOREM 9.4**    There is at least one triangle whose angle sum is less than the sum of two right angles.

**Proof**    Using the hyperbolic parallel postulate, we know that there is at least one line $\ell$ and at least one point $P$ not on $\ell$ such that at least two lines through $P$ are parallel to $\ell$. (See Figure 9.1.) From $P$, drop a perpendicular $PQ$ to line $\ell$. Construct line $m$ through $P$ perpendicular to $PQ$. Because $\ell$ and $m$ are both perpendicular to $PQ$, we know that $\ell \parallel m$. (You will have an opportunity to prove this in the exercises.) There is at least one more line through $P$ that is parallel to $\ell$; call this line $n$. Let $X$ and $Y$ be points on lines $m$ and $n$, respectively, on the same side of $PQ$. Because $m$ and $n$ are different lines, the measure of $\angle XPY$ is positive (even if very small). Furthermore, $\angle XPY + \angle QPY = 90°$ (or a right angle).

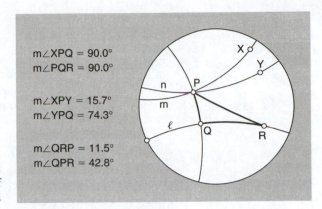

m∠XPQ = 90.0°
m∠PQR = 90.0°

m∠XPY = 15.7°
m∠YPQ = 74.3°

m∠QRP = 11.5°
m∠QPR = 42.8°

**FIGURE 9.1**
A Triangle with Angle Sum
Less Than the Sum of
Two Right Angles

There is a point $R$ on $\ell$ (on the same side of $PQ$ as $X$ and $Y$) so that points $P$, $Q$, and $R$ form $\triangle PQR$. Because $\overrightarrow{PR}$ intersects line $\ell$ and $\overrightarrow{PY}$ doesn't, $\overrightarrow{PR}$ must lie interior to $\angle QPY$, so that $\angle QPR < \angle QPY$. As $R$ is pushed farther and farther out on line $\ell$, making $QR$ longer, $\angle QRP$ gets smaller and smaller, approaching 0. Eventually, the measure of $\angle QRP$ will be smaller than the measure of $\angle XPY$.

So, $\triangle PQR$ has one right angle (at $Q$), and the other two angles add up to less than a right angle:

$$\angle QRP < \angle XPY \text{ and } \angle QPR < \angle QPY,$$

so that

$$\angle QRP + \angle QPR < \angle XPY + \angle QPY = 90°.$$

So, $\triangle PQR$ is a triangle whose angle sum is less than the sum of two right angles.

-------------------------------------------

The existence of a triangle whose angle sum is less than the sum of two right angles makes it possible to prove a remarkable fact: Rectangles do not exist in hyperbolic geometry! (You will have a chance to prove this in the exercises.) Because rectangles do not exist, we can strengthen the hyperbolic parallel postulate from an existence statement to a universal statement.

**THEOREM 9.5**  **Hyperbolic Parallel Theorem**  Given any line $\ell$ and any point $P$ not on $\ell$, there are at least two lines through $P$ that are parallel to $\ell$.

**Proof**  Consider line $\ell$ and point $P$ not on $\ell$. Drop a perpendicular from $P$ to $\ell$, and call the foot of $P$ in $\ell$ point $Q$. Construct a line through $P$ perpendicular to $PQ$, and call this line $n$. Choose point $R$ (different from $Q$) on line $\ell$, and construct line $t$ through $R$ perpendicular to $\ell$. From $P$, construct a perpendicular to line $t$, and call the foot of $P$ in $t$ point $S$. (See Figure 9.2.) Now we have quadrilateral $PQRS$, with $PQ \perp QR$, $QR \perp RS$, and $RS \perp SP$. Point $S$ cannot be on line $n$, for if it were, $PQRS$ would be a rectangle. Therefore, given any line $\ell$ and any point $P$ not on $\ell$, there are at least two lines through $P$ that are parallel to $\ell$ (Greenberg, 1980).

-------------------------------------------

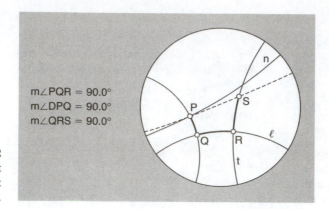

**FIGURE 9.2**
Lines *n* and *PS* are distinct
lines through *P* that do not
intersect line $\ell$.

$m\angle PQR = 90.0°$
$m\angle DPQ = 90.0°$
$m\angle QRS = 90.0°$

## PARALLEL LINES IN THE HYPERBOLIC PLANE

Two lines are *parallel* if they do not intersect. As you saw when you were working on Activity 6, there are different ways in which lines do not intersect. Two lines might seem to intersect at a point on the fundamental circle, but this point is not actually in the hyperbolic plane. These lines are parallel because they fail to intersect—but just barely. Such lines might be said to *intersect at a point at infinity*. We call such lines *limiting parallels*.

If you start with a hyperbolic line $\ell$ and a point $P$ not on $\ell$, as you did in Activity 6, you can construct two limiting parallel rays through $P$. Assume the $\Omega$-points are labeled so that $\overrightarrow{PO_1}$ is a limiting parallel ray to $\ell$ on the left, and $\overrightarrow{PO_2}$ is a limiting parallel ray on the right. These limiting parallel rays divide the family of all possible rays through $P$ into three sets. Here are two of them: There are rays through $P$ that lie to the left of the limiting parallel $\overrightarrow{PO_1}$, and there are rays through $P$ that lie to the right of $\overrightarrow{PO_2}$. None of the rays in either of these sets intersects $\ell$; that is, all of these rays are parallel to $\ell$. The third set of rays contains the rays through $P$ that lie between $\overrightarrow{PO_1}$ and $\overrightarrow{PO_2}$; all of these rays intersect line $\ell$. (As you reread this paragraph, you might find it helpful to interact with your Sketchpad worksheet for Activity 6, particularly part e.)

What about $\overrightarrow{PO_1}$ and $\overrightarrow{PO_2}$? These rays lie on the boundary between the rays that intersect $\ell$ and the rays that do not. Do $\overrightarrow{PO_1}$ and $\overrightarrow{PO_2}$ intersect $\ell$?

Points $O_1$ and $O_2$ are ideal points—points that lie at the ends of the line. By dragging $O_1$ and $O_2$ all the way to the ends of the line, we have dragged them off to infinity. Because a line is open, it does not contain these endpoints. As ideal points, $O_1$ and $O_2$ lie on the fundamental circle; they do not lie on line $\ell$. Thus, the limiting parallel rays $PO_1$ and $PO_2$ are parallel to $\ell$.

All the lines through $P$ that lie outside the angle formed by the two limiting parallel rays (there are infinitely many of them) are also parallel to $\ell$. Various authors refer to lines that are parallel in this way as *ultraparallel, superparallel,* or *hyperparallel* lines (Baragar, 2001; Coxeter, 1969; Greenberg, 1980).

As you worked on Activity 7, you started with a line $\ell$ and a point $P$ not on $\ell$. You constructed the point $Q$, which is the foot of the perpendicular from

$P$ to $\ell$. Then you constructed the *limiting parallel rays PR and PS*. You might have observed that $\overrightarrow{PQ}$ bisects $\angle RPS$ or that $\angle QPR$ and $\angle QPS$ are congruent. The angle between $PQ$ and either one of the limiting parallel rays $PR$ or $PS$ is called the *angle of parallelism*.

**THEOREM 9.6**  For a given line $\ell$ and a point $P$ not on $\ell$, the two angles of parallelism are congruent.

---

In doing Activity 8, you constructed $\overleftrightarrow{PQ}$ so that it would be perpendicular to line $a_1$. You constructed another line, $a_2$, perpendicular to segment $PQ$ at point $R$. Thus, $\overleftrightarrow{PQ}$ is a *common perpendicular* to lines $a_1$ and $a_2$. It can be proved that in the hyperbolic plane, if two lines $\ell$ and $m$ do not intersect, they will either be limiting parallels to each other or they will have a common perpendicular. It can also be proved that if $\overleftrightarrow{PQ}$ is a common perpendicular to lines $\ell$ and $m$, it is the only common perpendicular between these lines. Furthermore, segment $PQ$ is the shortest segment joining a point of $\ell$ to a point of $m$.

## QUADRILATERALS IN THE HYPERBOLIC PLANE

Activities 9, 10, and 11 invited you to investigate properties of quadrilaterals in the hyperbolic plane.

**DEFINITION 9.5**  A quadrilateral with three right angles is called a *Lambert quadrilateral*.

Did you observe that the fourth angle is acute? You can prove this result by triangulating the quadrilateral and using what you have observed about the angle sum of triangles in the hyperbolic plane.

**DEFINITION 9.6**  A quadrilateral with a pair of congruent sides that are both perpendicular to a third side is called a *Saccheri quadrilateral*.

If $ABCD$ is a Saccheri quadrilateral with right angles at $A$ and $B$ and congruent sides $AD$ and $BC$, the two right angles, $\angle A$ and $\angle B$, are called the *base angles* and side $AB$, which lies between the base angles, is called the *base* of the Saccheri quadrilateral. The angles above the base angles, $\angle C$ and $\angle D$, are the *summit angles,* and side $CD$ is the *summit* of the Saccheri quadrilateral. As you worked on Activity 9, you probably observed that the summit angles are congruent. To prove this, triangulate the Saccheri quadrilateral in two different ways and use congruent triangles to compare the angles. Because the angle sum of a triangle in the hyperbolic plane is always less than 180°, the summit angles of a Saccheri quadrilateral must be acute.

Rectangles and squares are so familiar to us that it is hard to believe that there could be any difficulty constructing them in the Poincaré disk. As you worked on Activities 9, 10, and 11, you may have been able to draw or construct quadrilaterals with one, two, or three right angles but not one with four right angles. In fact, rectangles cannot be constructed in the hyperbolic plane—they do not exist there! But this needs to be proved. Otherwise, we might think that we have not been successful because we have not been sufficiently clever. (You will have the opportunity to develop this proof in the exercises.)

## 9.5 EXERCISES

Give clear and complete answers to the exercises, expressing your explanations in complete sentences. Include diagrams whenever appropriate.

1. Use the underlying Euclidean constructions to prove that any two right angles in the Poincaré disk model are congruent.

2. Consider a circle, $c_1$, in the Euclidean plane.
   a. Prove that for any two points, $P$ and $Q$, that lie interior to $c_1$, there is *exactly one* circular arc through $P$ and $Q$ that is orthogonal to $c_1$.
   b. Explain how this construction verifies that Euclid's First Postulate holds in the Poincaré disk model of the hyperbolic plane.

3. Given a line $\ell$ and a point $P$ not on $\ell$ in the Poincaré disk model of the hyperbolic plane, construct a segment, $PQ$, perpendicular to $\ell$. Prove that if a line, $m$, through $P$ is perpendicular to $PQ$, then $m$ will not intersect $\ell$.

4. Given two distinct points, $A$ and $B$, in the Poincaré disk, prove that there is a unique hyperbolic line through $A$ and $B$. (*Note:* There are two things to prove here. You can use methods from Chapter 3 here, because you are trying to prove that from among all the circles in a certain infinite family of circles, there is only one that meets the criteria for being a line in the Poincaré disk.)

5. Given a line $\ell$ and a point $P$ not on $\ell$, prove that the angle of parallelism is the same in both directions.

6. Prove that in the hyperbolic plane, the angle of parallelism is acute.

7. What is the angle of parallelism in the Euclidean plane? Explain why.

8. Prove that the first two metric axioms hold for any two points in the Poincaré disk.

9. Construct an example in the Poincaré disk that illustrates the third metric axiom.

10. Complete the proof that the maximum length of a line in the Poincaré disk is $\infty$. (See page 212.)

11. Let $A$ be any point in the Poincaré disk. Prove that the distance from $A$ to the fundamental circle is infinite.

12. In the Poincaré disk model, prove that every triangle has positive defect.

13. In the hyperbolic plane, prove the following.
    a. At least one triangle has positive defect.
    b. Every triangle has positive defect.

14. Prove that the angle sum of any triangle in the hyperbolic plane is less than 180°.

15. Suppose that $R$ and $T$ are two points in the hyperbolic plane and $S$ is an *ideal point*. Prove that the measure of $\angle RST$ is 0°.

16. What is the angle sum of a triply asymptotic triangle? Explain.

17. Let $PQRS$ be a quadrilateral in the hyperbolic plane. Prove or disprove that the sum of the angle measures of $PQRS$ is always less than 360°.

18. Let $P_1P_2 \ldots P_n$ be an $n$-sided polygon in the hyperbolic plane. Find upper and lower bounds for the angle sum of this polygon. How do you know that these bounds are correct?

19. Under what conditions will the perpendicular bisectors of the sides of a triangle be concurrent? Under what conditions will they be parallel?

20. Prove that if the three perpendicular bisectors of the sides of $\triangle ABC$ in the Poincaré disk are concurrent at a point $O$, then the circle with center $O$ and radius $OA$ also contains the points $B$ and $C$. Prove that this circumcircle is unique.

21. Consider $\triangle ABC$ in the Poincaré disk.
    a. Prove that the bisectors of $\angle A$, $\angle B$, and $\angle C$ are concurrent. Denote the point of concurrency as $I$.
    b. Prove that $I$ is the center of the *unique* incircle.

22. Prove that the summit and the base of a Saccheri quadrilateral are hyperparallel to each other.

23. Prove or disprove that the Pythagorean Theorem holds in the hyperbolic plane.

24. In the hyperbolic plane, if the hypotenuse and a leg of one right triangle are congruent to the hypotenuse and a leg of a second right triangle, are the triangles congruent? Prove this or find a counterexample.

25. In the hyperbolic plane, prove that AAA is a criterion for congruent triangles—that is, if two triangles are similar, they are also congruent.

26. To find the distance from a point $P$ to a line $\ell$, we must find the point $X$ on $\ell$ that is as close as possible to $P$. Then the distance from $P$ to $\ell$ is defined as $d(P, \ell) = d(P, X)$.
    a. Given a line $\ell$ and a point $P$ in the hyperbolic plane, develop a procedure for locating the point $X$ on $\ell$ that is closest to $P$. Prove that your procedure is correct.
    b. Given lines $\ell$ and $m$ with $\ell \parallel m$, let $P$, $Q$, and $R$ be points on $\ell$. Prove or disprove that $d(P, m) = d(Q, m) = d(R, m)$.

27. In the hyperbolic plane, prove that parallel lines are not everywhere equidistant—that is, prove that for a given pair of parallel lines, they are closer together in some places than in other places.

28. Prove that there are two kinds of parallel lines in the hyperbolic plane. If lines $\ell$ and $m$ are parallel, show that either they have a common perpendicular or they meet at an ideal point.

29. Suppose that a quadrilateral in the hyperbolic plane has four congruent angles. Prove that the four angles must be acute.

30. Prove that the summit angles of a Saccheri quadrilateral must always be congruent acute angles.

31. Prove that rectangles do not exist in the hyperbolic plane.

32. Let $PQ$ be a common perpendicular to lines $\ell$ and $m$ in the hyperbolic plane. Prove that $PQ$ is the only common perpendicular.

33. Let $PQ$ be a common perpendicular to lines $\ell$ and $m$ in the hyperbolic plane. Prove that any hyperbolic segment joining a point of $\ell$ to a point of $m$ will be longer than $PQ$.

34. Prove that Playfair's Postulate is equivalent to Euclid's Fifth Postulate.

## The Upper-Half-Plane Model

The Poincaré disk is one model of the hyperbolic plane. Another interesting model to investigate is the upper-half-plane model. In this model, the points of the hyperbolic plane are represented by the points in the Euclidean plane that lie *above* the $x$-axis. (This is why it is called the *upper*-half-plane model.) The points that lie *on* the $x$-axis are ideal points. The lines of this model are of two types:

- *Type 1 lines:* vertical rays emanating from the $x$-axis
- *Type 2 lines:* semicircular arcs whose centers lie on the $x$-axis

For two lines, $\ell$ and $m$, to intersect in this model, at least one of them must be a Type 2 line. To measure an angle between any two lines in this model, consider the angle between the tangents to the semicircular arcs at the point of intersection, or consider the angle between the ray and the tangent to the semicircular arc at the point of intersection.

Do Exercises 35–38 in the upper-half-plane model of the hyperbolic plane.

35. Given any two points $A$ and $B$ in the upper-half-plane model, prove that there is a *unique* line through $A$ and $B$. (*Note:* There are two cases to consider.)

36. Verify that Euclid's first four postulates hold in the upper-half-plane model.

37. a. Verify that the hyperbolic parallel postulate holds in the upper-half-plane model.

b. Prove that the hyperbolic parallel theorem holds in this model.

38. Construct a rectangular coordinate system in a Sketchpad worksheet.

a. Construct examples of hyperbolic triangles in the upper-half-plane model. Calculate the angle sum of these triangles.

b. If triangles are classified as different types depending on the number of sides that are represented by Type 1 lines, how many different types of triangles are there? Explain.

c. Construct examples of right triangles in this model. Prove or disprove the Pythagorean Theorem in this model.

d. Construct an example of a triangle in the upper-half-plane model that has positive defect.

e. Prove that every triangle in this model has positive defect.

f. Construct an example of a Saccheri quadrilateral in the upper-half-plane model.

g. Construct an example of a Lambert quadrilateral in the upper-half-plane model.

h. Prove that it is impossible to construct a rectangle in this model.

Exercises 39 and 40 are especially for future teachers.

39. In the *Principles and Standards for School Mathematics,* the National Council of Teachers of Mathematics observes that "Being able to reason is essential to understanding mathematics. By developing ideas, exploring phenomena, justifying results, and using mathematical conjectures in all content areas and—with different expectations of sophistication—at all grade levels, students should see and expect that mathematics makes sense." Further, "Reasoning and proof cannot simply be taught in a single unit on logic, for example, by 'doing proofs' in geometry.... Reasoning and proof should be a consistent part of students' mathematical experience in prekindergarten through grade 12" (NCTM, 2000, 56). What does this mean for you and your future students?

a. Read the discussion on Reasoning and Proof in the *Principles and Standards for School Mathematics* (NCTM, 2000, 56–59). Then study the Geometry Standard for at least two grade bands (i.e., two of pre-K–2, 3–5, 6–8, or 9–12). What are the NCTM recommendations regarding the development of logical reasoning and mathematical proof? How is this focus on logical reasoning developed across several grade bands? Cite specific examples.

b. Find copies of school mathematics textbooks for the same grade levels as you studied for part a. How are the NCTM standards for reasoning and proof implemented in those textbooks? Again, cite specific examples.

c. Write a report in which you present and critique what you learn. Your report should include your responses to parts a and b.

40. Beginning in the earliest grades and developing through high school, students need to learn increasingly sophisticated ways of thinking, reasoning, and proving. Design several classroom activities involving logical reasoning and mathematical proof that would be appropriate for students in your future classroom. Write a short report explaining how the activities you design reflect both what you have learned in studying this chapter and the NCTM recommendations.

Reflect on what you have learned in this chapter.

41. Review the main ideas of this chapter. Describe, in your own words, the concepts you have studied and what you have learned about them. What are the important ideas? How do they fit together? Which concepts were easy for you? Which were hard?

42. Reflect on the learning environment for this course. Describe aspects of the learning environment that helped you understand the main ideas in this chapter. Which activities did you like? Dislike? Why?

In this chapter, you had an opportunity to explore a strange new universe. The hyperbolic plane has a lot in common with the Euclidean plane. Euclid's first four postulates still hold. Yet, everything looks quite strange!

The hyperbolic parallel postulate is one of two possible negations of Euclid's parallel postulate:

**Hyperbolic Parallel Postulate**   There is a line $\ell$ and there is a point $P$ not on $\ell$ such that more than one line through $P$ is parallel to $\ell$.

Because Euclid's parallel postulate is a universal statement, its negation will be an existence statement. (The negation of a universal statement is always an existence statement.) In this chapter, we have shown that if we accept the hyperbolic parallel postulate as an axiom (or postulate), we can prove the hyperbolic parallel theorem.

**Hyperbolic Parallel Theorem**   Given any line $\ell$ and any point $P$ not on $\ell$, there are at least two lines through $P$ that are parallel to $\ell$.

The hyperbolic parallel theorem is a stronger statement than the hyperbolic parallel postulate because it is a universal statement.

Throughout this chapter, we have focused on the Poincaré disk model of the hyperbolic plane. In this model, the *points* of the hyperbolic plane are represented by the points interior to the fundamental circle. The *lines* of the hyperbolic plane are represented by circular arcs that lie interior to the fundamental circle and are orthogonal to it. Using this model, we have investigated some of the phenomena of the hyperbolic plane:

- Triangles can be constructed with an angle sum less than 180°. In fact, all triangles have angle sum less than 180°. The difference between 180° and the actual angle sum of a triangle is called the *defect* of the triangle.

- Quadrilaterals can be constructed with one, two, or three right angles. But no quadrilateral can be constructed with four right angles. Therefore, rectangles do not exist in the hyperbolic plane.

Although we have illustrated these ideas using examples in the Poincaré disk model, we have proved them using the axioms and theorems of the hyperbolic plane. These phenomena are theorems and will hold in any model of the hyperbolic plane. We introduced a second model of the hyperbolic plane in the exercises. All of the examples we have explored in the Poincaré disk model can also be developed in the upper-half-plane model.

Euclid's first four postulates hold in any model of the hyperbolic plane. The implementation of these in the upper-half-plane model will look a bit different than the corresponding examples in the Poincaré disk. Yet these two models of the hyperbolic plane are *isomorphic*; that is, in a deep geometric sense, they are the same. (A proof of this fact is beyond the scope of this course.)

In earlier chapters of this text, we saw that an effective proof strategy for many theorems is to show that triangles are congruent. Because straightness of lines and apparent distances are visually distorted in models of the hyperbolic plane, we need to rely even more on axioms and theorems to prove whether a given pair of triangles is congruent. In Chapter 2, we accepted SAS as a valid criterion for congruent triangles, and without using Euclid's Fifth Postulate, we proved that the ASA, SSS, and AAS criteria for congruent triangles follow from this. So these criteria continue to hold in the hyperbolic plane. In addition, AAA is a valid criterion for congruent triangles in the hyperbolic plane. In other words, in the hyperbolic plane, similar triangles are congruent.

We saw once again the idea of a *metric*, a rule for measuring distance in a geometric space. As before, any metric must satisfy the three metric axioms:

**Metric Axiom 1**  If $A$ and $B$ are points, then $d(A, B) \geq 0$, and $d(A, B) = 0$ if and only if $A = B$.

**Metric Axiom 2**  If $A$ and $B$ are points, then $d(A, B) = d(B, A)$.

**Metric Axiom 3**  If $A$, $B$ and $C$ are points, then $d(A, B) + d(B, C) \geq d(A, C)$.

Many, but not all, of the theorems that hold in the Euclidean plane continue to hold in the hyperbolic plane. We are accustomed to thinking about and viewing the world from a Euclidean perspective. Once we introduce an alternative to Euclid's Fifth Postulate, we must reexamine what we know about familiar geometric figures, such as triangles, circles, and rectangles. Anything that we can prove using only Euclid's first four postulates continues to hold in the hyperbolic plane. Those theorems that require Euclid's Fifth Postulate will not hold in the hyperbolic plane.

For example, in this chapter, we took a second look at two of the concurrence theorems from Chapter 2. One of these holds in the hyperbolic plane:

**Theorem**  The three angle bisectors of a triangle are concurrent.

The proof of this theorem in the hyperbolic plane follows the same outline as the proof in the Euclidean plane. (In fact, you can use the exact same proof because the proof is based on results that follow from Euclid's first four postulates.) However, the other theorem needs to be modified. In the hyperbolic plane, the three perpendicular bisectors of the sides of a triangle may intersect or they may be parallel.

**Theorem**  If the three perpendicular bisectors of the sides of a triangle in the hyperbolic plane are concurrent at a point $O$, then the circle with center $O$ and radius $OA$ also contains the points $B$ and $C$.

Thus, while an incircle with center $I$ can be constructed for any hyperbolic triangle (see Exercise 21), a circumcircle can only be constructed for those hyperbolic triangles for which the point $O$ exists (see Exercise 20).

In the hyperbolic plane, there are two kinds of parallel lines:

- Some parallel lines could be said to meet at infinity. These are *limiting parallel lines*. If $\ell$ and $m$ are limiting parallels, they will not have a common perpendicular. Any line that is perpendicular to one of them will not be perpendicular to the other.

- Other parallel lines have a common perpendicular. If lines $\ell$ and $m$ are parallel and have a common perpendicular, they will not meet at infinity. Such parallel lines are called *ultraparallel, hyperparallel,* or *superparallel* lines.

In this chapter, we have only touched the surface of hyperbolic geometry. Indeed, an entire course could focus on properties of the hyperbolic plane. We hope that in studying this chapter, you have begun to develop an appreciation for the impact that a choice of axioms has on the shape of the objects in a geometric space.

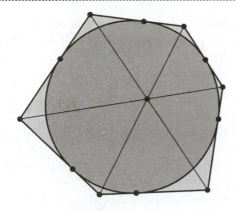

## CHAPTER TEN

# Projective Geometry

Projective geometry is a well-developed subject within geometry. Its mathematical roots go back to Pappus and even earlier to the ancient Greek mathematicians. It also has roots in art, through the theory of perspective drawing. The idea of a horizon line and the notion that parallel lines converge at the horizon brought a new realism to Renaissance painting, and many special techniques were invented for portraying objects in perspective. However, this shifted the focus of the geometry, for objects that were the same size in reality could no longer be drawn the same size in the painting. Instead, the objects should appear to shrink as they recede in the picture. The theory of perspective and projective geometry thus places more emphasis on incidence properties than on measurement.

In this chapter, we present three approaches to studying projective geometry. First, we examine an axiom system for a projective plane and prove some basic theorems. There are several interesting models for this axiom system, both a finite model and some infinite models. The notion of duality provides a powerful proof technique and leads to some important theorems. Second, we take an analytic approach by applying coordinate systems to these models. Third, we introduce projective transformations and give some of their properties, leading to the Fundamental Theorem of Projective Geometry.

Do the following activities, writing your explanations clearly in complete sentences. Include diagrams whenever appropriate. Save your work for each activity, as later work sometimes builds on earlier work. You will find it helpful to read ahead into the chapter as you work on these activities.

1. Create a coordinate system, and draw the horizontal line $y = 1$.
   a. Construct a point $P$ on this line, and construct line $\ell$ through $P$ and the origin. Measure the abscissa (the $x$-coordinate) of $P$, and measure the slope of $\ell$. Drag or animate $P$, and observe these values. How are these values related? Does every point $P$ have a slope value? Does every slope value have a point $P$?
   b. Does every point $P$ on $y = 1$ have a corresponding line $\ell$ from the origin? Does every line $\ell$ from the origin have a corresponding point $P$? Explain why or why not.
   c. Add to your sketch a unit circle centered at the origin. Construct the points where $\ell$ intersects this circle. Think of this situation as a function that maps a point on the circle to a corresponding point on the line $y = 1$. What is the domain of this function? Is it a one-to-one function? Is it onto? Does this function have an inverse function?

2. Create a line $\ell$, and construct segment $AB$ on this line. (You should be able to drag the segment without disturbing the line.) Label the endpoints of this segment.

   Now create another line $m$ and also a point $P$ not on $\ell$ or $m$. Construct the points $C = \overleftrightarrow{PA} \cap m$ and $D = \overleftrightarrow{PB} \cap m$. Then construct the segment $CD$. (Using several colors and thicknesses will make your sketch easier to follow.) Drag segment $AB$ along its line $\ell$, and observe what happens to segment $CD$. What changes? What stays the same?

   Now move point $P$ to another location. Again drag $AB$, and observe what happens. Try several locations for $P$. Do your observations still hold? Does anything change if you move line $m$?

3. Create three lines that are concurrent at a common point $P$. Construct points $A_1$ and $A_2$ on one of these lines, points $B_1$ and $B_2$ on a second line, and points $C_1$ and $C_2$ on the third line. Label these points.

   Construct segments to form $\triangle A_1 B_1 C_1$ and $\triangle A_2 B_2 C_2$, and construct the interiors of the triangles. (Using different colors will make your sketch easier to follow.) These triangles are said to be *perspective* from point $P$.

   Now extend side $A_1 B_1$ to a line. Then extend $A_2 B_2$ to a line, and construct the intersection point of these two new lines. Repeat this for sides $B_1 C_1$ and $B_2 C_2$ and for sides $C_1 A_1$ and $C_2 A_2$.

   You should now have three intersection points of corresponding sides of the triangles. How are these three points related to each other? Vary points $A_1$, $A_2$, $B_1$, $B_2$, $C_1$, and $C_2$. Does your observation still hold? Does your observation hold if the two triangles are on the same side of $P$? Does it hold if the triangles are on opposite sides of $P$? Does it hold if $P$ is interior to both triangles?

4. a. Create two lines $\ell_1$ and $\ell_2$. Construct three points $A_1$, $B_1$, $C_1$ on $\ell_1$ and three points $A_2$, $B_2$, $C_2$ on $\ell_2$. Then construct the *cross joins* $X = \overleftrightarrow{A_1B_2} \cap \overleftrightarrow{A_2B_1}$, $Y = \overleftrightarrow{B_1C_2} \cap \overleftrightarrow{B_2C_1}$, and $Z = \overleftrightarrow{C_1A_2} \cap \overleftrightarrow{C_2A_1}$. (Using different colors will help you understand this sketch.) What do you observe about the points $X$, $Y$, and $Z$? Drag some lines and points. Is your observation still valid?

   b. Now create a circle. Construct three points $A_1$, $B_1$, $C_1$ on one "side" of the circle and three points $A_2$, $B_2$, $C_2$ on the "other side." As in part a, construct the cross joins $X$, $Y$, $Z$. What do you observe about these points?

   In a general sense, the six points $A_1$, $B_2$, $C_1$, $A_2$, $B_1$, $C_2$, and the constructed lines form a hexagon inscribed in this circle. Select these six points, *in this order,* and construct the interior of the hexagon. Then drag vertices so that this is a convex hexagon. Does your observation still hold?

5. Construct four points, no three of which are collinear. Construct every possible line that uses two of these points. How many lines have you constructed? Now construct every possible point that uses two (or more) of these lines. (You might have to rearrange your diagram to see all of these points.) How many points have you constructed in all? How many of the constructed points are on each line? How many of the constructed lines are on each point?

6. a. Find at least five more points that lie on the line containing the origin $O$ and the point $P = (5, 2)$. How are the coordinates of these points related to each other?

   b. Find at least five points $X$ for which line $OX$ does not intersect the line $y = 1$. What do your answers have in common?

   c. Repeat these questions in $\Re^3$, using $P = (5, 2, 1)$ and the plane $z = 1$.

7. a. Create two lines $\ell_1$ and $\ell_2$ and a point $P_1$ not on either line. Construct a point $X$ on $\ell_1$, and construct the point $Y = \overleftrightarrow{P_1X} \cap \ell_2$. This creates a function $f(X) = Y$ with the points of $\ell_1$ as its source and the points of $\ell_2$ as the target. A function of this sort is called a *perspectivity* between the two lines. Does every input $X$ produce an output $f(X)$? That is, does every point $X$ of $\ell_1$ have a corresponding point $Y$ on $\ell_2$? Is every point of $\ell_2$ an output for some point from $\ell_1$? Vary $P_1$ and the two lines to see if this affects your answers.

   b. Now add to your diagram a third line, $\ell_3$, and another point, $P_2$, not on any of the three lines. Construct the point $Z = \overleftrightarrow{P_2Y} \cap \ell_3$. This is a composition of two perspectivity functions, which creates a new function $g(X) = Z$, with $\ell_1$ as its source and $\ell_3$ as its target. The new function $g$ is a *projectivity* between $\ell_1$ and $\ell_3$. Does every point of $\ell_1$ produce an output point on $\ell_3$? Does every point on $\ell_3$ have a corresponding point on $\ell_1$?

c. Do your answers to part b change if $\ell_3 = \ell_1$, that is, if the source and target are the same line? In this situation, can $g(X)$ ever equal $X$?

8. Create two lines $\ell_1$ and $\ell_2$. Construct six points: $A_1$, $B_1$, $C_1$ on $\ell_1$ and $A_2$, $B_2$, $C_2$ on $\ell_2$. Define a function from $\ell_1$ to $\ell_2$ as follows: Construct the points $P = \overleftrightarrow{A_1B_2} \cap \overleftrightarrow{A_2B_1}$ and $Q = \overleftrightarrow{A_1C_2} \cap \overleftrightarrow{A_2C_1}$. Construct $\overleftrightarrow{PQ}$. Also construct $\overleftrightarrow{A_1A_2}$ and point $R = \overleftrightarrow{A_1A_2} \cap \overleftrightarrow{PQ}$.

Let $X$ be a point on $\ell_1$. Construct point $T = \overleftrightarrow{A_2X} \cap \overleftrightarrow{PQ}$, and define $f(X) = \overleftrightarrow{A_1T} \cap \ell_2$. Drag $X$ along $\ell_1$, and observe the output $f(X)$.

a. Is this a well-defined function, that is, will every input give a unique output? If so, is it a one-to-one function? What is the domain of $f$? What is the range of $f$?

b. What are $f(A_1)$, $f(B_1)$, and $f(C_1)$?

c. Using the points and lines already in your diagram, find a perspectivity from $\ell_1$ to $\overleftrightarrow{PQ}$. (See Activity 7.) What point are you using for the center?

d. Again, using the points and lines in your diagram, find a perspectivity from $\overleftrightarrow{PQ}$ to $\ell_2$. Where is the center of this perspectivity?

## 10.2 DISCUSSION

## AN AXIOM SYSTEM

Axiom systems are an old idea in mathematics. The best-known example is Euclid's set of five axioms (postulates) for plane geometry, which we discussed in Chapter 1. From these five assumptions, Euclid and his contemporaries developed a long list of theorems. As long as his five axioms are accepted as true, all of these theorems must also be true. The metric axioms presented in Chapter 5 express the assumptions that we bring to measuring distance. Any rule for calculating distance must meet the requirements specified in those metric axioms. Another example of an axiom system is the set of axioms for a group, which we first presented in Chapter 6. This short list of assumptions yields a surprisingly rich set of theorems, and group theory is an active research area even today.

In Chapter 9, we explored some consequences of altering an axiom. The axioms for hyperbolic geometry presented in Chapter 9 are the same as the axioms for Euclidean geometry, except for one major change in Euclid's Fifth Postulate. As you saw, this one change led to a radically different, and sometimes surprising, set of theorems. Parallelism is a critical issue for hyperbolic geometry, and it will be an issue for projective geometry as well.

Many axiom systems have been developed for projective geometry. Some of these systems were written to describe only the real projective plane, an extended version of the usual Euclidean plane. Here we present a very general set of axioms (Batten, 1986, 23, 41–44). The various models will have properties beyond those that follow from these axioms.

**Axiom 1**   A line lies on at least two points.

**Axiom 2**   Any two distinct points have exactly one line in common.

**Axiom 3**   Any two distinct lines have at least one point in common.

**Axiom 4**   There is a set of four distinct points, no three of which are collinear.

It is Axiom 4 that guarantees that the geometry contains anything. This axiom says that there are points, and then Axiom 2 guarantees that there will be lines as well. Notice that these axioms say nothing about betweenness, order, continuity, or many other ideas. Incidence of points and lines is the only issue, at least in the axioms. Also notice that Axiom 3 specifically prohibits parallel lines in projective geometry. Any two lines in a projective plane will intersect.

We can prove something stronger than Axiom 1.

**THEOREM 10.1**   A line lies on at least three points.

**Proof**   The basic idea is shown in Figure 10.1. Given a line $\ell$, Axiom 1 guarantees that there are two points $P$ and $Q$ on $\ell$. By Axiom 4, there are two more points $R$ and $S$ not collinear with $P$ and $Q$. Axiom 3 says that $\overleftrightarrow{RS}$ must intersect $\ell$ at a point, $T$. Point $T$ must be a new point on $\ell$, for $R$, $S$, and $T$ are collinear and $P$ and $Q$ are not collinear with $R$ and $S$.

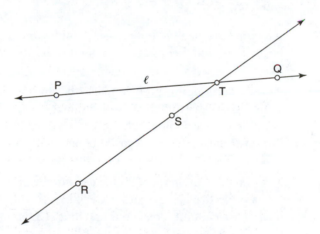

**FIGURE 10.1**
Proof That a Line Contains at Least Three Points

Axiom 3 can also be strengthened.

**THEOREM 10.2**   Any two distinct lines have exactly one point in common.

The proof of this is easy: Simply suppose two lines intersect twice, and apply Axiom 2.

Combining Axiom 3 with Theorem 10.2 has profound consequences. These statements mean that there are no parallel lines in projective geometry! Our experiences with perspective agree with this, for parallel lines, such as railroad tracks or highway lines, appear to intersect at the horizon. Of course, we can never reach those intersection points; they will always appear infinitely far away. (This idea will be important when we discuss models for projective geometry.)

In Activity 7a, you created a function between one line and another line, using one particular point to define the function. From Theorem 10.2, it is straightforward to prove a useful fact about this function. (You will be asked to prove this fact in the exercises.)

**THEOREM 10.3** The points on one line can be put into one-to-one correspondence with the points on any other line.

This function is known as a *perspectivity* between the two lines. We will work with perspectivities often in this chapter. The one-to-one correspondence shows that any two lines have the same number of points. For now, the correspondence can give us information about finite situations. (The correspondence still works in infinite situations, but it is trickier to explain what it means to have the same number of points.)

**COROLLARY 10.4** If a projective plane is finite (i.e., has a finite number of points), every line has the same number of points.

We can say something similar for the collection of lines on any given point. The proof uses another one-to-one correspondence, this time between points and lines.

**THEOREM 10.5** If a projective plane is finite, every point lies on the same number of lines. Further, this number of lines per point is the same as the number of points per line.

**Proof** Let $P$ be any point, and let $\ell$ be any line not containing $P$. (Which axioms guarantee that there is such a line?) Suppose line $\ell$ has $k + 1$ points. Each point on $\ell$ creates a line with $P$, so $P$ lies on at least $k + 1$ lines. In addition, any line on $P$ must intersect $\ell$, so $P$ lies on at most $k + 1$ lines.

The number $k$ in this proof is the *order* of the projective plane. The theorem says that a projective plane of order $k$ will have $k + 1$ points on every line and $k + 1$ lines on every point. This is our first hint of a symmetry between points and lines that will prove to be very important.

Knowing the number of points per line and the number of lines per point allows us to calculate the total number of points.

**COROLLARY 10.6** A finite projective plane of order $k$ will have $k^2 + k + 1$ points and will have $k^2 + k + 1$ lines.

**Proof** Consider a single point $P$. This point lies on $k + 1$ lines, each of which has $k + 1$ points, one of which is $P$. Every point of the projective plane is collinear with $P$, so every point is included on this set of lines. Thus, we have

(number of lines on $P$) × (number of points other than $P$ on a line) + ($P$ itself)

$$= (k + 1) \times k + 1.$$

A similar argument works if we start with a single line.

# MODELS FOR THE PROJECTIVE PLANE

A *model* of an axiom system is a situation in which all of the axioms hold true. For the axioms of the projective plane, there must be a set of things called *points* and another set of things called *lines*. Further, there must be some interpretation of what "lies on" means. It is very common to talk of points lying on a line, but it is less common to talk of lines lying on a point. Perhaps you have said "line through a point" or "line containing a point." Sometimes the term *incident* is used, as in "points incident on a line" or "lines incident on a point." The particular word to be used is not a critical issue, as long as you have an interpretation that agrees with the axioms.

The *undefined terms* of this particular axiom system are *point, line,* and *lies on.* In some models, the undefined terms will mean what you are accustomed to, but not always! If a model can be found, this shows that the axiom system is *consistent,* meaning that the axioms do not contradict each other. In this section, we present three very different models for the axioms of a projective plane.

## The Real Projective Plane

Activity 1 asked for a correspondence between the points on the line $y = 1$ and the lines through the origin. Most of the time this is pretty easy; a point lying on this horizontal line determines a line from the origin, and a line lying on the origin intersects this horizontal line to determine a point. However, one of the lines through the origin does not intersect $y = 1$, namely, the horizontal line $y = 0$. There are not enough points on $y = 1$ to account for all these lines in this way. Here, we can borrow an idea from artistic perspective. In a perspective drawing, parallel lines that recede into the picture intersect at the horizon line, which is imagined to be infinitely far away. The artist includes a point at infinity to mark the intersection of the parallel lines. This is mathematical nonsense—after all, parallel lines, by definition, do not intersect—but it suggests a way to extend the Euclidean plane and to make Axiom 3 work. The basic idea is to put in *points at infinity* where the parallels intersect. Then the infinite point on $y = 1$ will also be the infinite point on $y = 0$, and the two horizontal lines will intersect at the infinite point.

Let us do this carefully. For any line in the Euclidean plane, there will be a collection of lines that are parallel to it. The lines in this collection are *equivalent* to each other in a very strong way, for parallelism is an *equivalence relation.* This means three things: Any line $\ell$ is parallel to itself (the *reflexive* property); if $\ell$ is parallel to $m$, then $m$ is parallel to $\ell$ (the *symmetric* property); and if $\ell$ is parallel to $m$ and $m$ is parallel to $n$, then $\ell$ is parallel to $n$ (the *transitive* property). In the sense of parallelism, all these lines are essentially the same; they are equivalent.

Equivalence relations are useful tools in many areas of mathematics. The most important theorem about equivalence relations says that any such relation separates its set of objects into disjoint *equivalence classes.* Each equivalence class is a set containing all the objects that are related to each other. In our situation of Euclidean lines related by parallelism, one example of an equivalence class is the set of lines with slope 2. Every line with its slope equal to 2 belongs to this particular equivalence class, and lines with different slopes do not belong. The set of all vertical lines is another example of an equivalence class. Every Euclidean line belongs to exactly one of these equivalence classes.

Each equivalence class will be called an *ideal point*. Two lines that are parallel are in the same equivalence class, so they share a common ideal point. The set of ideal points forms the *ideal line*. The set of ordinary points in the Euclidean plane, together with the points of the ideal line, forms the *real projective plane*. A line is now one of two types: a Euclidean line plus its ideal point or the ideal line. To verify that Axiom 3 works in this model, we must examine two cases: when the two lines are both of the first type, or when one of the two lines is the ideal line. (You will get a chance to do this in the exercises.)

The real projective plane has properties beyond the axioms and the simple theorems mentioned earlier. For instance, something must be said about continuity. One approach is to add an axiom that for any line, all but one of the points can be put into one-to-one correspondence with the set of real numbers. Then the continuity of the real line $\Re^1$ will produce continuity in the real projective plane. (The classic reference for this model is *The Real Projective Plane* by Coxeter, 1949.)

Notice that the projective axioms say nothing about the ideal line. It is merely one line among many lines. This can be a great advantage when proving theorems. A proof that is based solely on the projective axioms avoids having to worry about special cases that involve parallel lines. For instance, here is a proof of the theorem from Activity 4a, which you first saw in Chapter 1 (see Activity 10, page 5). This proof uses Menelaus' Theorem, working on a triangle created by some of the lines in the theorem (Coxeter and Greitzer, 1967, 67–69).

**THEOREM 10.7**    **Pappus' Theorem**    Suppose that $A_1$, $B_1$, and $C_1$ are distinct points on one line, $\ell_1$, and that $A_2$, $B_2$, and $C_2$ are distinct points on a second line, $\ell_2$. Form the cross joins $X = \overleftrightarrow{A_1B_2} \cap \overleftrightarrow{A_2B_1}$, $Y = \overleftrightarrow{B_1C_2} \cap \overleftrightarrow{B_2C_1}$, and $Z = \overleftrightarrow{C_1A_2} \cap \overleftrightarrow{C_2A_1}$. Then the three points $X$, $Y$, and $Z$ are collinear.

**Proof**    See Figure 10.2. Suppose that $\overleftrightarrow{A_1B_2}$ and $\overleftrightarrow{B_1C_2}$ intersect at point $D$, that $\overleftrightarrow{B_1C_2}$ and $\overleftrightarrow{C_1A_2}$ intersect at $E$, and that $\overleftrightarrow{C_1A_2}$ and $\overleftrightarrow{A_1B_2}$ intersect at $F$. This uses

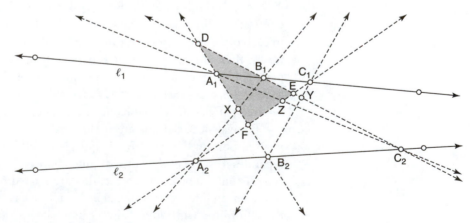

**FIGURE 10.2**
Proof of Pappus' Theorem

three of the eight lines given in the hypothesis, and the other five lines are transversals of $\triangle DEF$. From Menelaus' Theorem, each transversal creates a product.

From $A_1ZC_2$, we get

$$\frac{DC_2}{C_2E} \cdot \frac{EZ}{ZF} \cdot \frac{FA_1}{A_1D} = -1.$$

From $B_1XA_2$, we get

$$\frac{DB_1}{B_1E} \cdot \frac{EA_2}{A_2F} \cdot \frac{FX}{XD} = -1.$$

From $C_1YB_2$, we get

$$\frac{DY}{YE} \cdot \frac{EC_1}{C_1F} \cdot \frac{FB_2}{B_2D} = -1.$$

From $A_1B_1C_1$, we get

$$\frac{DB_1}{B_1E} \cdot \frac{EC_1}{C_1F} \cdot \frac{FA_1}{A_1D} = -1.$$

From $A_2B_2C_2$, we get

$$\frac{DC_2}{C_2E} \cdot \frac{EA_2}{A_2F} \cdot \frac{FB_2}{B_2D} = -1.$$

Now divide the product of the first three expressions by the product of the last two. After much cancellation, this produces

$$\frac{DY}{YE} \cdot \frac{EZ}{ZF} \cdot \frac{FX}{XD} = -1.$$

Therefore, the points $X$, $Y$, and $Z$ are collinear.

---

Pappus' Theorem has a very projective flavor to it, for it talks only about incidence of lines and points. The proof given here, of course, uses measurement concepts. Later we will see a proof based on projective transformations that avoids both measurement and parallelism, thus combining all possible cases into one short proof.

## A Finite Model: The Fano Plane

Theorem 10.1 showed that a line in a projective plane must have at least three points. Can we find a model in which a line will have exactly three points? This would be a projective plane of order 2. Activity 5 was an attempt to do this. The original four points should lead to six lines, and these six lines produce three additional points. This is a total of seven points, and $7 = 2^2 + 2 + 1$, as expected for order 2. Did you find a seventh line as well? This is tricky. Suppose we label the beginning four points as 1, 2, 3, 4. The lines {1, 2} and {3, 4} will intersect, say at 5. The lines {1, 3} and {2, 4} will intersect at 6. The lines {1, 4} and {2, 3} will intersect at 7. So far, the situation satisfies Axioms 1, 3, and 4. There is a problem, however, with Axiom 2, because points 5 and 6 are not collinear and neither are 5 and 7, nor 6 and 7. So add another line {5, 6, 7}. Here is our finite projective plane, known as the *Fano plane:*

$$\left\{\begin{matrix}1\\2\\5\end{matrix}\right\}, \left\{\begin{matrix}3\\4\\5\end{matrix}\right\}, \left\{\begin{matrix}1\\3\\6\end{matrix}\right\}, \left\{\begin{matrix}2\\4\\6\end{matrix}\right\}, \left\{\begin{matrix}1\\4\\7\end{matrix}\right\}, \left\{\begin{matrix}2\\3\\7\end{matrix}\right\}, \left\{\begin{matrix}5\\6\\7\end{matrix}\right\}.$$

All of the axioms are satisfied by this collection of lines. To draw this, however, requires us to bend one or more lines. Figure 10.3 shows one of many ways to draw the Fano plane. The Sketchpad command **Construct | Arc Through Three Points** may be helpful if you wish to try this yourself.

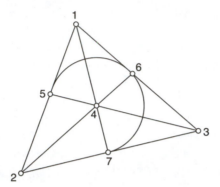

**FIGURE 10.3**
The Fano Plane

It is not hard to prove that this is the only possible way to create a projective plane of order 2. In fact, it has been proven that there are unique projective planes of order 2, 3, 4, 5, 7, and 8. You will be asked to draw the projective plane of order 3 in the exercises. How many points are needed? How many points lie on each line, and how many lines lie on each point? It may be helpful to draw the ordinary points first, then add the ideal points and the ideal line afterward. There is no projective plane of order 6, and there are at least four different projective planes of order 9. In 1991, a proof was announced that there is not a projective plane of order 10 (Lam, 1991). This proof involved checking hundreds of smaller questions and was assisted by a computer. The existence and uniqueness of projective planes having large order is still an unanswered question, and there is a general theorem waiting to be discovered.

## A Model on the Sphere

In Activity 1c, you looked for a correspondence between lines through the origin and points on the unit circle. Of course, each such line intersects the circle twice, so this is really a correspondence between a line and a pair of points, not just individual points. These diametrically opposed points are called *antipodes.* Lines through the origin also intersect the horizontal tangent line $y = 1$, so we can use these lines to create a correspondence between points on the horizontal tangent and pairs of antipodes on the unit circle. Each point on $y = 1$ corresponds to an antipodal pair on the circle, and each antipodal pair corresponds to a point on the tangent line.

There is, of course, one exception: The line along the horizontal axis does not intersect $y = 1$. In this case, the pair $(1, 0)$ and $(-1, 0)$ corresponds to the ideal point of the tangent line. If we consider each pair of antipodes as only one point, the unit circle acts as a projective line, including its ideal point.

Let's extend this idea to one higher dimension. In Chapter 9, we examined the sphere as a model of elliptic geometry. The lines in this model were the *great circles* on the sphere. It may be helpful to visualize a great circle as the intersection of the sphere with a plane that passes through its center. A great circle uses the center of the sphere as its center. It has the largest possible radius for a circle on the surface of the sphere. Earth's equator is perhaps the most familiar example of a great circle. If two great circles are drawn on the sphere, they will intersect twice, and these intersection points are diametrically opposed points—antipodes.

To use the sphere as a model of a projective plane, consider each antipodal pair of points to be a single point (Coxeter, 1969, 93–94). The lines will be the great circles. It is not difficult to see that the four axioms of a projective plane are satisfied in this model. A great circle has at least two points, that is, two antipodal pairs. Any two points lie on exactly one great circle. (This takes a little care. For instance, there are many great circles that contain both the "North Pole" and the "South Pole," which seems to violate Axiom 2. However, these are antipodes, so in this model, they are considered to be the same point.) Any two great circles intersect in a single antipodal pair. Finally, there are many possible choices for four distinct points, no three of which lie on a common great circle.

The spherical model is, in fact, isomorphic to the real projective plane. Think of the sphere as sitting on a plane, so that the plane is tangent to the sphere (see Figure 10.4). Imagine lines through the center of the sphere. Each such line intersects the sphere twice and intersects the tangent plane once. Thus, a point in the plane corresponds to an antipodal pair of points on the sphere, that is, to a single projective point. The exceptions are the points on the equator, for which the line from the center does not intersect the tangent plane. A pair of antipodal points on the equator corresponds to an ideal point for the tangent plane, and the equator corresponds to the ideal line.

This relationship between the sphere and the tangent plane is called *central projection*. A pair of points on the sphere is projected from its center to a point on the tangent plane, and any point on the tangent plane is the result of one of these projections. A great circle on the sphere is centrally projected to a line in the

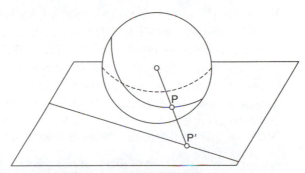

**FIGURE 10.4**
Central Projection

As point P traces a great circle on the sphere, P′ traces a line on the plane.

tangent plane. Do you see why? Imagine an arbitrary plane through the center of the sphere. This plane intersects the sphere to create a great circle and intersects the tangent plane to create a line. Of course, if this plane through the sphere's center is parallel to the tangent plane, it will intersect the sphere at the equator. This corresponds to the ideal line of the tangent plane.

With the spherical model, we encounter a disturbing fact about the real projective plane: It is a *nonorientable* surface. In the Euclidean plane, clockwise and counterclockwise are familiar concepts, and a clockwise rotation looks the same anywhere in the plane. This is not so in the real projective plane. Think of a small circle drawn in the upper half of a sphere, with arrows going clockwise around it. This corresponds to its antipodal circle in the lower half of the sphere. However, if you look at the antipodal circle from below the sphere, the arrows are going counterclockwise! Because these two circles are considered the same in the projective model, a clockwise/counterclockwise orientation is not possible.

## DUALITY

Look again at the axioms for a projective plane. Axioms 2 and 3 are very similar to each other. The resemblance is even stronger between Axiom 2 and Theorem 10.2. Axiom 2 says "Any two distinct points have exactly one line in common" and Theorem 10.2 says "Any two distinct lines have exactly one point in common." The pattern of the statements is the same. The only difference between the statements is that the words *point* and *line* have been switched. This is the idea of *duality;* if two major concepts are interchanged in a statement, the new statement also could be true. In the case of projective geometry, the two major concepts are *point* and *line,* two of the undefined terms. In this section, we will show that switching these terms in any theorem of projective geometry will indeed produce a new theorem; thus, projective geometry has the duality property. We will also examine two dual concepts of perspective and the surprising fact that they are equivalent.

Euclidean geometry does not have the duality property, nor does hyperbolic geometry. In both of those axiom systems, two distinct points determine a line. This is also true in projective geometry. However, both Euclidean and hyperbolic geometry include the notion of parallelism—two distinct lines do not necessarily determine a point, for the lines might not intersect. Projective geometry includes an axiom stating that two distinct lines will intersect, which is very different from Euclidean or hyperbolic geometry.

Here again are the axioms for a projective plane.

**Axiom 1**  A line lies on at least two points.

**Axiom 2**  Any two distinct points have exactly one line in common.

**Axiom 3**  Any two distinct lines have at least one point in common.

**Axiom 4**  There is a set of four distinct points, no three of which are collinear.

We have already looked at the dual of Axiom 2. What happens if the terms *point* and *line* are interchanged in all four of the axioms? Here is what we would get:

**Dual Axiom 1**   A point lies on at least two lines.

**Dual Axiom 2**   Any two distinct lines have exactly one point in common.

**Dual Axiom 3**   Any two distinct points have at least one line in common.

**Dual Axiom 4**   There is a set of four distinct lines, no three of which are concurrent.

Recall that *concurrent* means lying on a common point, while *collinear* means lying on a common line. So *concurrent* and *collinear* are dual concepts.

Are these dual statements true in a projective plane? Dual Axiom 2 is Theorem 10.2, and Dual Axiom 3 is a weaker form of Axiom 2, so they are both true statements. As for Dual Axiom 1, suppose we begin with a point $P$. By Axiom 4, there are at least three more points $A$, $B$, and $C$ such that no three of $P$, $A$, $B$, and $C$ are collinear. Pair $P$ with each of these points, and apply Axiom 2 to get at least two lines on point $P$. Dual Axiom 4 can also be proven from Axioms 4 and 2.

This shows that the projective plane has the *duality* property. If the terms *point* and *line* are interchanged in any true statement about a projective plane, we get another true statement. Of course, this interchange must occur for any term that is derived from the concepts of point and line, such as switching *concurrent* for *collinear*. Duality works for the axioms, as we have just shown, but it also works for any theorem of the projective plane. Suppose for some theorem we have a proof based on the four projective axioms. By writing the dual statements for every line of the proof, we get a proof of the dual theorem that is based on the dual axioms. Because we know that these dual axioms are true in a projective plane, we know that this dual proof is valid as well. So, anything derived from the dual axioms will also be true in a projective plane.

Here is an example of a dual theorem and its dual proof.

**THEOREM 10.8**   **Dual to Theorem 10.1**   A point lies on at least three lines.

**Proof**   Given a point $L$, Dual Axiom 1 guarantees that there are two lines $p$ and $q$ on $L$. By Dual Axiom 4, there are two more lines $r$ and $s$ not concurrent with $p$ and $q$. Dual Axiom 3 says that the point $r \cap s$ will join with $L$ to form a line, $t$. Line $t$ must be a new line on $L$, for $r$, $s$, and $t$ are concurrent and $p$ and $q$ are not concurrent with $r$ and $s$.

--------------------------------------------

Notice the similarity to Theorem 10.1. The terms *point* and *line* have been interchanged, and a few words have been altered to make it read more smoothly; otherwise, this is identical to the earlier proof. This proof illustrates the property of duality.

We have already seen some dual results. For instance, the order $k$ of a finite projective plane refers to both the number of points on a line and the number of lines on a point, with $k^2 + k + 1$ of each in the plane. Here is something more general.

**THEOREM 10.9    Dual to Theorem 10.3**    The lines on one point can be put into one-to-one correspondence with the lines on any other point.

As in the original theorem, the proof consists of defining a mapping (function) and verifying that it is one-to-one and onto. In the exercises, you will be asked to prove the original theorem and then to write the dual proof.

The set of lines lying on a single point is called a *pencil* of lines. The mapping created for the proof of this theorem is a *perspectivity* between the two pencils.

Duality can also be used when making definitions. For example, a *triangle* is defined as a set of three noncollinear points and the lines connecting them. (Remember that projective geometry deals only with incidence. Because betweenness is not available, we cannot discuss line segments.) The dual definition is the *trilateral*, a set of three nonconcurrent lines and the points connecting them, that is, the points where the lines intersect. But this is the same thing! The triangle is a *self-dual* concept.

What happens with more vertices? In Euclidean geometry, you have seen many objects created from four vertices and line segments of differing lengths. Because projective geometry does not deal with length, only incidence, there is not this variety. Here is the definition: A *quadrangle* is a set of four points, no three collinear, and the lines connecting them. You drew a quadrangle in Activity 5 with six lines on these four points. These lines also intersect at three additional points, which create the *diagonal triangle* of the quadrangle. In the Fano plane, the diagonal triangle will be degenerate, that is, the vertices will lie on a single line. In the real projective plane, this will be a legitimate triangle.

Now for the dual: A *quadrilateral* is a set of four lines, no three concurrent, and the points connecting them. This is sometimes called a *complete quadrilateral*. Try drawing this general quadrilateral. How many intersection points will there be from the four lines? Notice that a quadrilateral is not the same thing as a quadrangle. The numbers of points and lines are different, for instance. There are six points on the four lines of a quadrilateral. (Compare this with the quadrangle to see the duality in these numbers.) The six intersection points of the lines in a complete quadrilateral can be connected to make three additional lines. These lines form a diagonal triangle (trilateral) here as well.

We have already discussed two simple mappings, perspectivity between two lines and perspectivity between two pencils. These are dual concepts. Suppose we focus on these concepts more precisely.

**DEFINITION 10.1**    Two triangles, $\triangle A_1B_1C_1$ and $\triangle A_2B_2C_2$, are *perspective from a point* $P$ if the lines $\overleftrightarrow{A_1A_2}$, $\overleftrightarrow{B_1B_2}$, and $\overleftrightarrow{C_1C_2}$ are concurrent at $P$.

In the dual definition, we will use the notation $a_1a_2$ to represent the point of intersection for lines $a_1$ and $a_2$.

**DEFINITION 10.2**    Two trilaterals, $\triangle a_1b_1c_1$ and $\triangle a_2b_2c_2$, are *perspective from a line* $p$ if the points $a_1a_2$, $b_1b_2$, and $c_1c_2$ are collinear on $p$.

It may help you understand the first definition if you imagine yourself looking through a window with your eye located at the perspective point. Choose an object outside the window, and trace the image of this object onto the window, so that the outline on the glass matches the object outside. Buildings can be good choices for this, for most buildings have linear contours. The image and the object are in perspective from your eye's location. Figure 10.5 shows a famous etching by Albrecht Dürer in which an artist is looking through a grid at his subject and thus locating precisely where to draw his image.

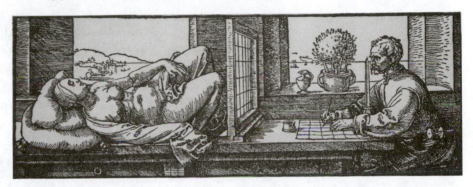

**FIGURE 10.5**
An Application of Perspectivity

In Activity 3, you constructed two triangles that were perspective from a point. As long as their corresponding vertices are collinear with the perspective point $P$, the triangles will be in perspective. This holds no matter where the corresponding vertices are with respect to $P$. The mathematical definition is a more general meaning of perspective than the artist's meaning.

Perspectivity from a line is a bit more difficult to visualize. Corresponding sides of two triangles will intersect, for every pair of lines intersects in the projective plane. With three corresponding pairs of sides, there are three intersection points. If these three points are collinear, the triangles are perspective from that line.

A remarkable fact is that in the real projective plane, these dual concepts of perspective are equivalent. This equivalence holds in many other projective planes as well, though there are some situations where it is not true.

**THEOREM 10.10**    **Desargues' Theorem**    In the real projective plane, if two triangles are perspective from a point, then they are perspective from a line.

In the Euclidean plane, when corresponding sides of the triangles are not parallel, Theorem 10.10 can be proven using Menelaus' Theorem. It is also possible to prove this theorem in the real projective plane as a consequence of Pappus' Theorem. In fact, many axiomatic treatments of the real projective plane use Pappus' Theorem as an axiom and then prove Desargues' Theorem and its converse. We give a proof that is interesting and unusual because it uses a three-dimensional result to prove the two-dimensional one (Graustein, 1930, 23–25).

**Proof**    Suppose that $\triangle A_1B_1C_1$ and $\triangle A_2B_2C_2$ are perspective from the point $P$. This means that points $P$, $A_1$, and $A_2$ lie on one line; points $P$, $B_1$, and $B_2$ lie on another line; and points $P$, $C_1$, and $C_2$ lie on a third line.

First, consider the case when these triangles are in different planes, $\Pi_1$ and $\Pi_2$, respectively. These planes intersect in a line $\ell$. We will show that any two corresponding sides of the perspective triangles will intersect on this line $\ell$.

The lines $PA_1A_2$ and $PB_1B_2$ lie in a common plane. The lines $A_1B_1$ and $A_2B_2$ also lie in this new plane, and they will intersect at a point $D$. Point $D$ will lie in the plane $\Pi_1$ because $D$ is on $A_1B_1$, which lies in $\Pi_1$. Because line $A_2B_2$ lies in $\Pi_2$, point $D$ will also lie in the plane $\Pi_2$. Point $D$ lies at the intersection of these three planes, and thus $D$ is on line $\ell$.

We can do the same argument for lines $B_1C_1$ and $B_2C_2$ intersecting at $E$, and again for lines $A_1C_1$ and $A_2C_2$ intersecting at $F$. All three points $D$, $E$, and $F$, where corresponding sides intersect, lie on the line $\ell$. Therefore, $\triangle A_1B_1C_1$ and $\triangle A_2B_2C_2$ are perspective from the line $\ell$. (It would be a good idea for you to create a drawing of this situation to make sure you understand where all these points and lines lie. Sketchpad can help, though you must visualize your diagram as three-dimensional.)

Now consider the case when $\triangle A_1B_1C_1$ and $\triangle A_2B_2C_2$ are in the same plane. The approach is to create a new triangle perspective to both original triangles, but in a new plane. We will then show that points $D$, $E$, and $F$ lie on the intersection of the two planes.

Let $R_1$ and $R_2$ be points collinear with $P$ but not in the plane of the two triangles. Construct the lines connecting $R_1$ to the vertices of $\triangle A_1B_1C_1$ and the lines connecting $R_2$ to the vertices of $\triangle A_2B_2C_2$.

The lines $PR_1R_2$ and $PA_1A_2$ lie in a common plane. In that plane, the lines $R_1A_1$ and $R_2A_2$ intersect at a point $A_3$. Similarly, the lines $R_1B_1$ and $R_2B_2$ intersect at a point $B_3$, and $R_1C_1$ and $R_2C_2$ intersect at a point $C_3$. This creates a triangle, $\triangle A_3B_3C_3$. This triangle will be in a different plane from the original two triangles. (Carefully examine Figure 10.6 to identify the three triangles and the perspectivities.)

We now have $\triangle A_1B_1C_1$ and $\triangle A_3B_3C_3$ perspective from $R_1$, because the lines $A_1A_3$, $B_1B_3$, and $C_1C_3$ are concurrent at $R_1$. Also, $\triangle A_2B_2C_2$ and $\triangle A_3B_3C_3$ are perspective from $R_2$.

Using the result from the three-dimensional case, this implies that $\triangle A_1B_1C_1$ and $\triangle A_3B_3C_3$ are perspective from the line $\ell$ where the two planes intersect. Furthermore, $\triangle A_2B_2C_2$ and $A_3B_3C_3$ are also perspective from $\ell$. This means that the lines $A_1B_1$ and $A_3B_3$ are concurrent with $\ell$ and that $A_2B_2$ and $A_3B_3$ are concurrent with $\ell$. Thus, we have four lines meeting at a single point: $A_1B_1$, $A_2B_2$, $A_3B_3$, and $\ell$. Of course, $\overleftrightarrow{A_1B_1}$ and $\overleftrightarrow{A_2B_2}$ are corresponding sides of the two original triangles, and they intersect on $\ell$.

Similar arguments show that the other corresponding sides also intersect on $\ell$, making $\ell$ the line of perspectivity for $\triangle A_1B_1C_1$ and $\triangle A_2B_2C_2$.

For the two forms of perspectivity to be equivalent, we must also have the reverse implication. To verify this reverse implication, we will use Desargues' Theorem in the proof.

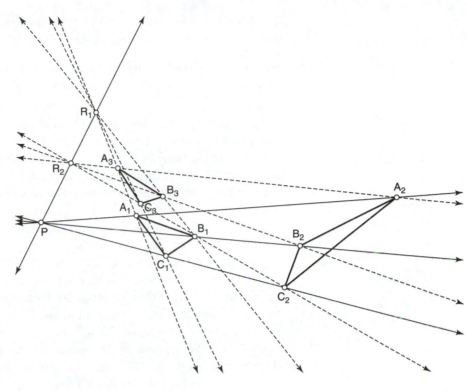

**FIGURE 10.6**
Proof of Desargues' Theorem

**THEOREM 10.11**  **The Dual (and Converse) of Desargues' Theorem**  In the real projective plane, if two triangles are perspective from a line, then they are perspective from a point.

**Proof**  Suppose $\triangle A_1 B_1 C_1$ and $\triangle A_2 B_2 C_2$ are perspective from line $\ell$. Let $D = \overleftrightarrow{A_1 B_1} \cap \overleftrightarrow{A_2 B_2}$, $E = \overleftrightarrow{B_1 C_1} \cap \overleftrightarrow{B_2 C_2}$, $F = \overleftrightarrow{C_1 A_1} \cap \overleftrightarrow{C_2 A_2}$, and $P = \overleftrightarrow{A_1 A_2} \cap \overleftrightarrow{B_1 B_2}$. Point $P$ will be the point of perspectivity; all we need to verify is that $P$ is collinear with $C_1$ and $C_2$.

Think about $\triangle F A_1 A_2$ and $\triangle E B_1 B_2$. These are perspective from point $D$. Hence, according to Desargues' Theorem, the triangles are perspective from a line. This line contains the points $\overleftrightarrow{F A_1} \cap \overleftrightarrow{E B_1} = C_1$, $\overleftrightarrow{A_1 A_2} \cap \overleftrightarrow{B_1 B_2} = P$, and $\overleftrightarrow{F A_2} \cap \overleftrightarrow{E B_2} = C_2$.

Desargues' Theorem and its converse hold in the Euclidean plane, though special treatment is needed when parallel lines are involved. The theorem and its converse are also true in the Fano plane, but it can be difficult to find two triangles that satisfy the hypothesis. (*Hint:* The six vertices do not have to be distinct points.) However, there are projective planes that do not have this property. An example will appear in the exercises.

In Activity 4a, you saw Pappus' Theorem, which was first seen in Chapter 1. What did you observe in part 4b? The situation in part 4b is very similar to Pappus' Theorem, and, in a sense we will discuss later, this is a more general version of Pappus' result. The result you observed in Activity 4b was first proven in 1640 by

Pascal, when he was sixteen years old! Pascal's original proof has been lost, though many other proofs have been given since then.

**THEOREM 10.12**     **Pascal's Theorem for Circles**    If a hexagon is inscribed in a circle, then the three points of intersection of pairs of opposite sides will be collinear.

------------------------------------------------

This statement may need some clarification. A *hexagon* is a set of six points, with no three successive points being collinear, and the six lines connecting them cyclically. (In the Euclidean plane, we would use line segments.) In more precise language, a hexagon is six points $A_1, \ldots, A_6$ and the lines $A_1A_2$, $A_2A_3$, $\ldots$, $A_6A_1$. Inscribing the hexagon in a circle means that all six vertices lie on a common circle. The hexagon does not have to be convex, and the sides could cross each other. In the real projective plane, opposite sides, like $\overleftrightarrow{A_1A_2}$ and $\overleftrightarrow{A_4A_5}$, will intersect; in the Euclidean plane, they could be parallel.

Pascal's Theorem in the Euclidean plane can be proven by Menelaus' Theorem or by Desargues' Theorem. It is necessary in the Euclidean case to assume that the intersections are ordinary points, not ideal points. It turns out that a much stronger version of Pascal's Theorem holds: The six points can be inscribed on any conic curve—a circle, an ellipse, a parabola, or a hyperbola—and the cross joins will be collinear. Even the degenerate conic consisting of two lines will work, which means that Pappus' Theorem is actually a special case of Pascal's Theorem.

A century and a half after Pascal's result, the dual version was proven.

**THEOREM 10.13**     **Brianchon's Theorem for Circles**    If a hexagon is circumscribed around a circle, then the three lines connecting pairs of opposite vertices are concurrent.

------------------------------------------------

This too can be strengthened to apply to any conic curve.

## COORDINATES FOR PROJECTIVE GEOMETRY

## Coordinates for a Projective Line

The idea of coordinates on a line is a familiar one; the set of real numbers matches up with the points of the line in a one-to-one correspondence. However, the real numbers are not enough for the projective line, for there is nothing to use for the ideal point. Something stronger is needed.

Consider the point $x = \frac{5}{2}$ on a number line. This coordinate can be written in many ways, such as

$$x = \frac{5}{2} = \frac{10}{4} = \frac{15}{6} = \frac{-20}{-8} = \cdots.$$

Any of these pairs of numbers represents the same point.

How are these fractions related to Activity 6a? A line through the origin has many possible direction vectors. The line through $(0, 0)$ and $(5, 2)$ has $(5, 2)$, $(10, 4)$, and $(-20, -8)$ as some of its direction vectors, as well as $(\frac{5}{2}, 1)$. All of these vectors represent the same direction, so they represent the same line through the origin. If we think of the horizontal line $y = 1$ as our desired number line, all of these vectors represent the same point on the number line.

This is the idea of *homogeneous coordinates*. A point on the line is represented by an ordered pair of numbers $(x_1, x_2)$, with the understanding that any nonzero multiple of this—$(ax_1, ax_2)$—represents the same point. The pair $(0, 0)$ does not fit this system, for if we were to allow multiplication by 0, every point would be the same. So, $(0, 0)$ is excluded from homogeneous coordinates. It is easy to convert homogeneous coordinates to Euclidean coordinates by using $x = \frac{x_1}{x_2}$.

Where does the point $(1, 0)$ lie on this number line? If we think of it as a direction vector in a plane, $(1, 0)$ is the horizontal direction. In Activity 6b, you may have listed this as one of your points on the horizontal line $y = 0$. This line is parallel to $y = 1$, so the pair $(1, 0)$ represents the ideal point on the projective line.

On the projective line, we now have $(0, 1)$ for the origin, $(1, 1)$ for the unit point, and $(1, 0)$ for the ideal point. Converting these to Euclidean coordinates gives precisely the values we want for the origin and for the unit point, namely, $x = 0$ and $x = 1$. The ideal point cannot be converted to a real number. This is not a problem, of course, because the ideal point does not belong in the Euclidean line.

## Coordinates for the Real Projective Plane

Homogeneous coordinates also work for a plane; however, with one additional dimension, we need a third coordinate (Smart, 1998, 244–249). A point is now represented by an ordered triple of numbers $(x_1, x_2, x_3)$, with the understandings that the triple $(ax_1, ax_2, ax_3)$ is the same point and that $(0, 0, 0)$ is excluded. To convert between Euclidean and homogeneous coordinates, use the substitutions

$$x = \frac{x_1}{x_3}, \; y = \frac{x_2}{x_3}.$$

As we did with the projective line, we can interpret this as a Euclidean system of one higher dimension. Think of our projective plane as being the plane $z = 1$ in $\mathfrak{R}^3$. Most lines through the origin will intersect $z = 1$, and a direction vector for such a line gives homogeneous coordinates for the intersection point of the line and the plane. The vector $(10, 4, 2)$, for instance, gives the direction of a line that intersects $z = 1$ at the point $(5, 2, 1)$; so, $(10, 4, 2)$ is one way to describe this point. Lines that do not intersect $z = 1$ have direction vectors ending in 0, such as $(1, 2, 0)$, and these correspond to ideal points for the projective plane. The set of ideal points makes up the ideal line.

In the real projective plane, there are three lines of special importance:

$x_1 = 0$,  the $y$-axis;

$x_2 = 0$,  the $x$-axis;

$x_3 = 0$,  the ideal line.

Do these equations make sense when you convert to Euclidean coordinates? In two cases, you should get a familiar equation; what happens in the third case? Together these three lines create the *fundamental triangle* in the real projective plane (see Figure 10.7, next page). Each region created by these lines has a distinctive pattern of positive and negative entries for its homogeneous coordinates. When we cross one of these lines, the sign of one coordinate will change. Notice,

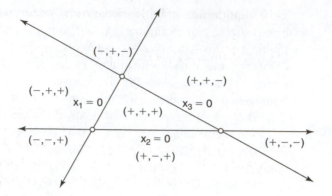

**FIGURE 10.7**
The Fundamental Triangle

however, that there are only four regions, despite the eight possible patterns of plus or minus. The homogeneity causes $(+, -, +)$ to be the same as $(-, +, -)$, because it is multiplication by $-1$. The other patterns each have a match as well.

In the Euclidean plane, the equation of a line can always be written in the form $ax + by + c = 0$, with $a$, $b$, $c$ not all zero. If we use the substitutions to convert this to homogeneous coordinates, this equation becomes

$$a\left(\frac{x_1}{x_3}\right) + b\left(\frac{x_2}{x_3}\right) + c = 0,$$

which simplifies to

$$ax_1 + bx_2 + cx_3 = 0.$$

Because every variable in this equation is raised to the same power and there are two variables in each term, the sum of the degrees in each term is 2. This is a *homogeneous equation* of degree 2.

For example, the equation $2x_1 - 5x_2 + x_3 = 0$ describes a line. So does the equation $4x_1 - 3x_3 = 0$. Where do these lines intersect? There is a difficulty in that there are three variables but only two equations. Thus, there is not a unique solution. We *can* solve this in homogeneous coordinates, however. An immediate observation is that

$$x_1 = \frac{3}{4}x_3.$$

Eliminating $x_1$ from the two given equations gives

$$x_2 = \frac{1}{2}x_3.$$

From these relationships, we get the coordinates $(\frac{3}{4}x_3, \frac{1}{2}x_3, x_3)$. This is indeed the solution; pick any nonzero value for $x_3$, and it produces a representation of the point. Another way to see this is to rewrite the coordinates as $x_3(\frac{3}{4}, \frac{1}{2}, 1)$. In this form, we can easily see that the solution is one point represented by all the multiples.

Let's look at this situation more closely. For convenience, choose $x_3 = 4$. Then the point is represented by $(3, 2, 4)$, which has only integer values for coordinates. This point lies on the line $2x_1 - 5x_2 + x_3 = 0$, so substituting the coordinates of the point will produce a valid equation:

$$2 \times 3 - 5 \times 2 + 4 = 0.$$

This equation can be written nicely in vector format:

$$[2, -2, 1] \cdot (3, 2, 4) = 0.$$

In general, the homogeneous linear equation $ax_1 + bx_2 + cx_3 = 0$ can be written as

$$[a, b, c] \cdot (x_1, x_2, x_3) = 0.$$

The triple $(x_1, x_2, x_3)$ gives the homogeneous coordinates of a point lying on this line. What role does the triple $[a, b, c]$ play in this equation? Think of duality. These are the *homogeneous coordinates of the line.* Just as points have coordinates in the real projective plane, lines too have coordinates.

Here is an example of line coordinates: Consider the Euclidean line $x = y$. In homogeneous coordinates, this becomes $x_1 = x_2$. Now write this in the form used above: $x_1 - x_2 + 0x_3 = 0$. Therefore, the coordinates of this line are $[1, -1, 0]$. The origin lies on this line, and a quick calculation confirms it: $[1, -1, 0] \cdot (0, 0, 1) = 0$. This calculation still works if other representations are used, such as $[-2, 2, 0] \cdot (0, 0, 15)$. We can also ask for the coordinates of the ideal point on this line. To find these coordinates, set $x_3 = 0$ and solve the equation $[1, -1, 0] \cdot (x_1, x_2, 0) = 0$.

Pappus' Theorem, Pascal's Theorem, and Desargues' Theorem are all concerned with three points being collinear. With homogeneous coordinates, there is a calculation that checks for collinearity.

**THEOREM 10.14**    Three projective points $(x_1, x_2, x_3)$, $(y_1, y_2, y_3)$, and $(z_1, z_2, z_3)$ are collinear if and only if

$$det \begin{bmatrix} x_1 & x_2 & x_3 \\ y_1 & y_2 & y_3 \\ z_1 & z_2 & z_3 \end{bmatrix} = 0.$$

(This computation is the determinant of the $3 \times 3$ matrix.)

**Proof**    Using the linear homogeneous equation $ax_1 + bx_2 + cx_3 = 0$, this theorem is routine but a bit messy to prove with basic algebra. Here is a more sophisticated proof using linear algebra.

The three points are collinear if and only if the linear system

$$\begin{cases} ax_1 + bx_2 + cx_3 = 0 \\ ay_1 + by_2 + cy_3 = 0 \\ az_1 + bz_2 + cz_3 = 0 \end{cases}$$

has a nontrivial solution for $[a, b, c]$. This solution exists if and only if the column vectors

$$\begin{bmatrix} x_1 \\ y_1 \\ z_1 \end{bmatrix}, \quad \begin{bmatrix} x_2 \\ y_2 \\ z_2 \end{bmatrix}, \quad \begin{bmatrix} x_3 \\ y_3 \\ z_3 \end{bmatrix}$$

are linearly dependent. This dependence occurs if and only if the given determinant is 0.

What is the dual version of this theorem? You will get a chance to state and prove this in the exercises.

The linear algebra approach in this proof gives us another useful fact. Because the determinant equals 0 if and only if the row vectors are dependent as well, we can say that three points are collinear if and only if any one of them can be written as a linear combination of the others. For example, consider the line $x_2 = 2x_1$ (or $y = 2x$) in the real projective plane. This line contains the origin, which is represented by $(0, 0, 1)$, and the point $(1, 2, 1)$. The ideal point $(1, 2, 0) = -1 \cdot (0, 0, 1) + 1 \cdot (1, 2, 1)$ thus belongs on this line.

## Homogeneous Coordinates for the Fano Plane

The Fano plane is the smallest projective plane, having merely seven points and seven lines. Because it is a finite system, we must have a finite list of possible coordinates. Unfortunately, using the real numbers creates an infinite set of possible coordinates. Therefore, we must restrict ourselves to a finite number system.

If you have studied abstract algebra, you have encountered modular arithmetic. The most familiar example of modular arithmetic is clock arithmetic, with the numbers $1, \ldots, 12$. If it is presently 5 o'clock and we add 10 hours, the answer is 3 o'clock. This is written as $5 + 10 \equiv 3 \pmod{12}$. Similarly, 8 o'clock plus 12 hours is 8 o'clock, so $8 + 12 \equiv 8 \pmod{12}$. This demonstrates that 12 is an identity for addition in this system. We also know that 6 o'clock plus 6 hours is 12 o'clock, so 6 is its own inverse for addition. However, it is more common in mathematics to say $6 + 6 \equiv 0 \pmod{12}$. It is written this way because we are accustomed to using 0 to represent the additive identity. Mathematicians would say we are using the modular numbers $0, \ldots, 11$.

For the coordinates of the Fano plane, we will use the modular number system with only two values, 0 and 1. This system is denoted by $Z_2$. Here are the addition and multiplication tables for $Z_2$.

| + | 0 | 1 |
|---|---|---|
| 0 | 0 | 1 |
| 1 | 1 | 0 |

| × | 0 | 1 |
|---|---|---|
| 0 | 0 | 0 |
| 1 | 0 | 1 |

This is an example of a *finite field*. It has (almost) all of the properties of the real numbers: closure, commutativity, associativity, distributivity, identities, and inverses. One major difference, however, is that the real numbers are *ordered* because we can clearly separate positive from negative. This cannot be done in $Z_2$, because $-1 + 1 = 0 = 1 + 1$, which would make $-1 = 1$.

To set up homogeneous coordinates for the Fano plane, we must have three components for the coordinates: one for each dimension and one to distinguish the ideal points. Suppose we choose point $A$ to be the origin $(0, 0, 1)$ and point $B$ to be the ideal point $(1, 0, 0)$. These two points lie on a common line that has a third point, $C = (z_1, z_2, z_3)$. According to Theorem 10.14,

$$det \begin{bmatrix} 0 & 0 & 1 \\ 1 & 0 & 0 \\ z_1 & z_2 & z_3 \end{bmatrix} = 0.$$

Expanding this determinant gives $z_2 = 0$. Point $C$ is not an ideal point, so $z_3 \neq 0$. Further, $C \neq A$, so the first coordinate cannot be 0. Thus, the coordinates of the third point must be $C = (1, 0, 1)$.

Actually, this is easier to do by linear combinations. In $Z_2$, we have only two coefficients to try, namely, 0 and 1. So there are very few linear combinations of $A$ and $B$, and the only new point generated is $1 \cdot A + 1 \cdot B = 1 \cdot (0, 0, 1) + 1 \cdot (1, 0, 0) = (1, 0, 1)$.

Now pick any point $D$ not on this line $ABC$. For instance, we could pick $D = (1, 1, 1)$. The determinant test shows that $D$ is not collinear with $A$, $B$. Another way to check this is to find the coordinates of $\overleftrightarrow{AB}$, which are $[0, 1, 0]$. (You should verify this.) Then $[0, 1, 0] \cdot (1, 1, 1) \neq 0$, so $D$ is not on this line. Now use $D$ to form $\overleftrightarrow{AD}$, $\overleftrightarrow{BD}$, and $\overleftrightarrow{CD}$. Each of these will have a third point, and it is not hard to find the coordinates of these new points. For instance, the third point on $AD$ is $1 \cdot A + 1 \cdot D = 1 \cdot (0, 0, 1) + 1 \cdot (1, 1, 1) = (1, 1, 0)$.

Figure 10.8 shows the Fano plane with its points labeled by their homogeneous coordinates. This is only one of many possible ways to label these points. Notice that the triple $(0, 0, 0)$ does not appear in this diagram. Where is the ideal line in this figure?

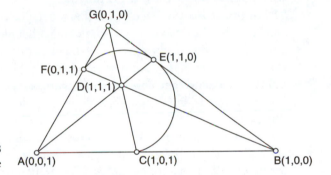

**FIGURE 10.8**
Coordinates in the Fano Plane

## PROJECTIVE TRANSFORMATIONS

In Activity 2, you looked at a relationship between two lines. Any point on the line $\ell$ is related to a corresponding point on the line $m$. How could you describe this correspondence?

**DEFINITION 10.3**  For two lines $\ell$ and $m$ in a projective plane and a point $P$ not on either of these lines, a *perspectivity* is a mapping between the sets of points on these lines defined by $f(X) = \overleftrightarrow{PX} \cap m$ for a point $X$ on $\ell$. The point $P$ is the *center* of the perspectivity. The fact that these lines correspond in this way is denoted $\ell \overline{\wedge} m$.

In Activity 2, you experimented with a perspectivity. Notice that the length of the initial segment $AB$ did not match the length of its image; in fact, if the lines $\ell$ and $m$ are not parallel, the two segments are not even proportional. The order of the points could reverse, depending on where $P$ is located. In the Euclidean plane, on the other hand, betweenness is preserved. If point $X$ is between $A$ and

$B$ on line $\ell$, then $f(X)$ will be between $f(A)$ and $f(B)$ on line $m$. Betweenness is an *invariant* of the mapping. Based on your observations, what are some other invariants for this perspectivity? Suppose you created a perspectivity in three-dimensional space between two planes. What properties do you think would be invariants in that situation?

The mapping in Activity 7a is also a perspectivity. This function is almost a one-to-one correspondence between the sets of points on the two lines, but there is a difficulty. It is possible to position point $X$ so that $P_1X$ is parallel to $\ell_2$. In this situation, we need the ideal point of $\ell_2$ to act as point $Y$. The ideal point of $\ell_1$ is also needed, otherwise there will be a point on $\ell_2$ that cannot be an output for this mapping. (Think about $P_1Y$ when it is parallel to $\ell_1$.) With the addition of these ideal points, this function $f$ truly becomes a one-to-one correspondence between the two projective lines.

This one-to-one correspondence was the important fact for proving that two lines in a finite projective plane have the same number of points. (See Corollary 10.4.) Notice that the inverse of a perspectivity uses the same center; it simply reverses the domain and range sets. In addition, any line is perspective to itself. To see this, pick any point $P$ not on line $\ell$ to use as the center. The mapping in this situation will be the identity function.

In projective space, we can define a perspectivity between two planes in a similar fashion, using a center point not on either plane. Lines in one plane will be perspective to lines in the other plane. Incidence of points and lines is preserved by this perspectivity. Thus, collinear points in one plane are perspective to collinear points in the other, and concurrent lines are perspective to concurrent lines. The image of a circle, however, might not be a circle; it could be an ellipse or some other curve.

A classic example of this is the cross sections of a cone. Figure 10.9 shows a cone sliced by a plane in various ways. There are a variety of curves that can be formed by the intersections: circles, ellipses, parabolas, and hyperbolas (or even a pair of intersecting lines). These curves, which are called the *conic sections*, have many interesting properties, one of which is that any two conic section curves are perspective from the point at the apex of the cone.

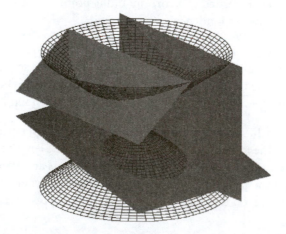

**FIGURE 10.9**
Sections of a Cone

This perspectivity property allows us to extend Pascal's Theorem. This theorem was originally stated for a hexagon inscribed in a circle, but it is, in fact, true for any conic curve. The conic curve is created by a plane intersecting the cone. A hexagon inscribed on the conic curve is a set of lines in this plane. Now imagine a second plane intersecting the cone to form a circle. The hexagon lines of the conic curve are perspective to other lines in the second plane, lines that form a hexagon on the circle. Incidence of lines is invariant, so the collinear cross joins for the hexagon in the circle correspond to collinear cross joins for the hexagon on the conic curve. Furthermore, the line common to the cross joins for the circle corresponds to a line common to the cross joins for the curve. (This is not really a proof, but it gives some idea why the more general theorem works.)

Because perspectivities are functions, they can be composed. This creates a new sort of mapping.

**DEFINITION 10.4**  For two lines $\ell_1$ and $\ell_2$, a *projectivity* is a mapping between the sets of points on the two lines created by composing a sequence of perspectivities. The fact that these lines correspond in this way is denoted $\ell_1 \barwedge \ell_2$.

In Activity 7b, you created a projectivity by composing two perspectivities. Euclidean betweenness is invariant for a projectivity, just as it is for the individual perspectivities. It is also possible for a line to be projective to itself. Because perspectivities have inverses, projectivities will have inverses as well. So the next theorem should not be a surprise.

**THEOREM 10.15**  The set of projectivities in a projective plane forms a group.

**Proof**  This theorem is routine to prove. Closure is easy: A composition of one sequence of perspectivities followed by another such sequence forms a single, longer sequence of perspectivities, which is a projectivity. Composition of functions is always associative. The function $f(X) = X$, a projectivity from any line to itself, serves as the identity for the group. Inverses are the major issue. Look first at a single perspectivity $f(X) = PX \cap \ell_2$ for points $X$ on line $\ell_1$. If we switch the domain and range, the same center will produce the inverse perspectivity $f^{-1}(Y) = \overleftrightarrow{PY} \cap \ell_1$ for points $Y$ on $\ell_2$. Because a projectivity is a sequence of perspectivities, the inverses of these perspectivities, composed in the reverse order, will be the inverse of the projectivity.

------------------------------------------------

Activity 8 asked you to create a projectivity, in a somewhat roundabout way. (Did you recognize Pappus' Theorem?) This mapping was deliberately designed to make $f(A_1) = A_2$, $f(B_1) = B_2$, and $f(C_1) = C_2$. This projectivity can actually be done as a sequence of two perspectivities. The first uses $A_2$ as the center of perspectivity between $\ell_1$ and $\overleftrightarrow{PQ}$, while the second uses $A_1$ as the center for a perspectivity between $\overleftrightarrow{PQ}$ and $\ell_2$. Because this procedure will always work, we can state a theorem.

**THEOREM 10.16**  Three distinct points on one line can be projectively related to three distinct points on another line by a sequence of exactly two perspectivities.

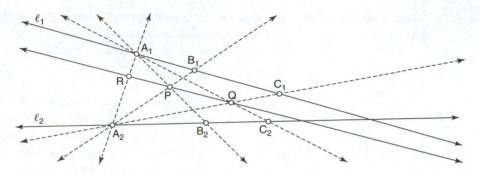

**FIGURE 10.10**
The Axis of Projectivity

**Proof of Theorem 10.16**  Follow the construction of Activity 8. (An example is shown in Figure 10.10.) With $A_2$ as the center of perspectivity, we have $A_1 B_1 C_1 \overline{\wedge} RPQ$. With $A_1$ as the center, we have $RPQ \overline{\wedge} A_2 B_2 C_2$. Therefore, $A_1 B_1 C_1 \wedge A_2 B_2 C_2$ in exactly two steps.

A detail to consider: What if $A_1 B_1 C_1$ are directly perspective to $A_2 B_2 C_2$? In this case, use the identity as the first perspectivity, and use the given perspectivity to complete the sequence.

--------

Notice the important role played by $\overleftrightarrow{RPQ}$ in this proof. This line is called the *axis of projectivity* for this particular mapping. Once constructed, it can be used to perform the projectivity, instead of using a pair of perspective centers. For any point $X$ on $\ell_1$, find $X_0 = \overleftrightarrow{A_2 X} \cap \overleftrightarrow{PQ}$, and then find $f(X) = \overleftrightarrow{A_1 X_0} \cap \ell_2$.

Theorem 10.16 does assume that the domain and range are different lines. As was noted earlier, however, a line can be projective to itself. Proving this requires only one additional step.

**COROLLARY 10.17**  Any three distinct collinear points are projective to any other three distinct collinear points by means of at most three perspectivities.

**Proof**  If the domain and range are different lines, Theorem 10.16 says that only two perspectivity steps are needed. If the domain and range are the same line $\ell_1$, create any second line $\ell_2$ and a point $P$ not on these two lines. Perform the perspectivity centered at $P$ from $\ell_1$ to $\ell_2$. This is one step. It then takes only two more steps to complete the projectivity from $\ell_2$ to the original range $\ell_1$.

--------

These results say that any three collinear points can be projectively related to any other three collinear points and that only a few perspectivity steps are needed to do so. However, a projectivity between lines usually has more than three points in its domain, except in the Fano plane. What happens to these other points? Can there be different projectivities that agree for three particular points but that disagree elsewhere? The answer is "no;" knowing what happens to three points completely determines the projectivity. This is the *Fundamental Theorem of Projective Geometry*.

The proof of the Fundamental Theorem builds on the preceding results but is somewhat long. We will state, but not prove, one final portion.

**LEMMA 10.18**   If a projectivity from a line $\ell$ to itself leaves three distinct points fixed, that projectivity must be the identity.

Now we are ready to prove the main result.

**THEOREM 10.19**   **Fundamental Theorem of Projective Geometry**   Given three distinct collinear points $A_1$, $B_1$, $C_1$ and three corresponding distinct collinear points $A_2$, $B_2$, $C_2$, there is exactly one projectivity for which

$$A_1B_1C_1 \barwedge A_2B_2C_2.$$

**Proof**   We already know that at least one projectivity can be found. Consider any other point $D_1$ on line $\ell_1$. If there are two different projectivities, then

$$A_1B_1C_1D_1 \barwedge A_2B_2C_2D_2 \quad \text{and} \quad A_1B_1C_1D_1 \barwedge A_2B_2C_2D_3$$

for points $D_2$ and $D_3$ on line $\ell_2$. Our task is to show that $D_2 = D_3$.

Use the inverse of the first projectivity to get

$$A_2B_2C_2D_2 \barwedge A_1B_1C_1D_1 \barwedge A_2B_2C_2D_3,$$

and hence

$$A_2B_2C_2D_2 \barwedge A_2B_2C_2D_3.$$

This composition is a projectivity from $\ell_2$ to itself that leaves three points fixed. Therefore, it is the identity projectivity and $D_2 = D_3$.

We have discussed two types of projective transformations, the perspectivities and the projectivities. Projectivities are defined as compositions of perspectivities, so they are more complicated functions. The following corollary gives an easy way to tell if a projectivity is actually the simpler mapping, a perspectivity.

This corollary is an "if and only if" statement. That means it is actually two statements together, an implication and its converse. There are two things to prove, so our proof will be in two parts.

**COROLLARY 10.20**   A projectivity between two distinct lines is a perspectivity if and only if the intersection of the lines is a fixed point of the projectivity.

**Proof**

*Only if*: If the projectivity consists of a single perspectivity, the center $P$ will not be on either line. The intersection point $A = \ell_1 \cap \ell_2$ will not be transformed because $\overleftrightarrow{PA} \cap \ell_2 = A$.

*If*: Let $A = \ell_1 \cap \ell_2$. Pick two more points $B_1$, $C_1$ on $\ell_1$, and find their images $B_2$, $C_2$ on $\ell_2$. Thus, we have

$$AB_1C_1 \barwedge AB_2C_2.$$

Let $P = \overleftrightarrow{B_1B_2} \cap \overleftrightarrow{C_1C_2}$. This point $P$ will be the center of a perspectivity that makes

$$AB_1C_1 \doublebarwedge AB_2C_2,$$

and therefore the projectivity equals the perspectivity from $P$.

This corollary is an interesting result, but its real value is that it allows us to give a more general proof of Pappus' Theorem. Because this proof uses only projective ideas, the theorem holds even when ideal points are involved.

**THEOREM 10.21**   **Pappus' Theorem**   Suppose $A_1$, $B_1$, $C_1$ are three distinct points on one line and $A_2$, $B_2$, $C_2$ are three distinct points on another line. Form the cross joins $X = \overleftrightarrow{A_1B_2} \cap \overleftrightarrow{A_2B_1}$, $Y = \overleftrightarrow{B_1C_2} \cap \overleftrightarrow{B_2C_1}$, and $Z = \overleftrightarrow{C_1A_2} \cap \overleftrightarrow{C_1A_2}$. Then the points $X$, $Y$, $Z$ are collinear.

**Proof**   (It will be helpful if you create a diagram of this proof.) We need a few more points. Let $P = \overleftrightarrow{A_1B_1} \cap \overleftrightarrow{A_2B_2}$, $Q = \overleftrightarrow{A_1C_2} \cap \overleftrightarrow{A_2B_1}$, and $R = \overleftrightarrow{B_1C_2} \cap \overleftrightarrow{A_2C_1}$. This gives

$$A_2XQB_1 \overline{\barwedge} A_2B_2C_2P \overline{\barwedge} RYC_2B_1,$$

where $A_1$ is the center of the first perspectivity and $C_1$ is the center of the second. The composition of these two perspectivities forms a projectivity between $\overleftrightarrow{A_2B_1}$ and $\overleftrightarrow{B_1C_2}$, which has $B_1$ as a fixed point. By the corollary, this projectivity can be done as a single perspectivity. The center of this perspectivity is

$$\overleftrightarrow{A_2R} \cap \overleftrightarrow{QC_2} = \overleftrightarrow{A_2C_1} \cap \overleftrightarrow{A_1C_2} = Z.$$

Point $X$ is perspective to $Y$ under this perspectivity, so $X$, $Y$, $Z$ are collinear.

----

It is worth noting that duality can be applied to the theorems and corollaries in this section to produce a new set of theorems.

This chapter has only been an introduction to projective geometry. Pappus' Theorem, Desargues' Theorem, and the Fundamental Theorem are interrelated in intricate ways that are worth studying. When using coordinates, Pappus' Theorem also has a deep connection to the commutativity of the number system (this was proven by Hilbert; see Robinson, 1946, 92–94). There are many other interesting topics, such as the cross ratio and the theory of conics. Furthermore, matrix equations can be used to represent projectivities, similar to what was done in Chapter 7. Thus, algebraic tools can be used to study projective geometry. In recent years, these matrix tools have found important applications in computer graphics.

## 10.3  EXERCISES

Give clear and complete answers to the exercises, expressing your explanations in complete sentences. Include diagrams whenever appropriate.

1. From the axioms, prove that two distinct lines have exactly one point in common.

2. Prove Dual Axiom 4, using the axioms of the projective plane.

3. a. Prove that the points on one line can be put in a one-to-one correspondence with the points on any other line.

   b. State the dual of part a. Then write the dual proof, and check that this proves the dual statement.

4. Draw the finite projective plane of order 3. Recall that the "lines" do not have to be straight in your

picture. It may help to list the points and lines before trying to make the drawing. (*A harder question:* Do this for order 4.)

5. In the Euclidean plane, prove that parallelism is an equivalence relation for lines.

6. Verify that the real projective plane satisfies the axioms for a projective plane. Include both possible cases for Axiom 3.

7. For the spherical model, give a careful proof that Axiom 2 holds.

8. Write the dual of Pappus' Theorem, and construct an example.

9. Consider Desargues' Theorem in the Euclidean plane. Use Menelaus' Theorem to prove the special case in which none of the lines in the configuration is parallel.

10. Construct examples of a quadrangle and a quadrilateral. Identify the diagonal triangle in each example.

11. Find an example of Desargues' Theorem in the Fano plane.

12. Construct an example of Brianchon's Theorem.

13. Convert the homogeneous equations $x_1 = 0$, $x_2 = 0$, and $x_3 = 0$ to Euclidean form. Do your answers make sense? Explain why or why not.

14. a. Find the homogeneous coordinates of the point where the lines $x_1 - 2x_2 + 3x_3 = 0$ and $5x_1 + 2x_2 - x_3 = 0$ intersect.
    b. Find the homogeneous coordinates of the line connecting the points $(\frac{1}{2}, -3, 0)$ and $(4, 4, 7)$.

15. See Figure 10.8. Find the coordinates of every line in this labeling of the Fano plane. Which is the ideal line?

16. Can point $(a, b, c)$ ever lie on line $[a, b, c]$? Either give an example in some model or explain why not.

17. See Figure 10.8. Relabel the points of the Fano plane so that the line that appears curved is the ideal line.

18. In the real projective plane, list all possible $+/-$ patterns for three coordinates. Then match these triples to the regions of the fundamental triangle. Also, locate the usual four Euclidean quadrants in the fundamental triangle.

19. Consider the real projective plane. What is $[1, 0, 0]$? What is $(1, 0, 0)$? What is $[0, 1, 0]$? What is $(0, 1, 0)$? What is $[0, 0, 1]$? What is $(0, 0, 1)$?

20. Identify the type of curve for each of the following Euclidean equations. Convert each to homogeneous coordinates. Then find any ideal points that lie on each curve.
    a. $y = px^2$
    b. $\dfrac{x^2}{a^2} + \dfrac{y^2}{b^2} = 1$
    c. $\dfrac{x^2}{a^2} - \dfrac{y^2}{b^2} = 1$

21. Use homogeneous coordinates to analytically prove Pappus' Theorem. (This will be simpler if you choose the first line carefully.)

22. State and prove the dual of Theorem 10.12.

23. Write the dual definitions for perspectivity and projectivity, and construct examples.

Exercises 24–26 are more advanced.

24. *Moulton's Geometry.* Another model of the projective plane was developed by Moulton (Robinson, 1946, 126–128). Start with the points of the Euclidean plane. The lines in this model are described in Euclidean terms. Suppose this plane has a rectangular coordinate system. All the horizontal, vertical, and Euclidean lines with negative slope are Moulton lines. If the Euclidean line $y = m(x - a)$ has positive slope, the corresponding Moulton line is given by

$$y = \begin{cases} m(x - a) & \text{for } x < a \\ \frac{m}{2}(x - a) & \text{for } x \geq a \end{cases}$$

(To visualize this, you can think of a beam of light refracting as it crosses a boundary.) To this collection of points and lines, add ideal points and the ideal line as before.
    a. Verify that Moulton's geometry satisfies the axioms for a projective plane.
    b. Draw some triangles, including some that cross the horizontal axis.

c. Find a pair of triangles for which Desargues' Theorem holds.

d. Find a pair of triangles for which Desargues' Theorem does not hold.

25. In Sketchpad, create five points $A_1$, $A_2$, $B_1$, $B_2$, and $C_1$ in general position (meaning that no three are collinear). Construct the point $P = \overleftrightarrow{A_1 B_2} \cap \overleftrightarrow{A_2 B_1}$, and create a line $\ell$ through $P$. Now construct the point

$$X = A_1(\ell \cap \overleftrightarrow{A_2 C_1}) \cap B_1(\ell \cap \overleftrightarrow{B_2 C_1}).$$

(Be careful with this statement. Is it clear when to form lines and when to form points?) Construct the locus of $X$ as $\ell$ varies. What curve is produced? Try this for different arrangements of the original five points. How is this exercise related to Pascal's Theorem?

26. a. Suppose you wish to create a perspective drawing of a floor that is tiled in a checkerboard pattern. If one edge of the floor is directly in front of you and parallel to the horizon, the lines moving away from it will appear to intersect at the horizon, at a *vanishing point*. This is *one-point perspective*. The lines parallel to the horizon are called *transversals*, and the lines moving away are called *orthogonals*. The challenge to constructing this picture is that corresponding diagonals of the checkerboard must appear parallel, so these diagonals also must intersect at the horizon. Construct this picture.

b. *For the ambitious:* If only one corner of the checkerboard, not an entire edge, is directly in front of you, there will be two vanishing points: one for the parallel lines going left and another for the parallel lines going right. This is *two-point perspective*. As before, the diagonals should appear parallel, so they must intersect on the horizon. Construct this picture. Once you are satisfied with your construction, draw the diagonals in the other direction to see if they also intersect on the horizon.

    (There is also a technique called *three-point perspective* that is used for objects seen from an unusual angle, such as a bird's-eye view of a tower.)

Exercises 27 and 28 are especially for future teachers.

27. In the *Principles and Standards for School Mathematics,* the National Council of Teachers of Mathematics observes that "Geometry has long been regarded as the place in the school mathematics curriculum where students learn to reason and to see the axiomatic structure of mathematics." Further, "Through the middle grades and into high school, . . . students should learn to use deductive reasoning and more formal proof techniques to solve problems and prove conjectures. At all levels, students should learn to formulate convincing explanations for their conjectures and solutions. . . . They should also be able to understand the role of definitions, axioms, and theorems and be able to construct their own proofs" (NCTM, 2000, 41–42). What does this mean for you and your future students?

a. Read the discussion on Geometry in the *Principles and Standards for School Mathematics* (NCTM, 2000, 41–43). Then study the Reasoning and Proof Standard for at least two grade bands (i.e., two of pre-K–2, 3–5, 6–8, or 9–12). What are the NCTM recommendations regarding the teacher's role in helping students recognize reasoning and proof as fundamental aspects of mathematics? How is this focus on reasoning and proof developed across several grade bands? Cite specific examples.

b. Find copies of school mathematics textbooks for the same grade levels as you studied for part a. How are the NCTM standards for reasoning and proof implemented in those textbooks? Again, cite specific examples.

c. Write a report in which you present and critique what you learn. Your report should include your responses to parts a and b.

28. In the *Principles and Standards for School Mathematics,* the National Council of Teachers of Mathematics recommends that "Instructional programs from prekindergarten through grade 12 should enable all students to . . . develop

and evaluate mathematical arguments and proofs" (NCTM, 2000, 56). Throughout this course, you have investigated several different kinds of axiom systems. For example, you worked with Euclid's Postulates, the axioms of metric geometry, the axioms of the hyperbolic plane, and the axioms of the projective plane.

a. Design several classroom activities involving logical reasoning and mathematical proof that would be appropriate for students in your future classroom.

b. Write a short report explaining how the activities you design reflect both what you have learned as you have worked with various axiom systems throughout this course and the NCTM recommendations.

Reflect on what you have learned in this chapter.

29. Review the main ideas of this chapter. Describe, in your own words, the concepts you have studied and what you have learned about them. What are the important ideas? How do they fit together? Which concepts were easy for you? Which were hard?

30. Reflect on the learning environment for this course. Describe aspects of the learning environment that helped you understand the main ideas in this chapter. Which activities did you like? Dislike? Why?

## 10.4 CHAPTER OVERVIEW

In Chapters 9 and 10, we presented two systems of geometry that are very different from Euclidean geometry. In both cases, parallelism is the key difference. Hyperbolic geometry allows the existence of many parallels to a given line, while in projective geometry, there are no parallel lines at all! The axiom system for a projective plane is very short and simple, but it leads to surprisingly strong results.

After a discussion of these axioms, a few basic theorems about points and lines were presented. The case in which the projective plane is finite got extra attention. Here we proved that every line contains the same number of points and that every point is on this same number of lines. From this information, we found a formula for the total number of points in a finite projective plane.

For these axioms and theorems to be meaningful, of course, there must be mathematical situations (models) to which they apply. Three models were shown. The first was the real projective plane, which begins with the Euclidean plane and adds an ideal point—a point at infinity—to each line. A major theorem of the real projective plane is Pappus' Theorem. The second model was a finite one, the Fano plane. This is the finite projective plane of smallest possible order. The third model shown was the spherical model, which is related to the real projective plane by central projection. To create a projective plane from the sphere, it is necessary to identify antipodal points, those points that lie diametrically across from each other.

A remarkable property of the projective plane is duality. The notions of point and line are interchangeable in the axioms and, hence, are interchangeable in any theorem proven from those axioms. Of particular interest were the ideas of perspective from a point or from a line. These led to Desargues' famous theorem and its converse, which together state that the two notions of perspective are equivalent. We also saw Pascal's Theorem about a hexagon inscribed in a circle, and its dual, Brianchon's Theorem about a hexagon circumscribed around a circle.

The introduction of homogeneous coordinates brought algebraic tools to the study of projective geometry. We saw that instead of four quadrants, the real projective plane has the fundamental triangle and its four regions. Because of duality, we can have coordinates for a line as well as for a point. The Fano plane can also have homogeneous coordinates, though this finite structure requires a finite number system, $Z_2$.

The final topic of this chapter was transformations in the projective plane. These come in two types: the *perspectivity*, a mapping between two lines that uses one center point; and the *projectivity*, which is a composition of perspectivities. Just as the Euclidean isometries did, the projectivities form a group. Several basic theorems about how these projectivities are constructed led to the Fundamental Theorem of Projective Geometry. The Fundamental Theorem tells us that knowing the correspondence between two sets of three points gives enough information to completely determine the projectivity. This led to a projective proof of Pappus' Theorem. In the Euclidean plane, the cases of Pappus' Theorem involving parallel lines each require special treatment. The projective approach covers all the cases in one proof that is both more general and more powerful.

**Desargues' Theorem**   In the real projective plane, if two triangles are perspective from a point, then they are perspective from a line.

**The Dual (and Converse) of Desargues' Theorem**   In the real projective plane, if two triangles are perspective from a line, then they are perspective from a point.

**Pascal's Theorem for Circles**   If a hexagon is inscribed in a circle, then the three points of intersection of pairs of opposite sides will be collinear.

**Brianchon's Theorem for Circles**   If a hexagon is circumscribed around a circle, then the three lines connecting pairs of opposite vertices are concurrent.

**Fundamental Theorem of Projective Geometry**   Given three distinct collinear points $A_1$, $B_1$, $C_1$ and three corresponding distinct collinear points $A_2$, $B_2$, $C_2$, there is exactly one projectivity for which

$$A_1B_1C_1 \barwedge A_2B_2C_2.$$

**Pappus' Theorem**   Suppose $A_1$, $B_1$, $C_1$ are distinct points on one line $\ell_1$ and $A_2$, $B_2$, $C_2$ are distinct points on a second line $\ell_2$. Form the cross joins $X = \overleftrightarrow{A_1B_2} \cap \overleftrightarrow{A_2B_1}$, $Y = \overleftrightarrow{B_1C_2} \cap \overleftrightarrow{B_2C_1}$, and $Z = \overleftrightarrow{C_1A_2} \cap \overleftrightarrow{C_2A_1}$. Then the three points $X$, $Y$, $Z$ are collinear.

# APPENDIX A: Trigonometry

We assume that you have studied trigonometry before. This appendix is not intended to be a thorough, detailed presentation of the subject. Rather, it will be another look at triangles and at the six trigonometric functions. We will provide different ways of developing some trigonometry facts, ways that emphasize their ties to geometry. (Many of the ideas in this appendix are based on Maor, 1998.)

Do the following activities, writing your explanations clearly in complete sentences. Include diagrams whenever appropriate. Save your work for each activity, as later work sometimes builds on earlier work. You will find it helpful to read ahead into this appendix as you work on these activities.

1. Figure A.1 shows a square and an equilateral triangle. Use these diagrams to calculate the exact values of $\sin(30°)$, $\cos(30°)$, $\tan(30°)$, $\sin(45°)$, $\cos(45°)$, $\tan(45°)$, $\sin(60°)$, $\cos(60°)$, and $\tan(60°)$.

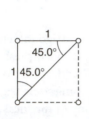

**FIGURE A.1**
Two Standard Triangles

2. Figure A.2 shows $\triangle ABC$ and its three altitudes. We can see that $\frac{AE}{AB} = \cos(\angle A)$, which makes $AE = AB\cos(\angle A)$. Write similar expressions for $AF$, $FB$, $BD$, $DC$, and $CE$, using only the sides and angles of $\triangle ABC$ in your answers.

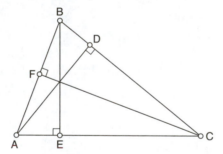

**FIGURE A.2**
Looking for Cosines

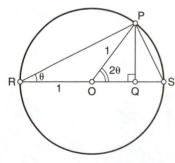

**FIGURE A.3**
Double Angle Formulas

3. Figure A.3 shows a unit circle centered at point $O$. (Recall that a unit circle has a radius of 1.)
   a. In $\triangle RPQ$, find expressions for $\sin(\theta)$ and $\cos(\theta)$.
   b. In $\triangle OPQ$, find expressions for $\sin(2\theta)$ and $\cos(2\theta)$.
   c. In $\triangle RPS$, find an expression for $\cos(\theta)$.
   d. Use the results from parts a, b, and c to prove the double angle formulas:
   $$\sin(2\theta) = 2\sin(\theta)\cos(\theta)$$
   $$\cos(2\theta) = 2\cos^2(\theta) - 1$$

4. In Figure A.4, a unit circle is drawn on a coordinate system with the origin at $A$. The lines tangent to the circle at $C$ and at $E$ are parallel to the $y$- and $x$-axes, respectively. $D$ is a point on the circle, and $\angle CAD = \theta$. Ray $AD$

intersects line $y = 1$ at $F$ and line $x = 1$ at $H$. (There is an interactive version of this diagram on the CD that accompanies this text.)

a. With $D$ in the first quadrant (as shown in Figure A.4), find segments equal to $\sin(\theta)$, $\cos(\theta)$, $\tan(\theta)$, $\sec(\theta)$, $\csc(\theta)$, and $\cot(\theta)$.

b. Repeat part a for the situation when $D$ has been moved to the second quadrant.

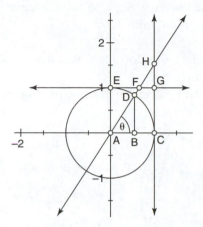

**FIGURE A.4**
Finding Trigonometric Functions

5. In Figure A.4, $\triangle ADB$, $\triangle AHC$, and $\triangle FAE$ are similar to each other.
   a. Draw each triangle separately and label the sides of each with the appropriate trigonometric functions of $\theta$.
   b. Use the similarity between these triangles to write $\tan(\theta)$ in terms of $\sin(\theta)$ and $\cos(\theta)$.
   c. Repeat part b for $\sec(\theta)$, $\csc(\theta)$, and $\cot(\theta)$.

6. Each of the following equations has several solutions in the interval $0 \leq t < 2\pi$. Find these solutions for each equation.
   a. $\sin(t)\tan(t) = \sin(t)$
   b. $4\sin(t)\cos(t) + 2\sin(t) - 2\cos(t) = 1$

7. Figure A.5 shows an angle $\alpha$ inscribed in a circle, its corresponding central angle, its corresponding chord $a$, and a segment perpendicular to that chord. Find the value of $\frac{\sin(\alpha)}{a}$. Draw similar diagrams, and find the values of $\frac{\sin(\beta)}{b}$ and $\frac{\sin(\gamma)}{c}$, where $\beta = \angle ABC$ and $\gamma = \angle BCA$. What do you observe?

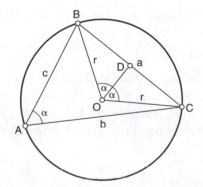

**FIGURE A.5**
Ratio of Sine to Side

8. Draw a circle and use it to construct a cyclic quadrilateral *ABCD*. Construct the diagonals *AC* and *BD* of this quadrilateral. Measure the lengths of these six segments, and calculate the products *AB · CD*, *BC · DA*, and *AC · BD*. What relationship do you observe among these three products? Make a conjecture.

9. Figure A.6 shows a cyclic quadrilateral for which the diagonal *AC* is a diameter of the circle. Suppose that the length of this diameter is 1.
   a. Use the result from Activity 7 to show that segment *BD* has length $\sin(\alpha + \beta)$.
   b. Label the four sides of the quadrilateral as $\sin(\alpha)$, $\cos(\alpha)$, $\sin(\beta)$, or $\cos(\beta)$.
   c. Use your observation from Activity 8 to write a formula for $\sin(\alpha + \beta)$.

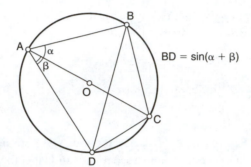

**FIGURE A.6**
Angle Sum

## A.2 DISCUSSION

## RIGHT TRIANGLE TRIGONOMETRY

From the three sides of a right triangle, there are six possible ways to form a ratio. The values of these ratios are affected by the size of the other angles in this triangle; change any one of the angles, and you change the six ratios. These ratios are the six trigonometric functions:

$$\sin(\theta) = \frac{\text{opposite side}}{\text{hypotenuse}} \qquad \tan(\theta) = \frac{\text{opposite side}}{\text{adjacent side}} \qquad \sec(\theta) = \frac{\text{hypotenuse}}{\text{adjacent side}}$$

$$\cos(\theta) = \frac{\text{adjacent side}}{\text{hypotenuse}} \qquad \cot(\theta) = \frac{\text{adjacent side}}{\text{opposite side}} \qquad \csc(\theta) = \frac{\text{hypotenuse}}{\text{opposite side}}$$

For example, consider the 60° angle of the 30°-60°-90° triangle shown in Figure A.1. This shows a triangle with a hypotenuse of 2 and a side of 1 adjacent to the 60° angle. By the Pythagorean Theorem, the opposite side to the 60° angle is $\sqrt{3}$. With these three numbers, all six functions can be computed. For example, $\sin(60°) = \frac{\sqrt{3}}{2}$.

How do we know that these functions are well-defined? Will a different right triangle with the same angles produce the same ratios? Consider a 30°-60°-90° triangle with a hypotenuse of 10. Because the angles are the same, this larger triangle is similar to our example triangle. Thus, the adjacent side to the 60° angle has length 5, and the opposite side is $5\sqrt{3}$. In this larger triangle, $\sin(60°) = \frac{5\sqrt{3}}{10}$, which reduces to $\frac{\sqrt{3}}{2}$. For similar triangles, ratios between corresponding sides will always be equal.

Activity 2 presents a triangle with its three altitudes. Each altitude creates two new triangles. In $\triangle ABE$, $AE$ is the side adjacent to $\angle A$, and $AB$ is the hypotenuse. Thus, $\cos(\angle A) = \frac{AE}{AB}$ and $AE = AB\cos(\angle A)$. In $\triangle CBE$, $CE$ is adjacent to $\angle C$, and $BC$ is the hypotenuse. Thus, $CE = BC\cos(\angle C)$. Each of the six segments listed in Activity 2 can be expressed as one of the sides of $\triangle ABC$ times the cosine of one of its angles.

Suppose we wanted to calculate the area of the $\triangle ABC$ from Activity 2. The usual formula for area of a triangle is

$$\text{area} = \frac{1}{2} \cdot \text{base} \cdot \text{altitude}.$$

For $\triangle ABC$, this area could be calculated as $\frac{1}{2} \cdot AC \cdot BE$. Notice, however, that $BE$ is the side opposite to $\angle A$ in $\triangle ABE$. Thus, $BE = AB\sin(\angle A)$, so that the area of this triangle can also be calculated using

$$\text{area} = \frac{1}{2} \cdot AC \cdot AB \cdot \sin(\angle A).$$

In the six definitions of the trigonometric functions, only three lengths were used. Consequently, there are many interconnections between these six functions. For instance, $\sec(\theta) = \frac{1}{\cos(\theta)}$. (There are two other reciprocal pairs; do you see them?) Another useful fact is

$$\tan(\theta) = \frac{\text{opposite}}{\text{adjacent}} = \frac{\left(\dfrac{\text{opposite}}{\text{hypotenuse}}\right)}{\left(\dfrac{\text{adjacent}}{\text{hypotenuse}}\right)} = \frac{\sin(\theta)}{\cos(\theta)}.$$

## UNIT CIRCLE TRIGONOMETRY

The definitions given on the preceding page for the various trigonometric functions work well for acute angles. But what about right angles or obtuse angles? Because these angles do not fit into a right triangle, these definitions cannot apply. It gets even worse: We can measure angles greater than 180° or less than 0°, but how do we compute trigonometric functions for these angles? We need stronger, more general definitions.

The solution to this difficulty is to think of the desired angle on a coordinate system. Put the vertex at the origin, and measure counterclockwise from the positive $x$-axis. (Clockwise angles are considered negative.) This allows you to draw angles of any size.

To define the trigonometric functions, imagine a circle of radius 1 centered at the origin. This is called the *standard unit circle*. The terminal side of an angle is a ray that will intersect the circle at a point $(x, y)$. The six functions are now defined as follows:

$$\sin(\theta) = \frac{y}{1} \quad \tan(\theta) = \frac{y}{x} \quad \sec(\theta) = \frac{1}{x}$$

$$\cos(\theta) = \frac{x}{1} \quad \cot(\theta) = \frac{x}{y} \quad \csc(\theta) = \frac{1}{y}$$

If the angle $\theta$ is acute, then these definitions are equivalent to those given on page 264. Figure A.4 allows us to compare the two sets of definitions. Point $D$ gives the $x$- and $y$-values for the definitions. In $\triangle ADB$, the opposite side $DB = y$, the adjacent side $AB = x$, and the hypotenuse $AD = 1$. So, the definitions agree.

When $D$ moves into the second quadrant, angle $\theta$ is between $90°$ and $180°$. The *reference triangle* for $\theta$, $\triangle ADB$, now lies in the second quadrant, with side $AB$ along the negative $x$-axis. The $x$-coordinate of $D$ is negative in this situation (the directed segment $AB$ is going in the negative direction), while the $y$-coordinate is positive. Thus, the $(x, y)$-coordinates of $D$ still provide the necessary values to compute the six trigonometric functions. The cosine, tangent, secant, and cotangent are negative for second quadrant angles, while the sine and cosecant are positive.

Because $x^2 + y^2 = 1$ on the unit circle, whether $x$ and $y$ are positive, negative, or zero, we have one of the *Pythagorean identities*:

$$\sin^2(x) + \cos^2(x) = 1.$$

With a little algebra, it is easy to prove the other Pythagorean identities:

$$\tan^2(x) + 1 = \sec^2(x),$$

$$1 + \cot^2(x) = \csc^2(x).$$

In Activity 4, you should have seen that $\tan(\theta) = CH$. Triangle $AHC$ is similar to $\triangle ADB$, and side $AC$ has length 1. In part 4b, the directed segment $CH$ is going downward, in the negative direction, which is another way to show that $\tan(\theta) < 0$ for this particular angle.

It is interesting to consider this segment $CH$ as the angle $\theta$ changes. (Notice that $CH$ is on a vertical line that is tangent to the unit circle. Might this be the source of the name *tangent*?) As $\theta$ approaches $90°$, point $H$ becomes arbitrarily high on its vertical line. At $90°$, line $AD$ is parallel to this vertical line, making $\tan(90°)$ undefined. Once $\theta$ turns into the second quadrant, $\tan(\theta)$ has negative values and approaches 0 as $\theta$ nears $180°$. Turning into the third quadrant, segment $CH$ is again positive and increasing. When $\theta$ reaches $270°$, $\tan(270°)$ is undefined for the same reason that $\tan(90°)$ is undefined. Fourth quadrant angles have $\tan(\theta) < 0$. Think about the graph of the function $\tan(\theta)$ and how its behavior is demonstrated by the length of segment $CH$.

## SOLVING TRIGONOMETRIC EQUATIONS

When solving trigonometric equations, many of the steps are the same as those used when solving algebraic equations. For instance, the first step in solving

$$4\sin(t)\cos(t) + 2\sin(t) - 2\cos(t) = 1$$

is to move all terms to one side, so you have an expression equal to 0:

$$4 \sin(t) \cos(t) + 2 \sin(t) - 2 \cos(t) - 1 = 0.$$

Now break this into factors. You might find it helpful to use abbreviations; visually, it may be easier to factor $4SC + 2S - 2C - 1 = 0$:

$$(2 \sin(t) - 1)(2 \cos(t) + 1) = 0.$$

Just as in algebra, a product equals 0 only when one or more of its factors equals 0. Thus, we have two smaller equations to solve:

$$2 \sin(t) - 1 = 0 \qquad \text{and} \qquad 2 \cos(t) + 1 = 0.$$

Let's consider these equations one at a time. The equation $2 \sin(t) - 1 = 0$ implies $\sin(t) = \frac{1}{2}$. Here is where trigonometry comes in. In Activity 1, you saw that $30°$ satisfies this equation; but $30°$ is not the only possibility. The sine function is also positive in the second quadrant, so there is a second quadrant angle that works. On the unit circle, this angle will also have a $y$-coordinate of $\frac{1}{2}$. Using a reference triangle of $30°$-$60°$-$90°$, we can see that $150°$ is another solution.

For $2 \cos(t) + 1 = 0$, we simplify to $\cos(t) = \frac{-1}{2}$. The $\frac{1}{2}$ should look familiar, for it is the cosine of $60°$. The challenge now is to place a reference triangle in the unit circle to produce a negative value for the cosine. The $x$-coordinate that represents the cosine is negative in the second and third quadrants, so that is where our reference triangle must go. Drawing the adjacent side along the horizontal axis and putting the hypotenuse at $60°$ to the horizontal gives angles of $120°$ and $240°$.

## DOUBLE ANGLE FORMULAS

Activity 3 concerns a unit circle (radius $= 1$) without a coordinate system. Figure A.3 shows an inscribed angle and its corresponding central angle, which is twice the size of the inscribed angle. Because segment $RS$ is a diameter, $\angle RPS$ is a right angle. Thus, we can say that $\cos(\theta) = \frac{RP}{RS} = \frac{RP}{2}$.

Other right triangles in this diagram lead to similar statements. By substituting, you should be able to prove the *double angle formulas*:

$$\sin(2\theta) = 2 \sin(\theta) \cos(\theta),$$
$$\cos(2\theta) = 2 \cos^2(\theta) - 1.$$

Using a Pythagorean identity, we can write two other forms of the double angle formula for cosines:

$$\cos(2\theta) = \cos^2(\theta) - \sin^2(\theta) = 1 - 2 \sin^2(\theta).$$

## ANGLE SUM FORMULAS

Activity 8 introduces a theorem that is not well known but that is surprisingly useful.

**THEOREM A.1**  **Ptolemy's Theorem**   In a cyclic quadrilateral, the product of the diagonals is equal to the sum of the products of the opposite sides.

In the notation of Activity 8, $AC \cdot BD = AB \cdot CD + BC \cdot DA$. The exercises ask you to develop this proof, which involves finding congruent angles and similar triangles.

How can this theorem be useful? First of all, suppose the cyclic quadrilateral is a rectangle. Then every vertex is a right angle, and $AB = CD$, $BC = DA$, and $AC = BD$. Ptolemy's Theorem gives us

$$(AC)^2 = (AB)^2 + (BC)^2,$$

which is the Pythagorean Theorem.

Now let $\angle BAC = \alpha$. Then $AB = AC \cdot \cos(\alpha)$ and $BC = AC \cdot \sin(\alpha)$. The Pythagorean Theorem then becomes

$$(AC)^2 = (AC \cdot \cos(\alpha))^2 + (AC \cdot \sin(\alpha))^2,$$

which simplifies to a familiar identity:

$$1 = \cos^2(\alpha) + \sin^2(\alpha).$$

Figure A.7 shows a special case of Ptolemy's Theorem in which the diagonal $AC$ is a diameter of the circle. In this diagram, the radius of the circle is $\frac{1}{2}$. From the result of Activity 7, we can say that $BD = \sin(\alpha + \beta)$. The labels on sides $AB$, $BC$, $CD$, and $DA$ come from basic right triangle trigonometry. Now when we apply Ptolemy's Theorem, we get

$$1 \cdot \sin(\alpha + \beta) = \sin(\alpha) \cdot \cos(\beta) + \sin(\beta) \cdot \cos(\alpha),$$

which is the *angle sum formula* for sine.

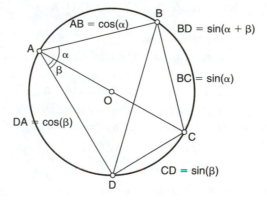

**FIGURE A.7**
A Special Case of Ptolemy's Theorem

We will not derive the angle sum formulas for cosine and tangent; instead, we merely state them:

$$\cos(\alpha + \beta) = \cos(\alpha) \cdot \cos(\beta) - \sin(\alpha) \cdot \sin(\beta),$$

$$\tan(\alpha + \beta) = \frac{\tan(\alpha) + \tan(\beta)}{1 - \tan(\alpha) \cdot \tan(\beta)}.$$

Notice that these formulas can be used for the difference of two angles, as long as we adjust for even versus odd functions:

$$\sin(\alpha - \beta) = \sin(\alpha + (-\beta))$$
$$= \sin(\alpha) \cdot \cos(-\beta) + \sin(-\beta) \cdot \cos(\alpha)$$
$$= \sin(\alpha) \cdot \cos(\beta) - \sin(\beta) \cdot \cos(\alpha).$$

Similarly,

$$\cos(\alpha - \beta) = \cos(\alpha) \cdot \cos(\beta) + \sin(\alpha) \cdot \sin(\beta)$$

and

$$\tan(\alpha - \beta) = \frac{\tan(\alpha) - \tan(\beta)}{1 + \tan(\alpha) \cdot \tan(\beta)}.$$

We can use the angle sum formulas to confirm the double angle formulas given earlier:

$$\sin(2\theta) = \sin(\theta + \theta) = \sin(\theta) \cdot \cos(\theta) + \sin(\theta) \cdot \cos(\theta) = 2\sin(\theta)\cos(\theta),$$

$$\cos(2\theta) = \cos(\theta + \theta) = \cos(\theta) \cdot \cos(\theta) + \sin(\theta) \cdot \sin(\theta) = \cos^2(\theta) - \sin^2(\theta).$$

From these, we can get one more double angle formula:

$$\tan(2\theta) = \frac{\sin(2\theta)}{\cos(2\theta)} = \frac{2\tan(\theta)}{1 - \tan^2(\theta)}.$$

## HALF-ANGLE FORMULAS

Suppose we let $\theta = \frac{\alpha}{2}$. With a great deal of algebra, the double angle formulas can be rearranged to give the *half-angle formulas*:

$$\sin\left(\frac{\alpha}{2}\right) = \pm\sqrt{\frac{1 - \cos(\alpha)}{2}},$$

$$\cos\left(\frac{\alpha}{2}\right) = \pm\sqrt{\frac{1 + \cos(\alpha)}{2}},$$

$$\tan\left(\frac{\alpha}{2}\right) = \pm\sqrt{\frac{1 - \cos(\alpha)}{1 + \cos(\alpha)}}.$$

A more useful form of the last formula is

$$\tan\left(\frac{\alpha}{2}\right) = \frac{1 - \cos(\alpha)}{\sin(\alpha)}.$$

## THE LAW OF SINES AND THE LAW OF COSINES

In Activity 7, you compared the sine of an inscribed angle $\alpha$ with the chord it subtends (the chord joining the endpoints of the angle). You can do this indirectly by looking at the corresponding central angle. Because the central angle is twice the inscribed angle, this central angle is $2\alpha$. Because $\triangle OBC$ is isosceles, the perpendicular to the chord bisects the central angle. Thus, we can find $\sin(\alpha)$ by examining $\triangle OBD$.

The answer to the question in Activity 7 is $\frac{\sin(\alpha)}{a} = \frac{1}{2r}$. The value of this ratio is determined solely by the radius of the circle. Furthermore, the diagram for $\beta = \angle ABC$ will show that $\frac{\sin(\beta)}{b}$ also equals $\frac{1}{2r}$. In fact,

$$\frac{\sin(\alpha)}{a} = \frac{\sin(\beta)}{b} = \frac{\sin(\gamma)}{c} = \frac{1}{2r},$$

which is the *Law of Sines*.

The *Law of Cosines,* which is a generalization of the Pythagorean Theorem, states that

$$c^2 = a^2 + b^2 - 2ab\cos(\gamma),$$

where $\gamma$ is the angle between sides $b$ and $c$. In a right triangle, $\gamma = 90°$ and the cosine term is 0, so that the Law of Cosines reduces to the Pythagorean Theorem.

The Law of Sines and the Law of Cosines are useful for solving triangles. This means that if you have enough information about sides and angles of a certain triangle, it is possible to calculate the other sides and angles using these formulas.

This appendix has been a very condensed presentation of trigonometry, and it is by no means complete. There are many more interesting, important facts and applications. Our intent is that this refreshes your memory and gives you helpful tools for the study of geometry.

## A.3 EXERCISES

Give clear and complete answers to the exercises, expressing your explanations in complete sentences. Include diagrams whenever appropriate.

1. We saw that the area of $\triangle ABC$ can be calculated by $\frac{1}{2} \cdot AB \cdot AC \cdot \sin(\angle A)$. Write corresponding formulas using $\angle B$ and $\angle C$ for the area of $\triangle ABC$.

2. Use the sum or difference formulas and facts about the two standard triangles to calculate the following.
   a. $\cos(15°)$ (Do this two ways.)
   b. $\sin(75°)$
   c. $\tan(105°)$
   d. $\cos(135°)$
   e. $\sin(150°)$ (Do this two ways.)
   f. $\cos(210°)$
   g. $\tan(225°)$

3. For which angles is $\sec(\theta) < 0$? Explain.

4. Use the standard unit circle to determine which of the trigonometric functions are negative for fourth quadrant angles. Explain your reasoning.

5. Use the standard unit circle to determine the sign of each trigonometric function for $\theta$ if

   $$\frac{13\pi}{2} \leq \theta \leq 7\pi.$$

   Explain your reasoning.

6. Use the angle sum formulas for sine and cosine to derive the angle sum formula for tangent.

7. Derive the following Pythagorean identities:
   a. $\tan^2(x) + 1 = \sec^2(x)$
   b. $1 + \cot^2(x) = \csc^2(x)$

8. Suppose that $a = 5$, $b = 10$, and $\angle C = 135°$ in $\triangle ABC$. Find the third side, $c$.

9. a. The base of an isosceles triangle measures 4 in., and the base angles are 52°. Find the altitude of this triangle.
   b. The base of an isosceles triangle measures $b$ units, and the base angles are $\theta$. Find an expression for the altitude of this triangle.

10. a. Find the area of an equilateral triangle whose sides measure 5 cm.
    b. Find an expression for the area of an equilateral triangle whose sides measure $a$ units.

11. Use the given information to find (if possible) the remaining sides and angles of $\triangle ABC$. If two solutions exist, find both. If no solution exists, explain how you know this.
    a. $A = 30°$, $B = 45°$, $a = 10$ units
    b. $A = 60°$, $B = 10°$, $b = 4.5$ cm
    c. $A = 150°$, $C = 20°$, $a = 200$ units
    d. $A = 100°$, $a = 125$ cm, $b = 10$ cm
    e. $C = 120°$, $a = 4$ in., $b = 6$ in.
    f. $A = 58°$, $a = 4.5$ cm, $b = 12.8$ cm
    g. $A = 58°$, $a = 4.5$ ft., $b = 5$ ft.
    h. $A = 110°$, $a = 125$ cm, $b = 200$ cm

**12.** Find the area of $\triangle ABC$ if $C = 120°$, $a = 4$ in., and $b = 6$ in.

**13.** Find the area of $\triangle ABC$ if $B = 130°$, $a = 62$ ft., and $c = 20$ ft.

**14.** Suppose that $\triangle ABC$ has $\angle A = 60°$ with the opposite side $a = 12$. Find the radius of the circumcircle of $\triangle ABC$.

**15.** Suppose that $\alpha$, $\beta$, $\gamma$ are the angles of a triangle. Prove that $\tan(\alpha) + \tan(\beta) + \tan(\gamma) = \tan(\alpha) \cdot \tan(\beta) \cdot \tan(\gamma)$.

**16.** Solve the following trigonometric equations. Give all solutions in the interval $0 \leq x < 2\pi$.
  a. $\sin(x) \cdot \tan(x) + 1 = \sin(x) + \tan(x)$
  b. $2\sin(x) = \sqrt{3}\tan(x)$
  c. $\sin^2(x) + \frac{1}{2}\sin(x) = \frac{1}{2}$
  d. $2\cos^2(x) - 2\sqrt{2}\cos(x) + 1 = 0$

**17.** Explain why $\sin(\theta) = \cos(90° - \theta)$.

**18.** Write out a careful detailed proof of the Law of Sines.

**19.** Prove the Law of Cosines.

**20.** Prove the double angle formulas.
  a. $\sin(2\theta) = 2\sin(\theta)\cos(\theta)$
  b. $\cos(2\theta) = 2\cos^2(\theta) - 1$

  c. $\cos(2\theta) = \cos^2(\theta) - \sin^2(\theta)$
  d. $\cos(2\theta) = 1 - 2\sin^2(\theta)$

**21.** Prove the angle sum formulas for cosine and tangent.
  a. $\cos(\alpha + \beta) = \cos(\alpha)\cos(\beta) - \sin(\alpha)\sin(\beta)$
  b. $\tan(\alpha + \beta) = \frac{\tan(\alpha) + \tan(\beta)}{1 - \tan(\alpha) \cdot \tan(\beta)}$

**22.** Prove the half-angle formulas.
  a. $\sin\left(\frac{\alpha}{2}\right) = \pm\sqrt{\frac{1 - \cos(\alpha)}{2}}$
  b. $\cos\left(\frac{\alpha}{2}\right) = \pm\sqrt{\frac{1 + \cos(\alpha)}{2}}$
  c. $\tan\left(\frac{\alpha}{2}\right) = \frac{1 - \cos(\alpha)}{\sin(\alpha)}$

**23.** Prove that $\tan(2\theta) = \frac{2\tan(\theta)}{1 - \tan^2(\theta)}$.

**24.** Prove Ptolemy's Theorem. (*Hint:* Construct a point, $E$, on the diagonal $AC$ so that $\angle ABE$ is congruent to $\angle DBC$. Then look for pairs of similar triangles. Remember that $AC = AE + EC$.)

**25.** Use Ptolemy's Theorem to derive an angle sum formula for the cosine.

Vectors and matrices are the fundamental objects of a course in linear algebra. There are many important and interesting facts about matrices, as well as many important applications. For the purposes of this book, however, we need only a few basic ideas: how to calculate linear combinations of vectors, how to calculate the dot product of two vectors, how to multiply matrices, what the inverse of a matrix means, and some facts about the determinant.

The activities of this appendix do not require Sketchpad. The emphasis here is strictly on computation, so that you have the tools needed for Chapters 7 and 10.

Do the following activities, writing your explanations clearly in complete sentences. Include diagrams whenever appropriate.

1. Fill in the missing values in the following statements.

   a. $\begin{bmatrix} ? \\ 2 \end{bmatrix} + 4 \begin{bmatrix} 4 \\ ? \end{bmatrix} = \begin{bmatrix} 21 \\ 26 \end{bmatrix}$

   b. $? \begin{bmatrix} 1 \\ -2 \end{bmatrix} + 5 \begin{bmatrix} -4 \\ 1 \end{bmatrix} = \begin{bmatrix} -16 \\ ? \end{bmatrix}$

   c. $10 \begin{bmatrix} 8 \\ 2 \end{bmatrix} + 3 \begin{bmatrix} ? \\ ? \end{bmatrix} = \begin{bmatrix} 50 \\ 50 \end{bmatrix}$

   d. $? \begin{bmatrix} 3 \\ 2 \end{bmatrix} + ? \begin{bmatrix} 5 \\ 6 \end{bmatrix} = \begin{bmatrix} 1 \\ ? \end{bmatrix}$

2. Fill in the missing values in the following statements.

   a. $(5, 6) \cdot (x, ?) = 5x + 6y$

   b. $(?, ?) \cdot (a, b) = 12a - 2b$

   c. $(1, 2, 4) \cdot (x, y, x) = ?$

   d. $(-2, 3) \cdot (6, 4) = ?$

   e. $(2x, -1, x) \cdot (x, x, 3) = ?$

3. Fill in the missing values in the following statements.

   a. $\begin{bmatrix} 3 & 1 \\ ? & ? \end{bmatrix} \begin{bmatrix} x \\ 2 \end{bmatrix} = \begin{bmatrix} 3x + 2 \\ 5x - 4 \end{bmatrix}$

   b. $\begin{bmatrix} 6 & 0 \\ -2 & 1 \end{bmatrix} \begin{bmatrix} ? \\ 2 \end{bmatrix} = \begin{bmatrix} ? \\ 0 \end{bmatrix}$

   c. $\begin{bmatrix} 0 & 7 \\ 1 & -2 \end{bmatrix} \begin{bmatrix} x \\ y \end{bmatrix} = \begin{bmatrix} ? \\ ? \end{bmatrix}$

4. Multiplying matrices is similar to what you did in Activity 3. For example,

   $$\begin{bmatrix} 1 & 2 \\ 3 & 4 \end{bmatrix} \begin{bmatrix} \cos(\theta) & -\sin(\theta) \\ \sin(\theta) & \cos(\theta) \end{bmatrix} = \begin{bmatrix} \cos(\theta) + 2\sin(\theta) & -\sin(\theta) + 2\cos(\theta) \\ 3\cos(\theta) + 4\sin(\theta) & -3\sin(\theta) + 4\cos(\theta) \end{bmatrix}.$$

   Calculate the following products. Simplify your answers.

   a. $\begin{bmatrix} \cos(\theta) & -\sin(\theta) \\ \sin(\theta) & \cos(\theta) \end{bmatrix} \begin{bmatrix} 1 & 2 \\ 3 & 4 \end{bmatrix}$

   b. $\begin{bmatrix} 1 & 0 \\ 0 & 1 \end{bmatrix} \begin{bmatrix} x & 4 \\ -1 & y \end{bmatrix}$

   c. $\begin{bmatrix} 1 & 2 \\ 3 & 4 \end{bmatrix} \begin{bmatrix} -1 & 2 \\ 3 & -4 \end{bmatrix}$

   d. $\begin{bmatrix} \cos(\theta) & -\sin(\theta) \\ \sin(\theta) & \cos(\theta) \end{bmatrix} \begin{bmatrix} \cos(\theta) & -\sin(\theta) \\ \sin(\theta) & \cos(\theta) \end{bmatrix}$

   (The product in part d will have a special meaning after you study rotations in Chapter 7.)

**5.** Fill in the missing values in the following statement:

$$\begin{bmatrix} \frac{-1}{2} & \frac{3}{2} \\ 1 & -2 \end{bmatrix} \begin{bmatrix} 4 & ? \\ ? & ? \end{bmatrix} = \begin{bmatrix} 1 & 0 \\ 0 & 1 \end{bmatrix}.$$

**6.** Any square matrix has a value called its *determinant*. Here are two examples of determinants for $2 \times 2$ matrices:

$$det \begin{bmatrix} x & 1 \\ y & 2 \end{bmatrix} = 2x - y,$$

$$det \begin{bmatrix} 7 & z \\ -4 & 3 \end{bmatrix} = 21 + 4z.$$

Explain how these determinants were computed. Then find the following determinants.

a. $det \begin{bmatrix} 1 & 3 \\ 2 & 6 \end{bmatrix}$

b. $det \begin{bmatrix} \frac{3}{5} & \frac{-4}{5} \\ \frac{4}{5} & \frac{3}{5} \end{bmatrix}$

**7.** a. Here are two matrices:

$$A = \begin{bmatrix} 4 & 3 \\ 5 & 4 \end{bmatrix}, \qquad B = \begin{bmatrix} 8 & 6 \\ 10 & 8 \end{bmatrix}.$$

How is $B$ related to $A$? How is $det(B)$ related to $det(A)$?

b. Compare the following matrices to $A$. Then predict $det(C)$ and $det(D)$. Check your answers.

$$C = \begin{bmatrix} 1 & 0.75 \\ 1.25 & 1 \end{bmatrix}, \qquad D = \begin{bmatrix} -20 & -15 \\ -25 & -20 \end{bmatrix}.$$

**B.2 DISCUSSION**

## LINEAR COMBINATIONS OF VECTORS

Conceptually, a vector is a joining of two ideas: direction and distance. This makes vectors a convenient method for describing motion. As you will see in Chapter 7, translations can be described very nicely in vector notation.

In Activity 1, the vectors were written as columns, such as $\begin{bmatrix} 1 \\ -2 \end{bmatrix}$. It is also possible to write vectors as rows, such as $(1, -2)$. Row vectors are more convenient for typing, but we will see that the difference in notation can be important. This particular vector says to move 1 unit to the right and 2 units down. The starting point for the vector can be anywhere, as long as the ending point is 1 unit to the right and 2 units down from it.

If we put a coefficient with this vector—say, $4 \begin{bmatrix} 1 \\ -2 \end{bmatrix}$—each entry of the vector will be multiplied by 4 to give $\begin{bmatrix} 4 \\ -8 \end{bmatrix}$. The coefficient 4 is called a *scalar*, and this process is called *scalar multiplication*.

Multiplying a vector by 4 means what you probably expect it to mean, that the vector is added to itself four times. Is it clear how to add vectors? The entries correspond in the natural way, and that is how we add; corresponding entries are added together. For instance, the first sum in Activity 1 can be separated into two computations:

$$? + 4 \cdot 4 = 21,$$
$$2 + 4 \cdot ? = 26.$$

These computations include both addition and scalar multiplication of vectors, as do the other sums in Activity 1. Such sums are called *linear combinations* of the vectors. Notice that only two operations are used, addition and multiplication by a constant, which are exactly the operations used to create linear equations.

## DOT PRODUCT OF VECTORS

Activity 2 shows examples of the *dot product* of two vectors. To perform a dot product, each vector must have the same number of components. The corresponding components are multiplied together, then these values are added. So a dot product produces a single real number as its result.

We were somewhat casual about the notation in Activity 2. The correct representation of the first product is

$$(5, 6) \cdot \begin{bmatrix} x \\ ? \end{bmatrix} = 5x + 6y,$$

so that we are multiplying a row vector by a column vector. However, it is often convenient to write the dot product with two row vectors, such as $(1, 2, 4) \cdot (x, y, x)$. Some authors use brackets instead of parentheses, such as [5, 6]; usually, the different notation does not matter. However, in Chapter 10, the two notations will mean different things. Vectors with parentheses, such as $(3, 2, 4)$, will represent points in a projective plane, while vectors with brackets, such as $[2, -1, 1]$, will represent lines. In projective geometry, a point lies on a line if and only if the dot product of the two vectors equals 0.

## MULTIPLYING A MATRIX BY A VECTOR

Activity 3 presents a different form of multiplication, a matrix times a column vector. For this multiplication to make sense, the sizes of the matrix and the vector must match properly. Consider the first product. The matrix is $2 \times 2$, meaning that it has two rows and two columns. The vector is $2 \times 1$, with two rows and only one column. The product looks like $(2 \times 2) \cdot (2 \times 1)$. Because the innermost values match—both are 2—the product can be done. The outermost values tell the size of the result, namely, $2 \times 1$. In this problem, the sizes match properly, and the product can be calculated. In the other calculations of Activity 3, the matrices and vectors are the same size as in part a, so these products make sense as well.

To do these products, it is best to work with one row of the matrix at a time. For instance, in the calculation

$$\begin{bmatrix} 3 & 1 \\ ? & ? \end{bmatrix} \begin{bmatrix} x \\ 2 \end{bmatrix} = \begin{bmatrix} 3x + 2 \\ 5x - 4 \end{bmatrix},$$

we can begin with the first row times the column vector. Written as a separate problem, this looks like

$$[3 \quad 1] \begin{bmatrix} x \\ 2 \end{bmatrix}.$$

Notice that this is a dot product! The answer is $3x + 2$, which is $3 \cdot x + 1 \cdot 2$.

The second entry in this product is another dot product:

$$[? \quad ?] \begin{bmatrix} x \\ 2 \end{bmatrix} = 5x - 4.$$

Do you see what values belong in the second row of the matrix? The multiplication follows the same pattern as before.

## MULTIPLYING TWO MATRICES

Activity 4 presents several problems involving the product of two matrices. The example in this activity is

$$\begin{bmatrix} 1 & 2 \\ 3 & 4 \end{bmatrix} \begin{bmatrix} \cos(\theta) & -\sin(\theta) \\ \sin(\theta) & \cos(\theta) \end{bmatrix} = \begin{bmatrix} \cos(\theta) + 2\sin(\theta) & -\sin(\theta) + 2\cos(\theta) \\ 3\cos(\theta) + 4\sin(\theta) & -3\sin(\theta) + 4\cos(\theta) \end{bmatrix}.$$

One way to think of this multiplication is as the left matrix times the first column of the second matrix, together with the left matrix times the second column of the second matrix. Thus, we have the following two products:

$$\begin{bmatrix} 1 & 2 \\ 3 & 4 \end{bmatrix} \begin{bmatrix} \cos(\theta) \\ \sin(\theta) \end{bmatrix} = \begin{bmatrix} \cos(\theta) + 2\sin(\theta) \\ 3\cos(\theta) + 4\sin(\theta) \end{bmatrix} \quad \text{and}$$

$$\begin{bmatrix} 1 & 2 \\ 3 & 4 \end{bmatrix} \begin{bmatrix} -\sin(\theta) \\ \cos(\theta) \end{bmatrix} = \begin{bmatrix} -\sin(\theta) + 2\cos(\theta) \\ -3\sin(\theta) + 4\cos(\theta) \end{bmatrix}.$$

(Notice that each calculation uses dot products.) These two products are put together to make the $2 \times 2$ matrix that is the answer.

It is not necessary to take the matrices apart to do this multiplication. We can use the following general pattern:

$$AB = \begin{bmatrix} \text{first row of } A \text{ times} & \text{first row of } A \text{ times} \\ \text{first column of } B & \text{second column of } B \\ \text{second row of } A \text{ times} & \text{second row of } A \text{ times} \\ \text{first column of } B & \text{second column of } B \end{bmatrix}.$$

As this shows, each entry of the matrix product is calculated by doing the product of a row and a column.

Multiplication of matrices is associative, as long as the sizes match correctly. (It is somewhat tedious to prove this, even in the $2 \times 2$ case.) However, multiplication of matrices is *not* commutative, as you may have noticed in Activity 4. In part a, which is the reverse of the product shown in the example, the result is very different.

The matrix $I = \begin{bmatrix} 1 & 0 \\ 0 & 1 \end{bmatrix}$ is called the *identity matrix*. As you saw in Activity 4b, $IX = X$ for any matrix $X$. This is also true in the reverse order; $XI = X$. Any matrix $X$ does not change when multiplied by the identity $I$. (The multiplication must make sense, of course, meaning that the sizes of the matrices must match properly. We have shown you only the $2 \times 2$ identity matrix. However, there is an identity matrix for any square size.)

The identity serves the same role that the number 1 does in normal multiplication. Just as in normal multiplication, we have the question of inverses. A reminder: Two numbers $x$, $y$ are *inverses* for multiplication if $xy = 1$ and $yx = 1$. In the same way, two matrices $X$, $Y$ are *inverses* if $XY = I$ and $YX = I$. In Activity 5, you were asked to complete the entries of the matrix that is the inverse of $\begin{bmatrix} \frac{-1}{2} & \frac{3}{2} \\ 1 & -2 \end{bmatrix}$. You must find the values that make the matrix product equal the identity.

A warning: Not every matrix will have an inverse. This is one major difference between multiplying numbers and multiplying matrices. For instance, $\begin{bmatrix} 1 & 3 \\ 2 & 6 \end{bmatrix}$ does not have an inverse. In the next section, you will see a quick way to determine whether a particular matrix has an inverse.

## THE DETERMINANT OF A MATRIX

A matrix has many numbers associated with it. This includes its entries, of course, but also various values that can be calculated from these entries. Many interesting theorems have been proven about these numerical descriptions of a matrix. One of the most useful is the *determinant*. This number combines all the entries of the matrix into a single value.

There are several ways to define the determinant, and there are also several ways to compute it. For the case of the small $2 \times 2$ matrices, the determinant is fairly simple:

$$det \begin{bmatrix} a & b \\ c & d \end{bmatrix} = ad - bc.$$

Does this agree with what you did in Activity 6?

In Chapter 10, it occasionally will be necessary to calculate the determinant of a $3 \times 3$ matrix, which is a square matrix with nine entries. There are several ways to proceed. Here is one fairly direct way:

$$det \begin{bmatrix} a_{11} & a_{12} & a_{13} \\ a_{21} & a_{22} & a_{23} \\ a_{31} & a_{32} & a_{33} \end{bmatrix} = a_{11} \cdot det \begin{bmatrix} a_{22} & a_{23} \\ a_{32} & a_{33} \end{bmatrix} - a_{12} \cdot det \begin{bmatrix} a_{21} & a_{23} \\ a_{31} & a_{33} \end{bmatrix}$$

$$+ a_{13} \cdot det \begin{bmatrix} a_{21} & a_{22} \\ a_{31} & a_{32} \end{bmatrix}.$$

For example,

$$det \begin{bmatrix} 1 & 2 & 3 \\ 4 & 5 & 6 \\ 7 & 8 & 9 \end{bmatrix} = 1 \cdot det \begin{bmatrix} 5 & 6 \\ 8 & 9 \end{bmatrix} - 2 \cdot det \begin{bmatrix} 4 & 6 \\ 7 & 9 \end{bmatrix} + 3 \cdot det \begin{bmatrix} 4 & 5 \\ 7 & 8 \end{bmatrix}$$

$$= 1 \cdot (-3) - 2 \cdot (-6) + 3 \cdot (-3) = 0.$$

This is only one of many ways to calculate a determinant for a $3 \times 3$ matrix. Larger matrices require more complicated methods, but we will not need those in this course.

Using the pattern for $2 \times 2$ matrices, you can quickly calculate the determinants in Activity 6. Notice that the determinant of $\begin{bmatrix} 1 & 3 \\ 2 & 6 \end{bmatrix}$ equals 0, even though none of the entries are 0. This will prove to be important.

Here are some basic facts about determinants:

**THEOREM B.1**   $det(AB) = det(A) \cdot det(B)$.

For $2 \times 2$ matrices, straightforward algebra will verify this theorem. For larger matrices, more advanced methods of proof are required. Consult any linear algebra text.

**THEOREM B.2**   The square matrix $A$ has an inverse if and only if $det(A) \neq 0$.

**Partial Proof**   Suppose $B$ is the inverse of $A$. Then $AB = I$. Because $det(I) = 1$, this gives us $det(B) = \frac{1}{det(A)}$. Thus $det(A)$ cannot be 0.

**THEOREM B.3**   The inverse of the matrix $A = \begin{bmatrix} a & b \\ c & d \end{bmatrix}$ is the matrix $A^{-1} = \frac{1}{ad-bc} \begin{bmatrix} d & -b \\ -c & a \end{bmatrix}$, provided that this matrix exists.

**THEOREM B.4**   For a $2 \times 2$ matrix $A$ and a scalar $s$, $det(sA) = s^2 det(A)$.

This final theorem is the key to Activity 7. The matrices $C$ and $D$ are multiples of matrix $A$, and $det(A) = 1$. Once you decide what scalar is involved, it is an easy calculation to determine the determinants of $C$ and $D$.

Give clear and complete answers to the exercises, expressing your explanations in complete sentences. Include diagrams whenever appropriate.

1. Let $(m \times n)$ denote a matrix of size $m$ by $n$. Which of the following matrix products are not defined? Explain why or why not for each situation.

   a. $(2 \times 2) \cdot (1 \times 2)$
   b. $(4 \times 2) \cdot (2 \times 1)$
   c. $(2 \times 3) \cdot (2 \times 1)$
   d. $(3 \times 2) \cdot (2 \times 2)$

2. For the vectors $\vec{a} = \begin{bmatrix} -3 \\ 1 \end{bmatrix}$ and $\vec{b} = \begin{bmatrix} 0 \\ 2 \end{bmatrix}$, compute the following.

   a. $-4\vec{a}$
   b. $2\vec{a} + \vec{b}$
   c. $5\vec{a} - 5\vec{b}$

3. a. Find scalars $a$, $b$ so that
$$a \begin{bmatrix} 2 \\ -5 \end{bmatrix} + b \begin{bmatrix} -6 \\ 15 \end{bmatrix} = \begin{bmatrix} 0 \\ 0 \end{bmatrix}.$$

   b. Find scalars $c$, $d$ so that
$$c \begin{bmatrix} 4 \\ 7 \end{bmatrix} + d \begin{bmatrix} 2 \\ -3 \end{bmatrix} = \begin{bmatrix} 0 \\ 0 \end{bmatrix}.$$

   c. How are these problems different? Explain the difference.

4. Calculate the following dot products.

   a. $[8, 3] \cdot [2, 5]$
   b. $\left(\frac{7}{2}, -1\right) \cdot \left(4, \frac{3}{2}\right)$
   c. $(7, -1, 4) \cdot (0, 4, 2)$
   d. $[-2, y, 5] \cdot [-2, z]$
   e. $(-1, 0, 1) \cdot \begin{bmatrix} 2 \\ 1 \\ 2 \end{bmatrix}$
   f. $[1, -2, 3, 4, 5] \cdot [10, 9, 8, -7, 6]$

5. Find at least three vectors $(a, b, c)$ so that
$$(4, -1, -4) \cdot (a, b, c) = 0.$$

6. Compute the following products.

   a. $\begin{bmatrix} 7 & \frac{1}{2} \\ 4 & 2 \end{bmatrix} \begin{bmatrix} 1 & -1 \\ -1 & 4 \end{bmatrix}$

   b. $\begin{bmatrix} 6 & 2 \\ 3 & -3 \end{bmatrix} \begin{bmatrix} 1 & 0 \\ 0 & 1 \end{bmatrix}$

   c. $\begin{bmatrix} 1 & -1 \\ -1 & 4 \end{bmatrix} \begin{bmatrix} 7 & \frac{1}{2} \\ 4 & 2 \end{bmatrix}$

   d. $\begin{bmatrix} 4 & 2 \\ -1 & \frac{-1}{2} \end{bmatrix} \begin{bmatrix} 1 & \frac{1}{2} \\ -2 & -1 \end{bmatrix}$

7. For the products in Exercise 6, calculate the determinant of each factor and the determinant of the product. Compare your answers with Theorem B.1.

8. Which of the following matrices has an inverse? For each invertible matrix, find its inverse.

   a. $\begin{bmatrix} 2 & 3 \\ 3 & 4 \end{bmatrix}$

   b. $\begin{bmatrix} 7 & -6 \\ 3 & -3 \end{bmatrix}$

   c. $\begin{bmatrix} 2 & -3 \\ -8 & 12 \end{bmatrix}$

9. Calculate the determinant of $\begin{bmatrix} 1 & -1 & 4 \\ 3 & 2 & -2 \\ 4 & 0 & 2 \end{bmatrix}$.

10. A square matrix is called *unimodular* if its determinant is 1, 0, or $-1$. The matrix is *strictly unimodular* if every square submatrix also has a determinant equal to 1, 0, or $-1$. (These matrices are important in the theory of linear programming.) For a $3 \times 3$ matrix, $M$, this means that the determinant of $M$ is 1, 0, or $-1$; that every $2 \times 2$ submatrix of $M$ is 1, 0, or $-1$; and that the determinant of every $1 \times 1$ submatrix is 1, 0, or $-1$. This last requirement means that every entry must be 1, 0, or $-1$. Find at least four examples of a strictly unimodular $3 \times 3$ matrix.

# BIBLIOGRAPHY

Baragar, Arthur. *A Survey of Classical and Modern Geometries.* Upper Saddle River, NJ: Prentice Hall, 2001.

Batten, Lynn Margaret. *Combinatorics of Finite Geometries.* Cambridge: Cambridge University Press, 1986.

Bix, Robert. *Topics in Geometry.* San Diego, CA: Academic Press, 1994.

Coxeter, H.S.M. *The Real Projective Plane.* New York: McGraw-Hill, 1949.

Coxeter, H.S.M. *Introduction to Geometry,* 2nd ed. New York: John Wiley & Sons, 1969.

Coxeter, H.S.M., and Greitzer, S.L. *Geometry Revisited.* New Mathematical Library. Washington, D.C.: The Mathematical Association of America, 1967.

Eves, Howard. *An Introduction to the History of Mathematics,* 4th ed. New York: Holt, Rinehardt and Winston, 1976.

Fenton, William E., and Ed Dubinsky. *Introduction to Discrete Mathematics with ISETL.* New York: Springer, 1996.

Graustein, William C. *Introduction to Higher Geometry.* New York: MacMillan, 1930.

Greenberg, Marvin Jay. *Euclidean and Non-Euclidean Geometries.* New York: W.H. Freeman, 1980.

Hartshorne, Robin. *Geometry: Euclid and Beyond.* New York: Springer-Verlag, 2000.

Heath, Sir Thomas, trans. *The Thirteen Books of Euclid's Elements.* New York: Dover, 1956.

Jacobson, Nathan. *Basic Algebra I.* San Francisco: W.H. Freeman, 1974.

Kay, David C. *College Geometry: A Discovery Approach.* New York: HarperCollins College, 1994.

Kimberling, Clark. *Geometry in Action.* Emeryville, CA: Key College Publishing, 2003.

Knight, Robert. *Using Laguerre Geometry to Discover Euclidean Theorems.* Unpublished diss., University of California–San Diego, 2000.

Krause, Eugene F. *Taxicab Geometry.* Menlo Park, CA: Addison-Wesley, 1975.

Lam, C.W.H. "The Search for a Finite Projective Plane of Order 10." *The American Mathematical Monthly* 98, no. 4 (April 1991): 305–318.

Lay, David C. *Linear Algebra and Its Applications,* 2nd ed. Reading, MA: Addison-Wesley Longman, 1997.

Maher, Richard J. "Step by Step Proofs and Small Group Work in Courses in Algebra and Analysis." *PRIMUS* 4, no. 3 (September 1994).

Maor, Eli. *Trigonometric Delights.* Princeton, NJ: Princeton University Press, 1998.

National Council of Teachers of Mathematics (NCTM). *Principles and Standards for School Mathematics.* Reston, VA: The National Council of Teachers of Mathematics, 2000.

Nelsen, Roger. "Proofs Without Words," *Mathematics Magazine,* April 2002.

Ogilvy, C. Stanley. *Excursions in Geometry.* New York: Dover Publications, Inc. 1990.

Reynolds, Barbara E. "Taxicab Geometry: An Example of Minkowski Space." Unpublished diss., Saint Louis University, 1979.

Reynolds, Barbara E. "Taxicab Geometry," *Pi Mu Epsilon Journal,* Spring 1980.

Robinson, Gilbert de B. *The Foundations of Geometry.* Toronto: University of Toronto Press, 1946.

Rogers, Elizabeth C., Barbara E. Reynolds, Neil A. Davidson, and Anthony D. Thomas, eds. *Cooperative Learning in Undergraduate Mathematics: Issues that Matter and Strategies that Work.* MAA Notes Series 55. Washington, D.C.: The Mathematical Association of America, 2001.

Shorlin, Kelly A., John R. de Bruyn, Malcolm Graham, and Stephen W. Morris, "Development and geometry of isotropic and directional shrinkage-crack patterns," *Phys. Rev. E.* 61, no. 6, 6950–6957 (June 2000), http://arxiv.org/pdf/patt-sol/9911003 (accessed March 29, 2005).

Sibley, Thomas Q. *The Geometric Viewpoint: A Survey of Geometries.* Reading, MA: Addison-Wesley Longman, 1998.

Smart, James R. *Modern Geometries,* 3rd ed. Pacific Grove, CA: Brooks/Cole, 1988 (also 5th ed., 1998).

Smith, Douglas, Maurice Eggen, and Richard St. Andre. *A Transition to Advanced Mathematics.* Pacific Grove, CA: Brooks Cole, 2001.

Sved, Marta. *Journey to Geometries.* Spectrum Series. Washington, D.C.: The Mathematical Association of America, 1991.

Thomas, David. *Modern Geometry.* Pacific Grove, CA: Brooks Cole, 2002.

Wallace, Edward C., and Stephen F. West. *Roads to Geometry.* Englewood Cliffs, NJ: Prentice-Hall, 1992.

Wells, David. *The Penguin Dictionary of Curious and Interesting Geometry.* London: Penguin Books, 1991.

Yaglom, I.M. *Geometric Transformations.* The New Mathematical Library. New York: The L.W. Singer Company (Random House), 1962.

conditional statements (implication)
  contrapositive. *See* contrapositive
  converse of, 35–36, 53
  defined, 34
  *If . . . then . . .* form, 20, 76
  proving. *See* logic; proofs
  symbol for, 34
  truth value of, 34–35
  *See also* biconditional statements
conformal function of the plane, 153
congruence
  defined, 14–15
  Hilbert's axioms of, 61, 62
  in hyperbolic geometry, 214
  isometries and, 144
  of triangles. *See* congruence of triangles
Congruence Axiom for Triangles (SAS), 21, 42, 214
congruence of triangles
  AAA criterion for, 214
  AAS criterion for, 50
  ASA criterion for, 42, 149–150
  axiom for (SAS), 21, 42, 214
  in hyperbolic plane, 214
  ratio of, 18
  similarity and, 18
  SSS criterion for, 50
conic sections
  Pascal's Theorem for Circles and, 246
  perspectivity and, 252–253
conjectures, 24
  as conditional statements, 37
  counterexamples as disproof of, 77
  defined, 6
  *If . . . then . . .* format for, 20
  proving. *See* proofs
  *See also* axioms; *specific conjectures*
conjunction
  negation of, 53
  operator for, 32, 33, 34
  symbol for, 34
constructible numbers, 91
construction
  of an arbelos, 59
  basics required, 25
  of circles, 62, 67, 118
  of circles that share a common chord, 69
  of circumcircle, 46–47
  drawing compared to, 16, 25
  of an equilateral triangle, 7, 16
  Euclid's postulates and, 13–14
  of fundamental circle, 204
  impossible, 17–18, 23

inscribed regular pentagon in a circle, 23
  language of proofs referring to, 63
  of a locus, 118, 119
  of midpoints, 6, 11–12, 23
  of number line, 90–91
  of parallel lines, 10, 62
  of perpendicular lines, 10, 16–17, 62
  of perspectivities, 231
  of Poincaré disk, 204–206, 215–216
  of projectivities, 231
  of a quadrilateral, 6
  robust, 7, 62, 63, 76
  of a salinon, 74
  of a tangent, 68
  of taxicab metric, 119–122
  tiling, 182
  as visual proof, 62, 63
  *See also* drawing
continuity, projective geometry and, 236
contradiction, proof by, 84
  *See also* negation
contrapositive
  converse compared to, 53
  defined, 36
  indirect proofs and, 40, 77, 84
converse of a statement, 35–36, 78
  contrapositive compared to, 53
  separate proof required for, 78
convex figures, 9
convex polygons, 9
coordinate plane, symbol for, 91
coordinate systems. *See* analytic geometry; coordinate system, rectangular; homogeneous coordinates; origin; polar coordinates; skew coordinate system
coordinate system, rectangular
  distance and, 91–93, 97–100
  homogeneous coordinates converted to, 247–248
  origin of, 91
  polar coordinates, relationship to, 104, 105
  proofs using, 100–102
  scale in, 91
corollary, defined, 24
corresponding angles, 26
Cosines, Law of, 91–92, 270
cosine ratio
  angle sum formula, 268
  defined, 264
  difference of two angles, 269
  double angle formulas, 267
  on unit circle, 266
  *See also* trigonometry

distributive property
    of matrix multiplication, 169
    *See also* isometries
domain of a function, 139
dot product of vectors, 276, 277
double angle formulas, 267, 269
drawing
    construction compared to, 16, 25
    perspective, 229, 235, 258
    *See also* construction
duality
    axioms for, 240–241
    compared to other geometries, 240
    defined, 240
    examples of, 241–246
    homogeneous coordinates and, 249

Earth, spherical geometry and, 217,
        218, 239
edge to edge, in tiling, 194
elementary tilings, 195–196
*Elements* (Euclid), 12, 60–61, 63
ellipses
    defined, 126
    in Euclidean metric, 126–127
    in taxicab metric, 127
elliptic geometry, 217–218, 239
Elliptic Parallel Postulate, 217
Elliptic Parallel Theorem, 218
equations
    of a circle, 97–98, 103–104
    linear. *See* linear equations
    matrix. *See* matrices
    trigonometric, 266–267
equator, 217, 218, 239
equilateral triangles
    construction of, 7, 16
    defined, 18
equivalence relations
    classes of (ideal point), 235–236
    defined, 235
    properties of, 235
Escher, M. C., and tilings, 194, 195
Euclid, 5, 10, 60–61, 63
Euclidean geometry
    betweenness as preserved in, 251
    distance and, 91, 113, 117, 124–125
    homogeneous coordinates converted to,
        247–248
    impossible constructions of,
        17–18, 23
    projective geometry compared to, 240
    *See also* non-Euclidean geometries
Euclidean motions. *See* isometries

Euclid's postulates
    assumption of, in writing proofs, 76
    defined, 11, 12–13
    development of, 10–11, 12–13
    in hyperbolic geometry, 208–209
    as implicit hypotheses, 13
    Sketchpad and, 13–14
Euclid's First Postulate
    defined, 12
    in hyperbolic geometry, 208
    straightedge as representation of, 17
Euclid's Second Postulate
    defined, 12
    in hyperbolic geometry, 208, 211, 212
    straightedge as representation of, 17
Euclid's Third Postulate
    circle constructions and, 62
    compass as representation of, 17
    defined, 13
    in hyperbolic geometry, 208–209
Euclid's Fourth Postulate
    defined, 13
    in hyperbolic geometry, 209
    perpendicular-line constructions
        and, 62
Euclid's Fifth Postulate
    acceptance of, as implicit hypothesis, 13
    angle sum of a triangle and, 64
    Clavius' Axiom and, 11
    defined, 11, 13
    hyperbolic geometry and replacement
        of, 218
    illustration of, 25–26
    parallel-line constructions and, 62
    Playfair's Postulate as equivalent to, 11,
        13, 14
    Sketchpad and, 14
    *See also* Playfair's Postulate
Euler line, 45
*every,* as universal quantifier, 40
excircle, 51, 60, 65–66, 153–155
existential quantifier
    defined, 39
    key words indicating, 40
    negation and, 40, 217
    symbol for, 39
exterior angles
    activity introducing, 4
    alternate, 10, 26
    of a cyclic quadrilateral, 68
    on same side of the transversal,
        10, 26
    sum of, 18–19
    of a triangle, 18–19, 51, 65–66

logarithmic spiral, 104

logic

    *modus ponens*, 38, 76

    *modus tollens*, 38, 77

    operators for, 32–34

    statements. *See* statements

    syllogism, 38

    symbols and symbolic form of, 34

    *See also* proofs

logos, symmetry and, 190

longitudinal lines, 217, 218

mathematical models

    defined, 235

    set-up requirements of, 207

    *See also* axioms

matrices

    as associative, multiplication of, 168

    determinant of. *See* determinant

    as distributive, multiplication of, 169

    identity, 278

    inverse, 278, 279

    multiplication of, 277–278

    multiplied with a vector, 168, 276–277

    as noncommutative, multiplication

        of, 169

    projective geometry and, 249–251, 256

    similarity, 174–175

    unimodular, 280

    *See also* matrices and isometries

matrices and isometries

    composition of isometries and, 168–170,

        173–174

    determinant, and direct or opposite

        isometry, 170–172

    as distance preserving, 172–173

    fixed points proved through, 173

    general form of representation by, 170–172

    glide reflections represented by, 169–170

    order for, 169

    pattern of, 170

    proofs utilizing, 172–174

    reflections represented by, 166–168,

        169, 170

    rotations represented by, 165–166,

        168, 170

    translations represented by, 169, 170

medians of a triangle, 30, 43–44

    as Cevian, 47

    *See also* centroid

Menelaus' Theorem

    defined, 48

    proof of Desargues' Theorem

        and, 243

    proof of Pappus' Theorem and, 236, 237

    proof of Pascal's Theorem and, 246

metric

    axioms for, 124, 125, 130, 211–212

    defined, 117

    *See also* distance formulas

midline of frieze pattern, 191–192

midpoint of a line segment

    construction of, 11–12

    formula for, 93

midpoint of a side, construction of, 6

mirror line (line of reflection), 140, 142,

    166–167

models. *See* mathematical models

modular arithmetic, 250

*modus ponens*, 38, 77

*modus tollens*, 38, 77

Moulton, F. R., 257

Moulton's geometry, 257–258

Nagel point, 51–52

nature, symmetry and, 190

navigation, ocean, 89

negation

    indirect proofs and, 77, 84

    key words for, 40

    operator for, 33, 34

    of Playfair's Postulate, 41

    of quantified statements, 40–41

    symbol for, 34

    of universal statements, as

        existential, 217

negative numbers

    clockwise rotation signifying, 103

    direction of, on number line, 91

*n*-gon. *See* polygons

nine-point circle

    algebraic proof of, 105–110

    defined, 47

    history of, 75

    inversion and, 153–155

    radius of, 108

*none*, negation and, 40

non-Euclidean geometries

    elliptic, 217–218

    hyperbolic. *See* hyperbolic geometry

    *See also* projective geometry; taxicab

        geometry

nonorientable surface, 240

nonstatements, 32

*not* (logical operator)

    rules for, 33–34

    *See also* negation

*not all*, as negation, 40

quantified statements, 38–39
  existential. *See* existential quantifier
  negation of, 40–41
  universal. *See* universal quantifier
  Venn diagrams and, 39

radians, 10
radius
  of a circle, 20, 46
  of nine-point circle, 108
range of a function, 139
rays, polar coordinate system and, 103
real projective plane
  continuity and, 236
  coordinates for, 247–250
  defined, 236
  diagonal triangle of, 242
  dual concepts of perspective in, 243–246
  ideal line and, 236
  as model, 235–237
  as nonorientable surface, 240
  spherical model as isomorphic to, 239
reasoning
  deductive. *See* logic
  inductive. *See* inductive reasoning
reciprocal trigonometric functions, 265
rectangles, as nonexistent in hyperbolic
    plane, 219, 221–222
reference triangle, 266
reflection
  composition of isometries and, 144–145,
    146, 168, 169
  composition of symmetries and,
    185–186
  defined, 140
  finite symmetry groups and, 186–190
  fixed points and, 142
  frieze patterns and, 191–193
  line of (mirror line), 140, 142, 166–167
  matrices to represent, 166–168,
    169, 170
  not a subgroup, 149
  as opposite isometry, 141–142, 170
  through origin, 167–168
  proof of, 146–148
  in space, 151
  vectors and, 167
  *See also* glide reflection
reflection in a point, 151
reflexive property, equivalence and, 235
regular polygons
  defined, 182
  tiling with, 196–197
regular tiling, 196

right angles
  congruence of, 209
  Euclidean definition of, 61
  measure of, 10
right triangles
  defined, 18
  properties of. *See* Pythagorean Theorem;
    trigonometry
rigid motions. *See* isometries
rise over the run, 94
robust constructions
  defined, 7
  proof development and, 62, 63, 76
robust proofs
  goals of, 37
  robust constructions leading to, 62,
    63, 76
rose curves, 105
rotary reflection isometry, 151
rotation
  angle of, 165, 166, 193
  center of, 142, 165, 166, 173
  composition of isometries and, 146,
    166, 168
  composition of symmetries and, 185–186
  defined, 141
  as direct isometry, 141, 170
  finite symmetry groups and, 186–190
  fixed point of, 142, 173
  frieze patterns and, 191–193
  inverse, 166
  matrices to represent, 165–166, 168, 170
  polar coordinates and, 141, 165
  in space, 151
  as subgroup of isometries, 149
  wallpaper symmetry and, 193

Saccheri quadrilateral, 216, 221
salinon, 73, 74
salt cellar. *See* salinon
SAS Congruence Axiom, 21, 42, 214
scalar multiplication, 276
scalars, 276
scale, in coordinate systems, 91
scalene triangle, 18
scientific method, 21, 82–83
screw isometry, 151
secant ratio
  defined, 264
  on unit circle, 266
  *See also* trigonometry
sector of a circle, 67
segments
  of a circle, 67

line. *See* line segments
in taxicab metric, 128
self-duality, 242
semicircles
angle inscribed in, 20
Pythagorean Theorem and, 67
semiperimeter, 81
semiregular tiling, 197
shoemaker's knife. *See* arbelos
side-angle-side (SAS), 21, 42, 214
side-side-side (SSS), 50
similarity
congruence and, 18
defined, 15
function of, 174–175
matrix representation of, 174–175
slope and, 94
of triangles, 18
trigonometry and, 265
similarity function, 174–175
Simson line, 52
Sines, Law of, 269, 270
sine wave, 105
sine ratio
angle sum formula for, 268
defined, 264
difference of two angles, 268
double angle formulas, 267
on unit circle, 266
*See also* trigonometry
size change, similarities and,
174–175
skew coordinate system, 91–93
slope
angle of inclination and, 95–96
collinearity verified through, 152
defined, 94
equivalence classes and, 235–236
formula, 94
for horizontal lines, 94
of parallel lines, 96
of perpendicular lines, 96–97
point-slope form, 93, 95
similarity and, 94
slope-intercept form, 93, 95
symbol for, 93
taxicab geometry and, 128
for vertical lines, as undefined, 95
as well defined, 94
Sketchpad. *See* The Geometer's Sketchpad
slope-intercept form, 93, 95
*some do not*, as negation, 40
*some*, as existential quantifier, 40
source of a function, 139

space
isometries in, 150–151
projective, 252–253
symmetry in, 193, 194
spheres
elliptic geometry and, 217–218, 239
as projective geometry model, 238–240
spherical geometry, 217–218, 239
spirals, polar coordinates and equations
of, 104
square root notation, 98
squares
defined, 9
self-congruence of, 183–184
as subgroup of isometries, 149
tiling with, 197
SSS criterion for triangle congruence, 50
straight line, Euclidean definition of, 61
standard unit circle, 265
statements
closed, 32, 38–39
compound, 32–33, 34
conditional. *See* biconditional statements;
conditional statements
defined, 31–32
logical operators for, 32–34
nonstatements, 32
open, 32, 38–39
quantified. *See* quantified statements
step-by-step proofs
angle bisectors of a triangle as concurrent,
64–65
angle sum of a triangle, 63–64
process of, 62–63, 76
straight angles, 17, 20
straight edge
Greek philosophers allowing, 5
representing Euclid's First and Second
Postulates, 17
strictly unimodular matrices, 280
substitution, closing a statement with, 38
sum of angles
of exterior angles, 18–19
formulas for, 267–269
in hyperbolic geometry, 209–210
of quadrilateral, 210
of triangle, 63–64, 209–210, 218–219
triangulation of the figure and, 210
summit, 221
summit angles, 221
superparallel lines. *See* limiting
parallel rays
Sydney Opera House, 190
syllogism, defined, 38

similarity and, 174–175

stretching or shrinking the plane, 158–159

of symmetries. *See* composition of symmetries

transitive property of equivalence, 235

translation

and composition of isometries, 144–145, 146, 169

defined, 141

as direct isometry, 141, 170

fixed points, lack of, 142

frieze patterns and, 191–193

inverse of, 148, 191

matrices to represent, 169, 170

proof of, 146–148

in space, 150

as subgroup of isometries, 149

symmetry and, 190–197

tiling and, 194–197

vectors and, 141, 146, 148, 164–165, 191

wallpaper patterns and, 193

transversal lines, 10, 13, 25–26, 258

trapezium, 49

trapezoids, defined, 49

triangle inequality

defined, 22, 124

in hyperbolic geometry, 211, 212

isometries preserving betweenness and, 143–144

triangles

altitudes of. *See* altitudes of a triangle

angle bisectors of, 44–45, 64–66, 213–214

area of, 54, 265

asymptotic, 210

centroid of, 43–44, 45

circumcenter. *See* circumcenter

concurrence of, 43–45, 64–65

congruence of. *See* congruence of triangles

as convex, 9

as cyclic, 46–47

defect of, 210

diagonal triangle, 242

equilateral, 7, 16, 18

Euler line of, 45

excircle, 51, 60, 65–66, 153–155

exterior angles of, 18–19, 51, 65–66

in hyperbolic geometry, 209–210, 214–215, 218–219

incenter, 44–45, 64–65, 213–214

incircle. *See* incircle

interior angles of, 18–19, 63–64

isosceles, 18

medians of, 30, 43–44, 47

midpoint of, 23

nine-point circle of. *See* nine-point circle

orthocenter. *See* orthocenter

pedal, 47, 106

perpendicular bisectors of, 45, 46–47, 107–108, 213

perspective from a line, 52

perspective from a point, 52

points of concurrency of. *See* points of concurrency

in projective geometry, 242–245

reference, 266

right, 18

scalene, 18

similarity of, 18

sum of angles of, 63–64, 209–210, 218–219

tiling with, 196

vertices, labeling, 141

triangulation of a figure, 210

trigonometry, 261–270

activities exploring, 262–264

area and, 265

double angle formulas, 267, 269

half-angle formulas, 269

Law of Cosines, 91–92

reciprocal functions, 265

rotation and, 166

solving equations, 266–267

unit circle, 265–266

as well-defined functions, 265

trilateral, 8, 242

*See also* triangles

truth tables, for conditional statements, 34

truth value

conditional statements and, 34–35

converse statements and, 35–36

determination of, 32

logical operators and, 33–34

nonstatements as lacking, 32

quantifiers and determination of, 38–39

twist isometry, 151

two-point perspective, 258

ultraparallel lines. *See* limiting parallel rays

unbounded vs. bounded figures, 190–191

unimodular matrices, 280

unit circle trigonometry, 265–266

unit length, 90, 91

universal quantifier

defined, 39

key words indicating, 40

negation and, 40, 217

symbol for, 39

upper-half-plane model, 223–224, 225